COMPARTMENTAL ANALYSIS OF ECOSYSTEM MODELS

STATISTICAL ECOLOGY
Volume 10

a publication from the
satellite program in statistical ecology
international statistical ecology program

Statistical Ecology Series

General Editor: G. P. Patil

*For these first three volumes, contact: For all of the remaining volumes, contact:
The Pennsylvania State University Press   International Co-operative Publishing House
University Park, PA 16802 USA   P.O. Box 245
                                              Burtonsville, MD 20730 USA

COMPARTMENTAL ANALYSIS OF ECOSYSTEM MODELS

edited by

JAMES H. MATIS
Institute of Statistics
Texas A&M University
College Station, Texas

BERNARD C. PATTEN
Institute of Ecology
University of Georgia
Athens, Georgia

GARY C. WHITE
Environmental Science Group
Los Alamos Scientific Laboratory
Los Alamos, New Mexico

International Co-operative Publishing House
Fairland, Maryland USA

Mathematical ecology is moving out of its classical phase carrying with it untold promise for the future, but, as H.A.L. Fisher, the historian, remarks, progress is not a law of nature. Without enlightenment and eternal vigilance on the part of both ecologists and mathematicians there always lurks the danger that mathematical ecology might enter a dark age of barren formalism, fostered by an excessive faith in the magic of mathematics, blind acceptance of methodological dogma and worship of the new electronic gods. It is up to all of us to ensure that this does not happen.

J. G. SKELLAM (1972)
in Mathematical Models in Ecology
edited by J. N. R. Jeffers
Blackwell Scientific Publications
Oxford, England

For top management and general public policy development, monitoring data must be shaped into easy-to-understand indices that aggregate data into understandable forms. I am convinced that much greater effort must be placed on the development of better monitoring systems and indices than we have in the past. Failure to do so will result in sub-optimum achievement of goals at much greater expense.

R. E. TRAIN (1973)
National Conference on Managing the Environment
The United States Environmental Protection Agency
Washington, D.C., USA

International Statistical Ecology Program

ADVISORY BOARD
(Chairman: G. P. Patil)

Berthet, P.
Cairns, J., Jr.
Chapman, D. G.
Cormack, R. M.
Cox, D. R.
Holling, C. S.

Iwao, S.
Matern, B.
Matis, J. H.
Ord, J. K.
Pielou, E. C.
Rao, C. R.

Robson, D. S.
Rossi, O.
Simberloff, D. S.
Taillie, C.
Warren, W. G.
Waters, W. E.

SATELLITE PROGRAM IN STATISTICAL ECOLOGY
(1977-1978)
Director: G. P. Patil   Managing Editor: C. Taillie

COORDINATORS AND EDITORS

Artuz, M. I.
Cairns, J., Jr.
Chapman, D. G.
Cormack, R. M.
Gallucci, V.
Grassle, J. F.
Holling, C. S.
Innis, G. S.
Matis, J. H.

O'Neill, R. V.
Ord, J. K.
Orloci, L.
Patil, G. P.
Patten, B. C.
Rao, C. R.
Robson, D. S.
Rosenzweig, M. L.
Shoemaker, C.

Simberloff, D. S.
Smith, W. K.
Solomon, D. L.
Stiteler, W. M.
Taillie, C.
Usher, M. B.
Waters, W. E.
White, G. C.
Williams, F. M.

HOSTS

Matis, J. H.
Gates, C. E.

Waters, W. E.
Noy-Meir, I.

Rossi, O.
Zanni, R.

ADVISORS

Berthet, P.
Engen, S.
Glass, N.

Hennemuth, R.
Iwao, S.
Knox, G. A.

Matern, B.
Seber, G.
Warren, W. G.

SPONSORS

NATO Advanced Study Institutes Program
NATO Ecosciences Program

The Pennsylvania State University
The Texas A&M University
The University of California at Berkeley

National Marine Fisheries Service, USA
Environmental Protection Agency, USA
Fish and Wildlife Service, USA
Army Research Office, USA

Comunita Economica Europea
Universita degli Studi, Parma, Italy

Consiglio Nazionale delle Ricerche, Italy
Ministero dei Lavori Pubblici,
Affari Esteri, e Pubblica Istruzione, Italy
Societa Italiana di Statistica, Italy
Societa Italiana di Ecologia, Italy

The Participants and
Their Home Institutions and Organizations

PARTICIPANTS

SATELLITE A: COLLEGE STATION AND BERKELEY   July 18-August 13, 1977

Andrews, P. L., Montana
Anthony, R. G., Oregon
Artuz, M. I., Turkey
Bagiatis, K., Greece
Bajusz, B. A., Pennsylvania
Baumgaertner, J., California
Bell, E., Washington
Bellefleur, P., Canada
Berthet, P., Belgium
Beyer, J., Denmark
Bingham, R., Texas
Boswell, M. T., Pennsylvania
Braswell, J. H., Georgia
Braumann, C. A., Portugal
Brennan, J. A., Massachusetts
Bruhn, J. N., California
Cairns, J. Jr., Virginia
Callahan, C. A., Oregon
Cancela da Fonseca, J. P., France
Caraco, T. B., Arizona
Chapman, D. G., Washington
Cho, A., Pennsylvania
Colwell, R. California
Cormack, R., Scotland
Coulson, R. N., Texas
DeMars, C. J., California
Dennis, B., Pennsylvania
Derr, J., Iowa
deVries, P. G., Netherlands
Doucet, P. G., Netherlands
Elterman, A. L., California
Engen, S., Norway
Ernsting, G., Netherlands
Fiadeiro, P. M., Portugal
Flores, R. G., Brazil
Flynn, T. S., California
Folse, L. J., Texas
Ford, R. G., California
Gallucci, V. F., Washington
Gates, C. E., Texas
Gautier, C., France
Gerald, K. B., Texas

Giles, R. H., Virginia
Gokhale, D. V., California
Grant, W. G., Texas
Guardans, R. C., Spain
Hart, D., Virginia
Hazard, J. W., Oregon
Hendrickson, J. A., Pennsylvania
Hennemuth, R., Massachusetts
Hogg, D. B., Mississippi
Innis, G. S., Utah
Janardan, K. G., Illinois
Johnson, D., Texas
Johnson, D. H., North Dakota
Jolly, G. M., Scotland
Kester, T., Belgium
Kie, J. G., California
Kobayashi, S., Japan
Kubicek, F., Czechoslovakia
Labovitz, M. L., Pennsylvania
Lamberti, G. A., California
Lasebikan, B. A., Nigeria
Lindahl, K. Q., California
Laurence, G. C., Rhode Island
Livingston, G. P., Texas
Ludwig, J. A., New Mexico
Ma, J. C. W., Texas
Macken, C. A., Minnesota
Marsden, M. A., Montana
Mason, R., Oregon
Matis, J. H., Texas
Matthews, G. A., Texas
Minello, T. J., Texas
Mizell, R. F., Mississippi
Monserud, R. A., Idaho
Myers, C. C., Illinois
Myers, R. A., Canada
Naveh, Z., Israel
Nebeker, T. E., Mississippi
Neyman, J., California
Norick, N. X., California
O'Neill, R. V., Tennessee
Ord, J. K., England

Overton, S., Oregon
Patil, G. P., Pennsylvania
Pennington, M. R., Massachusetts
Poole, R. W., Rhode Island
Pulley, P. E., Texas
Quinn, T. J., Washington
Rawson, C. B., Washington
Reynolds, J. F., North Carolina
Riggs, L. A., California
Robson, D. S., New York
Roman, J. R., New York
Rosenzweig, M. L., Arizona
Roughgarden, J., California
Roux, J. J. J., South Africa
Sanders, F., Tennessee
Sen, A. R., Canada
Serchuk, F. M., Massachusetts
Shoemaker, C., New York
Singh, K. P., India
Smith, W. K., Massachusetts
Solomon, D. L., New York
Southward, G. M., New Mexico
Stafford, S. G., New York
Steinhorst, R. K., Texas
Stenseth, N. C., Norway
Stiteler, W. M., New York
Stout, M. L., California
Stromberg, L. P., California
Taillie, C., Pennsylvania
Tracy, D. S., Canada
Usher, M. B., England
vanBiezen, J. B., Netherlands
Walter, G. G., Wisconsin
Waters, W. E., California
Wensel, L. C., California
West, I. F., New Zealand
Wiegert, R. G., Georgia
Williams, F. M., Pennsylvania
Wright, J. R., Alabama
Wu, Y. C., California
Yandell, B. S., California
Zweifel, J. R., California

SATELLITE B: PARMA   July 31-September 5, 1978

Arditi, R., France
Azzarita F., Italy
Balchen, J. G., Norway
Bargmann, R. E., Georgia
Barlow, N. D., England
Baxter, M. B., Australia
Behrens, J., Denmark
Beran, H. G., Austria
Berman, M., Maryland
Berryman, A. A., Washington
Boswell, M. T., Pennsylvania
Brambilla, C., Italy
Braumann, C. A., New York
Breitenecker, M., Austria
Buerk, R., West Germany
Callahan, C. A., Oregon
Cancela da Fonseca, J. P., France
Chieppa, M., Italy

Clark, W. G., Italy
Cobelli, C., Italy
Cooper, C., California
Curry, G. L., Texas
DeMichele, D. W., Texas
Derr, J., Iowa
Diggle, P. J., England
Drakides, C., France
Ebenhöh, W., West Germany
Engen, S., Norway
Feoli, E., Italy
Fiadeiro, P. M., Portugal
Fischlin, A., Switzerland
Framstad, E. B., Norway
Frohberg, K., Austria
Gallucci, V. F., Washington
Garcia-Moya, E., Mexico
Gatto, M., Italy

Geri, C., France
Giavelli, G., Italy
Ginzburg, L. R., New York
Gokhale, D. V., California
Goldstein, R. A., California
Granero Porati, M. I., Italy
Greve, W., West Germany
Grosslein, M. D., Massachusetts
Grümm, H. R., Austria
Gulland, J. A., Italy
Gutierrez, A. P., California
Gydesen, H., Denmark
Hanski, I., England
Hanson, B. J., Utah
Hau, B., West Germany
Helgason, T., Iceland
Hendrickson, J. A., Pennsylvania
Hengeveld, R., Netherlands

Hennemuth, R. C., Massachusetts
Hoff, J. M., Norway
Holling, C. S., Canada
Hotz, M. C. B., Belgium
Jancey, R. C., Canada
Kooijman, S., Netherlands
Lamont, B. B., Australia
Levi, D., Italy
Liu, C. J., Kentucky
Marshall, W., Canada
Martin, F. W., Maryland
Matis, J. H., Texas
Menozzi, P., Italy
Meyer, J. A., France
Mohn, R. K., Canada
Mosimann, J., Maryland
Naveh, Z., Israel
Noy-Meir, I., Israel
Olivieri-Barra, S. T., Belgium
Ord, J. K., England
Orloci, L., Canada
Pacchetti, G., Italy
Pagani, L., Italy
Patil, G. P., Pennsylvania
Patten, B. C., Georgia

Pennington, M. R., Massachusetts
Policello, G. E., Ohio
Pospahala, R. S., Maryland
Purdue, P., Kentucky
Radler, K., West Germany
Ramsey, F. L., Oregon
Reyment, R. A., Sweden
Reyna Robles, R., Mexico
Rinaldi, S., Italy
Robson, D. S., New York
Rossi, O., Italy
Russek, E., Washington
Russo, A. R., Hawaii
Sadasivan, G., India
Schaefer, R., France
Shoemaker, C. A., New York
Show, I. T., California
Shuter, B. J., Canada
Simberloff, D., Florida
Slagstad, D., Norway
Smetacek, V. S., West Germany
Smith, W. K., Massachusetts
Sokal, R. R., New York
Soliani, L., Italy
Solomon, D. L., New York

Spremann, K., West Germany
Steinhorst, R. K., Idaho
Stenseth, N. C., Norway
Stiteler, W. M., New York
Subrahmanyam, C. B., Florida
Szöcs, Z., Hungary
Taillie, C., Pennsylvania
terBraak, C. J. F., Netherlands
Torrez, W. C., New Mexico
Tracy, D. S., Canada
Tursi, A., Italy
Vale, C., Portugal
van Biezen, J. B., Netherlands
Vazzana, C., Italy
Wahl, E., West Germany
Walker, B. H., England
Walter, G. G., Wisconsin
Walters, C. J., Canada
Warren, W. G., Canada
Waters, W. E., California
White, G. C., New Mexico
Wise, M. E., Netherlands
Zanni, R., Italy

SATELLITE C: JERUSALEM   September 7-September 15, 1978

Austin, M. P., Australia
Baxter, M. B., Australia
Berthet, P., Belgium
Carleton, T. J., Canada
Eisen, P. A., New York
Galluci, V., Washington
Goldstein, R. A., California
Goodman, D., California
Hanski, I., Finland
Hendrickson, J. A., Pennsylvania
Hengeveld, R., Netherlands

Hennemuth, R. C., Massachusetts
Hirsch, A., Washington, DC
Kempton, R. A., England
Naveh, Z., Israel
Noy-Meir, I., Israel
Odum, H. T., Florida
O'Neill, R. V., Tennessee
Patil, G. P., Pennsylvania
Quinn, T. J., Washington
Resh, V. H., California
Rossi, O., Italy

Rosenzweig, M. L., Arizona
Safriel, U., Israel
Sheshinski, R., Israel
Simberloff, D. S., Florida
Sokal, R. R., New York
Solem, J. O., Norway
Stenseth, N. C., Norway
Subrahmanyam, C. B., Florida
Torrez, W., New Mexico
Waters, W. E., California
Whittaker, R. H., New York

AUTHORS NOT LISTED ABOVE

Adams, J. E., Texas
Barber, M. C., Georgia
Barreto, M., Brazil
Batcheler, C. L., New Zealand
Beuter, K. J., West Germany
Bitz, D. W., Rhode Island
Brown, B. E., Massachusetts
Brown, G. C., Kentucky
Brown-Leger, L. S., Massachusetts
Chaim, S., Israel
Chardy, P. France
Clark, G. M., Ohio
Condra, C., California
Connor, E. F., Florida
Costa, H., Brazil
Dale, M., Australia
De LaSalle, P., France
Duek, J. L., Arizona
Elwood, J. W., Tennessee
Ferris, J. M., Indiana
Ferris, V. R., Indiana
Finn, J. T., Massachusetts
Foltz, J. L., Texas
Gibson, V. R., Massachusetts
Giddings, J. M., Tennessee

Gittins, R., Australia
Godron, M., France
Grassle, J. F., Massachusetts
Green, R., California
Halbach, U., West Germany
Halfon, E., Canada
Helthshe, J. F., Rhode Island
Hildebrand, S. G., Tennessee
Hogeweg, P., Netherlands
Iwao, S., Japan
Jacur, G. R., Italy
Kaesler, R. L., Kansas
Kravitz, D., Massachusetts
Kruczynski, W. L., Florida
Lackey, R., Virginia
Laurec, A., France
Lepschy, A., Italy
Malley, J. D., Maryland
Marcus, A. H., Washington
Matern, B., Sweden
McCune, E. D., Texas
Moncreiff, R., California
Moroni, A., Italy
Nichols, J. D., Maryland
O'Connor, J. S., New York

Paloheimo, J. E., Canada
Perrin, S., California
Pickford, S. G., Washington
Plowright, R. C., Canada
Podani, J., Hungary
Ratnaparkhi, M. V., Pennsylvania
Rescigno, A., Canada
Rickaert, M., France
Rohde, C., Maryland
Rosen, R., Canada
Scott, E., California
Scott, J. M., Hawaii
Seber, G. A. F., New Zealand
Siri, E., Italy
Skalski, J. R., Washington
Stehman, S., Pennsylvania
Steinberger, E. H., Israel
Swartzman, G., Washington
Taylor, C. E., California
Taylor, L. R., England
Tiwari, J. L., California
Watson, R. M., Scotland
Wehrly, T. E., Texas
Wigley, R. L., Massachusetts
Wissel, C., West Germany
Yang, M., Florida

Foreword

The Second International Congress of Ecology was held in Jerusalem during September 1978. In this connection, a Satellite Program in Statistical Ecology was organized by the International Statistical Ecology Program (ISEP) during 1977 and 1978. The emphasis was on research, review, and exposition concerned with the interface between quantitative ecology and relevant quantitative methods. Both theory and application of ecology and ecometrics received attention. The program consisted of instructional coursework, seminar series, thematic research conferences, and collaborative research workshops.

The 1977 and 1978 Satellite Program consisted of NATO Advanced Study Institutes at College Station in Texas, Berkeley in California, and Parma in Italy; NATO Advanced Research Institute at Parma; ISEP Research Conferences, Seminars, and Workshops at College Station, Berkeley, Parma, and Jerusalem; and a Research Conference at Jerusalem.

The Satellite Program has been supported by NATO Advanced Study Institutes Program; NATO Ecosciences Program; National Marine Fisheries Service, USA; Environmental Protection Agency, USA; Fish and Wildlife Service, USA; Army Research Office, USA; The Pennsylvania State University; The Texas A&M University; The University of California at Berkeley; Universita degli Studi, Parma; Consiglio Nazionale delle Ricerche, Italy; Ministero dei Lavori Pubblici, Affari Esteri, e Pubblica Istruzione, Italy; Societa Italiana di Statistica, Italy; Societa Italiana di Ecologia, Italy; Communita Economica Europea; and the participants and their home institutions and organizations.

Research papers and research-review-expositions were specially prepared for the program by concerned experts and expositors. These materials have been refereed and revised, and are now available in a series of ten edited volumes.

HISTORICAL BACKGROUND

The First International Symposium on Statistical Ecology was held in 1969 at Yale University with support from the Ford Foundation and the US Forest Service. The three symposium co-chairmen (G. P. Patil, E. C. Pielou, and W. E. Waters) represented the fields of statistics, theoretical ecology, and applied ecology. The program was well attended, and it provided a broad picture of where statistics and ecology stood relative to each other. While effort was apparent, communication between the two disciplines was inadequate.

It was clear that a focal forum was necessary to discuss and develop a constructive interface between quantifiable problems in ecology and relevant quantitative methods. As a partial solution to fill this need at professional organizations' level, the director of the symposium (G. P. Patil) made certain recommendations to the Presidents of the International Association for Ecology (A. D. Hasler), the International Statistical Institute (W. G. Cochran), and the International Biometric Society (P. Armitage). The International Association for Ecology (INTECOL) took a timely step in creating a section in the organization, namely, the statistical ecology section. The three societies together took a timely step in setting up a liaison committee on statistical ecology. The INTECOL Section and the Liaison Committee together developed the International Statistical Ecology Program. Since its inception in 1970, ISEP (as it has come to be known) has put emphasis on identifying the interdisciplinary needs of statistics and ecology at advanced instructional levels, and also at research conference and workshop levels.

The First Advanced Institute on Statistical Ecology in the United States was organized at Penn State for six weeks in 1972 with support from the US National Science Foundation, the US Forest Service, and the Mathematical Social Sciences Board. The participants of the Institute are all enjoying the benefits of their fruitful participation. With support from the UNESCO program of Man and Biosphere, a six month program was held in Venezuela for participants from Latin America in 1974 under the direction of Jorge Rabinovich, himself a participant in the 1972 Institute. With some initiatives from ISEP, special statistical ecology sessions have been held at the international conferences of the International Statistical Institute and the Biometric Society.

While plans were being made for the Second International Congress of Ecology, the then Secretary General and current President of INTECOL (G. A. Knox), and the ISEP chairman (G. P. Patil) discussed the need and the timeliness of a program in statistical ecology. The Satellite Program in Statistical Ecology took its final shape from this beginning under the care and concern of its director, advisors, coordinators, and sponsors.

SCIENTIFIC BACKGROUND AND PURPOSE

The perceptions of Skellam and Train quoted on page v in this volume recapitulate the cautions, inspirations, and objectives responsible for the Satellite Program. The rigorous formulation of a quantitative scientific concept requires and in a sense creates empirically measurable quantities. Conversely, the scientific validity of the concept is totally dependent upon the measured values of those quantities. This is a capsule version of the feedback process known as the "scientific method." In crude modern terms, we might label the first procedure "modeling" and the second "curve-fitting." For reasons of complexity, historical accident, or whatever, these mutually dependent, complementary components have never become firmly integrated within ecology and its application to environmental studies.

Both procedures involve forms of mathematics: In the "modeling" process, the mathematics is used *relationally* — as a system of logic to ensure rigor and clarity of reasoning. This is in the historical tradition of Volterra and Lotka. Validation is most often based on *qualitative* agreement: the right trend, or the correct shape of the curve. This is as it should be, especially for broad general theories: it is the ideas and understanding that count, not so much the quantitative detail.

In the "curve-fitting" tradition, the mathematics is used *numerically* — as a system for precise measurement and prediction. Validation is most often based on *quantitative* agreement: n-place accuracy, or minimum uncertainty. This is also as it should be, especially for application and management: it is the forecast and ability to act confidently that count, not so much the underlying concept.

It need not be taken as a sign of "physics envy" to assert that as a science matures, these two processes must converge — more quantification must be used in concept validation and more concepts must be incorporated into the methodology of quantification.

The purpose of the Satellite Program is to encourage that convergence within the science of ecology and promote its application in the study of the environment and environmental stress. Monitoring and assessment activities to be meaningful and defensible need: (i) a conceptual and philosophical basis, (ii) a theoretical framework, (iii) methodological support, (iv) a technological toolbox, and (v) administrative management. The ultimate purpose of the program is to help identify and integrate the specifics of these important factors responsible for protecting the environment.

We take as our theme the better melding of fundamental ecological concepts with rigorous empirical quantification. The overall result should be progress toward a stronger body of general ecological theory and practice.

PLANNING AND ORGANIZATION

The realization of any program of this nature and dimension often fails to fully meet the initial expectations and objectives of the organizers. Factors that are both logistic and psychological in nature tend to contribute to this general experience. Logistic difficulties include optimality problems for time and location. Other difficulties which must be attended to involve conflicting attitudes towards the importance of individual contributions to the proceedings.

We tried to cope with these problems by seeking active advice from a number of participants and special advisors. The advice we received was immensely helpful in guiding our selection of the best experts in the field to achieve as representative and balanced a coverage as possible. Simultaneously, the editors together with the referees took a rather critical and constructive attitude from initial to final stages of preparation of papers by offering specific suggestions concerning the suitability, and also the structure, content and size. These efforts of coordination and revision were intensified through editorial sessions at the program itself as a necessary step for the benefit of both the general readership and the participants. It is our pleasure to record with appreciation the spontaneous cooperation of the participants. Everyone went by scientific interests often at the expense of personal preferences. The program atmosphere became truly creative and friendly, and this remarkable development contributed to the maximal cohesion of the program and its proceedings within the limited time period available.

In retrospect, our goals were perhaps ambitious! We had close to 350 lectures and discussions during 50 days in the middle of the summer seasons of 1977 and 1978. For several reasons, we decided that an overworked program was to be preferred to a leisurely one. First of all, gatherings of such dimension are possible only every 5-10 years. Secondly, the previous meetings of this nature occurred some 5-10 years back, and the subject area of statistical ecology had witnessed substantial growth in this time. Thirdly, but most importantly, was the overwhelming response from potential participants, many of whom were to come across the continents!

Satellite A at College Station, Texas, and at Berkeley, California covered a four week period during July 18-August 13, 1977 and had 125 participants. Satellite B at

Parma, Italy had 130 participants spread over a six week period during July 31-September 5, 1978. Satellite C at Jerusalem, Israel had 35 participants. Approximately, one-third of the participants were graduate students, one-half were university faculty, and one-third were agency scientists. Approximately, one-third of the participants had an affiliation with one mathematical science or another, one-half an affiliation with one environmental science or the other, and one-quarter had an affiliation with one environmental management program or another. Thus the group was a good mix of great variety contributing to the effectiveness of the program. Not only what one heard was enlightening, but what one over-heard was equally enlightening!

Professors G. P. Patil, Paul Berthet, J. K. Ord, and Charles Taillie served as scientific directors of the program with Professor Patil assuming the responsibility of its direction from its conception to its conclusion. The inaugural speakers were: Professor H. O. Hartley at College Station, Professor J. Neyman at Berkeley, Professor C. S. Holling at Parma, and Professor G. P. Patil at Jerusalem.

SCIENTIFIC CONTENT AND PUBLICATION

The following summary information on the subjects and corresponding coordinators of the satellite program may be of some interest. The details of each subject and its publication volume are reported elsewhere.

Statistical Distributions in Ecological Work: J. K. Ord, G. P. Patil, and C. Taillie.

Spatial and Temporal Analysis in Ecology: R. M. Cormack and J. K. Ord.

Quantitative Population Dynamics: D. G. Chapman and V. Gallucci.

Sampling Biological Populations: R. M. Cormack, G. P. Patil, and D. S. Robson.

Ecological Diversity in Theory and Practice: J. F. Grassle, G. P. Patil, W. K. Smith, and C. Taillie.

Multivariate Methods in Ecological Work: L. Orloci, C. R. Rao, and W. M. Stiteler.

Systems Analysis of Ecosystems: G. S. Innis and R. V. O'Neill.

Compartmental Analysis of Ecosystem Models: J. H. Matis, B. C. Patten, and G. C. White.

Environmental Biomonitoring, Assessment, Prediction, and Management-Certain Case Studies and Related Quantitative Issues: J. Cairns, Jr., G. P. Patil and W. E. Waters.

Contemporary Quantitative Ecology and Related Ecometrics: G. P. Patil and M. L. Rosenzweig.

Three more subjects were organized in the program.

Scientific Modeling and Quantitative Thinking with Examples in Ecology: G. P. Patil, D. S. Simberloff, and D. L. Solomon.

Conceptual Foundations of Ecological Theory and Applications: M. B. Usher and F. M. Williams.

Optimizations in Ecological Theory and Management: C. S. Holling and C. Shoemaker.

It would be fruitful to reorganize these subjects and add a few more for the next satellite program when it occurs.

It should be mentioned here that the close coordination and cooperation between the coordinators and the authors/speakers of potential contributions to the Proceedings (which are particularly intensive during the satellite program) paid themselves handsomely when the editors were confronted with the technical work related to the publications after the close of the program at Jerusalem. It is therefore very satisfying to report that the edited research papers and research-review-expositions prepared for the program are ready for distribution within 12 months of the conclusion of the program. For purposes of convenience, the contributions are organized in ten volumes in the Statistical Ecology Series published by the International Co-operative Publishing House. Altogether, they consist of an estimated 4,000 pages of research, review, and exposition, in addition to this common foreword in each followed by individual volume introductions. Subject and author indexes are also prepared at the end. Every effort has been made to keep the coverage of the volumes close to their individual titles. May this ten volume set in its own modest way provide an example of synergism.

FUTURE DIRECTIONS

We wish there was no need for a program of this nature and dimension. It would be ideal if the needs of an interdisciplinary program were satisfactorily met in the existing institutions. Unfortunately, universities and governmental agencies have not been able to find effective ways to foster healthy interdisciplinary programs. The individuals attempting to do something in this direction tend to feel disheartened or disillusioned.

The satellite-like-programs help create and sustain enthusiasm, inward strength, and working efficiency of those who desire to meet a contemporary social need in the form of some interdisciplinary work. It should be only proper and rewarding for everyone involved that such programs are planned from time to time.

Plans are being made for a satellite program in conjunction with the next Biennial Conference of the International Statistical Institute and with the next International Congress of Ecology. Care should be exercised that the next program not become a mere replica of the present one, however successful it has been. Instead, the next program should be organized so that it helps further the evolution of statistical ecology as a productive field.

The next program is being discussed in terms of subject area groups. Each subject group is to have a coordinator assisted by small committees, such as a program committee, a research committee, an annual review committee, a journal committee, and an education committee. This approach is expected to respond to the need for a journal on statistical ecology, and also to the need of bringing out well planned annual review volumes. The education committee would formulate plans for timely manuals, modules, and monographs. Interested readers may feel free to communicate their ideas and interests to those involved in planning the next program. With mutual goodwill and support, we shall have met a timely need for today's science, technology, and society.

July 1979   G. P. Patil

Program Acknowledgments

For any program to be successful, mutual understanding and support among all participants are essential in directions ranging from critical to constructive and from cautious to enthusiastic. The present program is grateful to the members of the ISEP Advisory Board, and to the referees, editors, coordinators, advisors, sponsors and the participants for their timely advice and support.

The success of the program was due, in no small measure, to the endeavors of the Local Arrangements Chairmen: J. H. Matis and C. E. Gates at College Station, W. E. Waters at Berkeley, O. Rossi and R. Zanni at Parma, and I. Noy-Meir at Jerusalem. We thank them for their hospitality and support.

And finally those who have assisted with the arduous task of preparing the materials for publication. Barbara Alles has been an ever cheerful and industrious secretary in the face of every adversity. Charles Taillie managed both scientific and non-scientific aspects. Bharat Kapur copyedited and proofread. Bonnie Burris, Bonnie Henninger, and Sandy Rothrock prepared the final versions of the manuscripts. Marllyn Boswell helped with the subject and author indexes. So did Bharat Kapur, Satish Patil, and Rani Venkataramani.

All of these nice people have done a fine job indeed. To all of them, our sincere thanks.

July 1979   G. P. Patil

Reviewers of Manuscripts

With appreciation and gratitude, the program acknowledges the valuable services of the following referees who have served as reviewers of manuscripts submitted to the program for possible publication. The editors thank the reviewers for their critical and constructive reviews.

J. Balchen
University of Trondheim

R. E. Bargmann
University of Georgia

C. Chatfield
University of Bath

R. Colwell
University of California

A. Berryman
Washington State University

J. Beyer
Danish Institute of Fisheries and Marine Research

M. T. Boswell
Pennsylvania State University

C. Braumann
Instituto Universitario de Evora

M. A. Buzas
Smithsonian Institution

J. Cairns, Jr.
Virginia Polytechnic Institute and State University

C. Cobelli
Laboratorio per Ricerche di Dinamica dei Sistemi e di Bioingegneria

R. Cormack
University of St. Andrews

B. Coull
University of South Carolina

M. B. Dale
Commonwealth Scientific and Industrial Research Organization

A. P. Dawid
The City University, London

B. Dennis
Pennsylvania State University

F. Diemer
Office of Naval Research

P. Diggle
University of Newcastle upon Type

P. G. Doucet
Vrije University

J. E. Dunn
University of Arkansas

L. Eberhardt
Battelle Pacific Northwest Laboratories

S. Engen
University of Trondheim

R. G. Flores
Fundacao Instituto Brasileiro de Geografia e Estatistica

E. D. Ford
Institute of Terrestrial Ecology

C. E. Gates
Texas A&M University

K. B. Gerald
Rockwell International

L. Ginzburg
State University of New York

J. C. Gittins
University of Oxford

N. Glass
Environmental Protection Agency

D. V. Gokhale
University of California

R. Goldstein
Electric Power Research Institute

M. Granero-Porati
Istituto di Fisica, Parma

J. F. Grassle
Woods Hole Oceanographic Institution

J. C. Griffiths
Pennsylvania State University

I. Hanski
University of Oxford

W. Harkness
Pennsylvania State University

T. Helgason
University of Iceland

J. Hendrickson
Academy of Natural Sciences

R. C. Hennemuth
National Marine Fisheries Service

D. Hildebrand
University of Pennsylvania

A. Hirsch
Fish and Wildlife Service

P. Holgate
Birkbeck College

H. Horn
Princeton University

G. Innis
Utah State University

I. James
Commonwealth Scientific and Industrial Research Organization

K. G. Janardan
Sangamon State University

G. M. Jolly
University of Edinburgh

C. Jones
Woods Hole Oceanographic Institution

R. Kaesler
University of Kansas

R. A. Kempton
Plant Breeding Institute, Cambridge

J. Kirkley
National Marine Fisheries Service

G. Knott
National Institutes of Health

S. Kotz
Temple University

A. M. Kshirsagar
University of Michigan

S. Kullback
The George Washington University

R. C. Lewontin
Harvard University

B. F. J. Manly
University of Otago

A. H. Marcus
Washington State University

F. Martin
Patuxent Wildlife Research Center

J. H. Matis
Texas A&M University

J. R. McBride
University of California

D. Mollison
Heriot-Wett University

R. V. O'Neill
Oak Ridge National Laboratory

J. Newton
University of St. Andrews

J. K. Ord
University of Warwick

L. Orloci
University of Western Ontario

G. P. Patil
Pennsylvania State University

B. C. Patten
University of Georgia

M. Pennington
National Marine Fisheries Service

S. Pimm
Texas Tech University

K. H. Pollock
University of Reading

R. W. Poole
Brown University

R. S. Pospahala
Patuxent Wildlife Research Center

E. Preston
Environmental Protection Agency

F. Preston
Preston Laboratories

P. Purdue
University of Kentucky

F. L. Ramsey
Oregon State University

C. R. Rao
Indian Statistical Institute

P. A. Rauch
University of California

E. Renshaw
University of Edinburgh

R. Reyment
Uppsala University

D. S. Robson
Cornell University

M. L. Rosenzweig
University of Arizona

O. Rossi
University of Parma

W. E. Schaaf
National Marine Fisheries Service

T. Schopf
University of Chicago

H. T. Schreuder
Forest Service

G. A. F. Seber
University of Auckland

J. Sepkoski
University of Rochester

I. T. Show, Jr.
Science Applications, Inc.

D. Simberloff
Florida State University

D. B. Siniff
University of Minnesota

W. K. Smith
Woods Hole Oceanographic Institution

R. R. Sokal
State University of New York

G. M. Southward
New Mexico State University

S. Stehman
Pennsylvania State University

R. K. Steinhorst
University of Idaho

W. M. Stiteler
Syracuse University

P. Switzer
Stanford University

C. Taillie
International Statistical Ecology Program

C. E. Taylor
University of California

C. Tsokos
University of South Florida

E. Ursin
Danish Institute of Fisheries and Marine Research

D. Vaughan
Oak Ridge National Laboratory

G. G. Walter
University of Wisconsin

W. G. Warren
Western Forest Products Laboratory

W. E. Waters
University of California

S. D. Webb
University of Florida

G. C. White
Los Alamos Scientific Laboratory

R. H. Whittaker
Cornell University

M. E. Wise
Leiden University

S. Zahl
University of Connecticut

J. Zweifel
National Marine Fisheries Service

Contents of Edited Volumes

Determination of Plant Species Diversity in Mediterranean Shrub and Woodland Along Environmental Gradients. V. RESH, Biomonitoring, Species Diversity Indices, and Taxonomy. J. SOLEM, A Comparison of Species Diversity Indices in Trichoptera Communities. W. K. SMITH, V. R. GIBSON, L. S. BROWN-LEGER, and J. F. GRASSLE, Diversity as an Indicator of Pollution: Cautionary Results from Microcosm Experiments. C. B. SUBRAHMANYAM and W. L. KRUCZYNSKI, Colonization of Polychaetous Annelids in the Intertidal Zone of a Dredged Material Island in North Florida. C. E. TAYLOR and C. CONDRA, Competitor Diversity and Chromosomal Variation in Drosophia Pseudoobscura. B. DENNIS and O. ROSSI, Community Composition and Diversity Analysis in a Marine Zooplankton Survey.

Bibliography: B. DENNIS, G. P. PATIL, O. ROSSI, S. STEHMAN, and C. TAILLIE, A Bibliography of Literature on Ecological Diversity and Related Methodology.

MULTIVARIATE METHODS IN ECOLOGICAL WORK
L. Orloci, C. R. Rao, and W. M. Stiteler (editors)   **400 pp. approx.**
R. BARGMANN, Structural Analysis of Singular Matrices Using Union Intersection Statistics with Applications in Ecology. M. DALE, On Linguistic Approaches to Ecosystems and Their Classification. D. V. GOKHALE, Analysis of Ecological Frequency Data: Certain Case Studies. J. HENDRICKSON, Examples of Discrete Multivariate Methods in Ecological Work. R. HENGEVELD and P. HOGEWEG, Cluster Analysis of the Distribution Patterns of Dutch Carabid Species. R. JANCEY, Species Weighting. A. LAUREC, P. CHARDY, P. DE LASALLE, and M. RICKAERT, Use of Dual Structures in Inertia Analysis: Ecological Implications. J. MOSIMANN and J. D. MALLEY, Size and Shape Analysis. L. ORLOCI, Non-Linear Data Structure and Their Description. J. PODANI, A Generalized Strategy of Homogeneity-Optimizing Hierarchical Classificatory Methods. R. REYMENT, Multivariate Analysis in Statistical Paleoecology. E. SCOTT, Spurious Correlation. W. K. SMITH, D. KRAVITZ, and J. F. GRASSLE, Confidence Intervals for Similarity Measures Using the Two Sample Jackknife. R. K. STEINHORST, Analysis of Niche Overlap. W. M. STITELER, Multivariate Statistics with Applications in Statistical Ecology. Z. SZOCS, New Computer Oriented Methods for Structural Investigation of Natural and Simulated Vegetation Patterns. B. LAMONT and K. J. GRANT, A Comparison of Twenty Measures of Site Dissimilarity. R. GITTINS, Ecological Applications of Canonical Analysis.

SPATIAL AND TEMPORAL ANALYSIS IN ECOLOGY
R. M. Cormack and J. K. Ord (editors)   **400 pp. approx.**
J. K. ORD, Time-Series and Spatial Patterns in Ecology. P. J. DIGGLE, Statistical Methods for Spatial Point Patterns in Ecology. R. M. CORMACK, Spatial Aspects of Competition Between Individuals. R. W. POOLE, The Statistical Prediction of the Fluctuations in Abundance in Nicholson's Sheep Blowfly Experiments. W. G. WARREN and C. L. BATCHELER, The Density of Spatial Patterns: Robust Estimation Through Distance Methods. B. MATERN, The Analysis of Ecological Maps as Mosaics. J. A. LUDWIG, A Test of Different Quadrat Variance Methods for the Analysis of Spatial Pattern. S. A. L. M. KOOIJMAN, The Description of Point Patterns. R. HENGEVELD, The Analysis of Spatial Patterns of Some Ground Beetles (Col. Carabidae).

SYSTEMS ANALYSIS OF ECOSYSTEMS
G. S. Innis and R. V. O'Neill (editors)   **425 pp. approx.**
R. K. STEINHORST, Stochastic Difference Equation Models of Biological Systems. R. V. O'NEILL, Natural Variability as a Source of Error in Model Predictions. R. K. STEINHORST, Parameter Identifiability, Validation and Sensitivity Analysis of Large System Models. R. V. O'NEILL, Transmutation Across Hierarchical Levels. R. V. O'NEILL, J. W. ELWOOD, and S. G. HILDEBRAND, Theoretical Implications of Spatial Heterogeneity in Stream Ecosystems. R. V. O'NEILL and J. M. GIDDINGS, Population Interactions and Ecosystem Function: Plankton Competition and Community Production. J. P. CANCELA DA FONSECA, Species Colonization Models of Temporary Ecosystems Habitats. E. HALFON, Computer-Based Development of Large Scale Ecological Models: Problems and Prospects. G. S. INNIS, A Spiral Approach to Ecosystem Simulation.

COMPARTMENTAL ANALYSIS OF ECOSYSTEM MODELS
J. H. Matis, B. C. Patten, and G. C. White (editors)   **400 pp. approx.**
Applications of Compartmental Analysis to Ecosystem Modeling: R. V. O'NEILL, A Review of Linear Compartmental Analysis in Ecosystem Science. G. G. WALTER, A Compartmental Model of a Marine Ecosystem. M. C. BARBER, B. C. PATTEN, and J. T. FINN, Review and Evaluation of Input-Output Flow Analysis for Ecological Applications. I. T. SHOW, JR., An Application of Compartmental Models to Meso-scale Marine Ecosystems.

Identifiability and Statistical Estimation of Parameters in Compartmental Models: C. COBELLI, A. LEPSCHY, G. R. JACUR, Identification Experiments and Identifiability Criteria for Compartmental Systems. M. BERMAN, Simulation, Data Analysis, and Modeling with the SAAM Computer Program. G. C. WHITE and G. M. CLARK. Estimation of Parameters for Stochastic Compartment Models. R. E. BARGMANN, Statistical Estimation and Computational Algorithms in Compartmental Analysis for Incomplete Sets of Observations.

Stochastic Approaches to the Compartmental Modeling of Ecosystems: J. L. TIWARI, A Modeling Approach Based on Stochastic Differential Equations, the Principle of Maximum Entropy, and Bayesian Inference for Parameters. J. H. MATIS and T. E. WEHRLY, An Approach to a Compartmental Model with Multiple Sources of Stochasticity for Modeling Ecological Systems. P. PURDUE, Stochastic Compartmental Models: A Review of the Mathematical Theory with Ecological Applications. A. H. MARCUS, Semi-Markov Compartmental Models in Ecology and Environmental Health. M. E. WISE, The Need for Rethinking on Both Compartments and Modeling.

Mathematical Analysis of Compartmental Structures: G. G. WALTER, Compartmental Models, Digraphs, and Markov Chains. K. B. GERALD and J. H. MATIS, On the Cumulants of Some Stochastic Compartmental Models Applied to Ecological Systems. A. RESCIGNO, The Two-variable Operational Calculus in the Construction of Compartmental Ecological Models.

ENVIRONMENTAL BIOMONITORING, ASSESSMENT, PREDICTION, AND MANAGEMENT — CERTAIN CASE STUDIES AND RELATED QUANTITATIVE ISSUES
J. Cairns, Jr., G. P. Patil, and W. E. Waters (editors) 450 pp. approx.
Biomonitoring: J. CAIRNS, JR., Biological Monitoring — Concept and Scope. Z. NAVEH, E. H. STEINBERGER, and S. CHAIM, Use of Bio-Indicators for Monitoring of Air Pollution by Fluor, Ozone and Sulfur Dioxide.

Environmental assessment and prediction: G. P. PATIL, C. TAILLIE, and R. L. WIGLEY, Transect Sampling Methods and Their Application to the Deep-Sea Red Crab. W. E. WATERS, Biomonitoring, Assessment, and Prediction in Forest Pest Management Systems. C. A. CALLAHAN, V. R. FERRIS, and J. M. FERRIS, The Ordination of Aquatic Nematode Communities as Affected by Stream Water Quality. R. A. GOLDSTEIN, Development and Implementation of a Research Program on Ecological Assessment of the Impact of Thermal Power Plant Cooling Systems on Aquatic Environments.

Environmental Management: A. HIRSCH, Ecological Information and Technology Transfer, R. H. GILES, JR., Modeling Decisions or Ecological Systems. R. LACKEY, Appliction of Renewable Natural Resource Modeling in Public Decision-Making Process. F. MARTIN, R. S. POSPAHALA, and J. D. NICHOLS, Assessment and Population Management of North American Migratory Birds. J. E. PALOHEIMO and R. C. PLOWRIGHT, Bioenergetics, Population Growth and Fisheries Management. Z. NAVEH, A Model of Multiple-Use Management Strategies of Marginal and Untillable Mediterranean Upland Ecosystems.

Case Studies and Quantitative Issues: M. D. GROSSLEIN, R. C. HENNEMUTH, and B. E. BROWN, Research, Assessment, and Management of a Marine Ecosystem in the Northwest Atlantic — A Case Study. J. NEYMAN, Two Interesting Ecological Problems Demanding Statistical Treatment. B. DENNIS, G. P. PATIL, and O. ROSSI, The Sensitivity of Ecological Diversity Indices to the Presence of Pollutants in Aquatic Communities. D. SIMBERLOFF, Constraints on Community Structure During Colonization.

CONTEMPORARY QUANTITATIVE ECOLOGY AND RELATED ECOMETRICS
G. P. Patil and M. L. Rosenzweig (editors) 725 pp. approx.
Community Structure and Diversity: R. A. KEMPTON and L. R. TAYLOR, Some Observations on the Yearly Variability of Species Abundance at a Site and the Consistency of Measures of Diversity. G. P. PATIL and C. TAILLIE, A Study of Diversity Profiles and Orderings for a Bird Community in the Vicinity of Colstrip, Montana. M. L. ROSENZWEIG, Three Probable Evolutionary Causes for Habitat Selection. O. ROSSI, G. GIAVELLI, A. MORONI, and E. SIRI, Statistical Analysis of the Zooplankton Species Diversity of Lakes Placed Along a Gradient. S. KOBAYASHI, Another Model of the Species Rank-Abundance Relation for a Delimited Community. M. L. ROSENZWEIG and J. L. DUEK, Species Diversity and Turnover in an Ordovician Marine Invertebrate Assemblage.

Patterns and Interpretations: D. S. SIMBERLOFF and E. F. CONNOR, Q-Mode and R-Mode Analyses of Biogeographic Distributions: Null Hypotheses Based on Random Colonization. D. GOODMAN,

Applications of Eigenvector Analysis in the Resolution of Spectral Pattern in Spatial and Temporal Ecological Sequences. R. H. WHITTAKER and Z. NAVEH, Analysis of Two-Phase Patterns. R. R. SOKAL, Ecological Parameters Infered From Spatial Correlograms. M. P. AUSTIN, Current Approaches to the Non-Linearity Problem in Vegetation Analysis. M. GODRON, A Probabilistic Computation for the Research of "Optimal Cuts" in Vegetation Studies. S. IWAO, The m*-m Method for Analyzing Distribution Patterns of Single- and Mixed-Species Populations.

Modeling and Ecosystems Modeling: D. L. SOLOMON, On a Paradigm for Mathematical Modeling. R. W. POOLE, Ecological Models and Stochastic-Deterministic Question. R. ROSEN, On the Role of Time and Interaction in Ecosystem Modelling. E. HALFON, On the Parameter Structure of a Large Scale Ecological Model. G. SWARTZMAN, Evaluation of Ecological Simulation Models. R. WIEGERT, Modeling Coastal, Estuarine and Marsh Ecosystems: State-of-the-Art.

Statistical Methodology and Sampling: J. DERR and J. K. ORD, Field Estimates of Insect Colonization, II. J. A. HENDRICKSON, JR., Analyses of Species Occurrences in Community, Continuum and Biomonitoring Studies. R. SHESHINSKI, Interpolation in the Plane. The Robustness to Misspecified Correlation Models and Different Trend Functions. W. C. TORREZ, The Effect of Random Selective Intensities on Fixation Probabilities. R. GREEN, A Graph Theoretical Test to Detect Interference in Selecting Nest Sites. I. NOY-MEIR, Graphical Models and Methods in Ecology. T. J. QUINN, The Effects of School Structure on Line Transect Estimators of Abundance. J. W. HAZARD and S. G. PICKFORD, Line Intersect Sampling of Forest Residue.

Applied Statistical Ecology: R. C. HENNEMUTH, Man as Predator. V. F. GALLUCCI, On Assessing Population Characteristics of Migratory Marine Animals. P. A. EISEN and J. S. O'CONNOR, MESA Contributions to Sampling in Marine Environments. W. E. WATERS and V. H. RESH, Ecological and Statistical Features of Sampling Insect Populations in Forest and Aquatic Environments. R. L. KAESLER, Statistical Paleoecology: Problems and Perspectives. S. A. L. M. KOOIJMAN and R. HENGEVELD, The Description of Non-Linear Relationship Between Some Carabid Beetles and Environmental Factors. P. E. PULLEY, R. N. COULSON, and J. L. FOLTZ, Sampling Bark Beetle Populations for Abundance.

A Bibliography: B. DENNIS, G. P. PATIL, M. V. RATNAPARKHI, and S. STEHMAN, A Bibliography of Selected Books on Quantitative Ecology and Related Ecometrics.

QUANTITATIVE POPULATION DYNAMICS
D. G. Chapman and V. F. Gallucci, editors   **300 pp. approx.**

D. G. CHAPMAN and V. F. GALLUCCI, Population Dynamics Models and Applications. J. G. BALCHEN, Mathematical and Numerical Modeling of Physical and Biological Processes in the Barents Sea. A. BERRYMAN and G. C. BROWN, The Habitat Equation: A Fundamental Concept in Population Dynamics. C. A. BRAUMANN, Population Adaptation to a "Noisy" Environment: Stochastic Analogs of Some Deterministic Models. L. GINZBERG, Genetic Adaptation and Models of Population Dynamics. M. I. GRANERO-PORATI, Stability of Model Systems Describing Prey-predator Communities. K. J. BEUTER, C. WISSEL, and U. HALBACH, Correlation and Spectral Analysis of Population Dynamics in the Rotifer *Brachionus Calyciflorus Pallas*. G. G. WALTER, Surplus Yield Models of Fisheries Management. SOME MORE PAPERS IN PREPARATION.

Contributors to This Volume

Barber, M. Craig
Department of Zoology
University of Georgia

Bargmann, Rolf E.
Department of Computer Science and Statistics
University of Georgia

Berman, Mones
Laboratory of Theoretical Biology
National Institutes of Health

Clark, G. M.
Industrial and Systems Engineering
The Ohio State University

Cobelli, C.
Laboratorio per Ricerche di Dinamica dei
    Sistemi e di Bioingegneria, Padova

Finn, John T.
Department of Forestry and Wildlife
University of Massachusetts

Gerald, K. B.
Institute of Statistics
Texas A&M University

Jacur, G. Romanin
Laboratorio per Ricerche di Dinamica dei
    Sistemi e di Bioingegneria, Padova

Lepschy, A.
Istituto di Elettrotecnica e di Elettronica
Universita di Padova

Marcus, Allan H.
Department of Applied Mathematics
Washington State University

Matis, J. H.
Institute of Statistics
Texas A&M University

O'Neill, R. V.
Environmental Sciences Division
Oak Ridge National Laboratory

Patten, Bernard C.
Department of Zoology
University of Georgia

Purdue, P.
Department of Statistics
University of Kentucky

Rescigno, Aldo
The Bragg Creek Institute
    for Natural Philosophy

Show, Ivan T.
Science Applications, Inc.
La Jolla, California

Tiwari, J. L.
Department of Surgery
University of California

Walter, G. G.
Department of Mathematics
The Univeristy of Wisconsin

Wehrly, T. E.
Institute of Statistics
Texas A&M University

White, G. C.
Environmental Science Group
Los Alamos Scientific Laboratory

Wise, M. E.
Physiology Laboratory
Leiden University

PREFACE TO THE VOLUME

Ecosystem modeling, simulation, and analysis began in the late Fifties and early Sixties with the use of compartment models to mimic storages and flows of radiotracers, stable materials, and energy in food chains and food webs. The development of simple, low order models gave impetus to more ambitious efforts in the U.S. International Biological Program (IBP) during the late Sixties to mid Seventies to represent ecosystems with large scale computer models. Most of these approaches, perhaps 95% or more, were compartment model oriented, although with much innovation as new technological and quantitative tools and more refined concepts became available. The size and scope of ecological modeling throughout the Seventies has generated a rich diversity in the types of conceptual and mathematical models which have been employed, with corresponding symbolic embellishments such as box-and-arrow diagrams, energy circuit diagrams of H. T. Odum, feedback dynamics diagrams of J. W. Forrester, mathematical graphs, and control diagrams from engineering.

For the most part, the compartmental analysis of ecosystem models during the Sixties and Seventies has focused on large scale, deterministic simulation models with some lesser attention to small scale, deterministic analytical models. The statistical inferences associated with the validation, hypothesis testing, and parameter estimation aspects of these models have been largely ignored, and stochasticity in ecosystem models has, with some few exceptions, been absent. However, in the last few years stochastic modeling of and statistical inference in compartmental models, both of which are topics of natural interest in statistical ecology, have witnessed remarkable development among a relatively small but growing number of investigators. Recent research in these areas has not only broadened the theoretical foundation of compartment modeling but, at the same time, has also led to new promising practical applications.

In recognition of these developments, the International

Statistical Ecology Program (ISEP) invited a session on
Compartmental Analysis of Ecosystem Models as part of its
Satellite Program in Statistical Ecology to precede the Second
International Congress of Ecology in Jerusalem. This session
was held at the University of Parma in Parma, Italy, August 21
to 25, 1978. The meeting, endowed with a stimulating local
setting as well as the conducive overall ISEP sponsorship, and
a group of participants representing broad backgrounds and
technical expertise, was exceptionally successful. The
week was full with presentations and workshop interactions
throughout the days and with discussion sessions late into
the nights. The ample time and opportunity for exchanges in both
depth and breadth concerning this one general modeling area
created a new excitement for the work and a unique spirit of the
meetings. The papers presented were subsequently reviewed,
refined and integrated, and are collected together here in
the hope of preserving and disseminating both the content and
flavor of what occurred at Parma.

The interface between classical compartmental analysis of
ecosystem models on the one hand and new statistical ecology
research in stochastic modeling and statistical inference on
the other is immediately evidenced by some differences in
notation among the papers. This, of course, is characteristic
of a new emerging field and occurs in something as basic,
for example, as arrangement of the subscripts in a rate coefficient where, by convention, classical compartmental analysis
orders them one way but stochastic processes literature orders
them the reverse way. An editorial decision was made to not
attempt to compel a common notation but rather to preserve the
inherent linkage between the individual papers in this volume
and the respective authors' previous contributions in the
published literature.

The papers are divided into four natural subtopics on the
basis of both their immediate applicability and level of
abstraction. The first section of four papers (O'Neill; Walter;
Barber, Patten, and Finn; and Show) discusses present and
developing applications of compartmental analysis to ecosystem
modeling, and several have particular reference to the use
of stochastic models. The second section of four papers
(Cobelli, Lepschy, and Jacur; Berman; White and Clark; and
Bargmann) develops certain tools which are necessary for
identification and estimation of the applied models. The third
section of five papers (Tiwari; Matis and Wehrly; Purdue; Marcus;
and Wise) outlines several recent approaches to stochastic
modeling which have already been applied to the field and which
show promise for further practical application. The fourth
section of three papers (Walter; Gerald and Matis; and Rescigno)

contains some recent mathematical developments which are more abstract than the other sections, and yet which not only support present applications but also point the way to new areas in need of further research. *In toto*, we feel that the volume summarizes many recent developments in stochastic modeling and statistical inference aspects of the compartmental analysis of ecosystem models, and our hope is that it will serve as a useful stimulus to further research in these areas of statistical ecology.

We trace the origin of this volume back to the invitation from the ISEP Board to organize the meeting in Parma, and hence we express our appreciation to them. We are also particularly grateful and want to acknowledge the contribution of the Board Chairman, G. P. Patil, who devoted countless hours both before the meeting, in organizing the logistical and financial support, and during the meeting, through his continued organizational assistance and his scientific participation. His tireless efforts were an inspiration and his enthusiasm was contagious. We also appreciate the local hosts, O. Rossi, R. Zanni, and staff, who are in large measure responsible for making Parma such a delight to remember.

April 1979

James H. Matis
Bernard C. Patten
Gary C. White

ACKNOWLEDGMENTS

For permission to reproduce materials in this volume, thanks are due to Academic Press for Figure 1 on page 274 and for Figures 2 and 3 on page 275.

TABLE OF CONTENTS

COMPARTMENTAL
ANALYSIS
OF ECOSYSTEM MODELS

APPLICATIONS OF
COMPARTMENTAL ANALYSIS
TO
ECOSYSTEM MODELING

INTRODUCTION TO SECTION I

The four papers in this section emphasize uses of compartment models. A natural question is, why employ a model? Different types of models are useful in the following ways: 1) Models provide a means of conceptualizing the system being studied, i.e., of helping the investigator think about and formulate the system's structure and function. 2) Models provide a means of summarizing complex data sets into a much smaller set of numbers. 3) Models provide a means of predicting the future, or forecasting. 4) Models provide a management tool. Management decisions can be influenced by results of model manipulations. This list of applications is roughly in order of increasing data requirements. System conceptualization could be accomplished with only qualitative data, while management decisions would hopefully require a well quantified model.

The first paper by R. V. O'Neill reviews the applications of linear constant coefficient compartment models in ecology. The author argues that these types of models are not right for all circumstances, but are definitely useful for some applications.

G. G. Walter then illustrates the application of compartment modeling to the Georges Bank fishery. He demonstrates the usefulness of translating a compartment model into a Markov chain formulation to determine the ultimate energy distribution in top predator compartments. Also, the model is manipulated to study results of removing two forage fishes.

The paper by M. C. Barber, B. C. Patten, and J. T. Finn reviews input-output flow analysis for compartment models. Various indices are derived that show potential for measuring environmental impact of chronic disturbances on ecosystems. Models developed from post-disturbance data may show significant differences in some of these measures when compared to a model developed from pre-disturbance data on the same system. For example, the loss of nutrients by an ecosystem due to an environmental impact may be detected with a cycling measure.

In order to assess impacts of offshore oil wells on marine communities, I. T. Show in the final paper develops a compartment model incorporating physical, chemical, and biological features of the problem. Spatial variability is included in this model through horizontal flows. This paper illustrates what will probably become a significant approach in the future: utilizing compartment models in environmental impact analysis.

In all, the four papers of this section illustrate present applications of compartment models in ecology and, in addition, are suggestive of further likely lines of application in the future.

J. H. Matis, B. C. Patten, and G. C. White, (eds.),
Compartmental Analysis of Ecosystem Models, pp. 3-28. All rights reserved.
Copyright ©1979 by International Co-operative Publishing House, Fairland, Maryland.

A REVIEW OF LINEAR COMPARTMENTAL ANALYSIS IN ECOSYSTEM SCIENCE

R. V. O'NEILL

Environmental Sciences Division
Oak Ridge National Laboratory
Oak Ridge, Tennessee 37830 USA

SUMMARY. The linear, constant coefficient, compartment model
has a large number of applications in ecosystem science. In this
paper, an attempt is made to outline the problem areas and
objectives for which this type of model has particular advantages.
The areas identified are: (1) an adequate model of tracer movement through an undisturbed but non-equilibrium ecosystem,
(2) an adequate model of the movement of material in greater than
tracer quantity through an ecosystem near steady state, (3) a
minimal model based on limited data, (4) a tool for extrapolating
past trends, (5) a framework for the summarization of large data
sets, and (6) a theoretical tool for exploring and comparing
limited aspects of ecosystem dynamics. The review is set in an
historical perspective which helps explain why these models
were adopted in ecology. References are also provided to
literature which documents available mathematical techniques in
an ecological context.

KEY WORDS. model, systems analysis, compartment model, tracer
dynamics, linear analysis.

1. INTRODUCTION

Linear, constant coefficient compartment modeling has
played an important role in the development of systems ecology
since the inception of the field (Neel and Olson, 1962; Olson,
1963a,b, 1964, 1965; Patten, 1964). Some authors (Bledsoe and
Jameson, 1970; Clymer and Bledsoe, 1970; Jameson, 1970; Wiegert,
1975) have suggested that this form of model is mainly of

historic or heuristic value. Nevertheless, linear models continue
to find a significant number of applications in ecology (Emanuel
and Mulholland, 1976; Gist and Crossley, 1975; Mitchell *et al.*,
1975; Patten, 1975; Patten *et al.*, 1975a,b; Webster, 1975;
Webster *et al.*, 1975; Shugart *et al.*, 1976; Garten *et al.*, in
press). The continuing use of this model form provides the
motivation for examining the advantages and limitations of these
models for future applications.

For the purposes of this review, I will consider models
describing transfer of some entity (e.g. energy, nutrient,
pollutant, radio-nuclide) through an ecological system. In the
compartment modeling approach, the system is first divided into
a number of compartments (e.g. plants, animals, soil). Material
or energy transport is described by fluxes between these compartments. The model is developed as a set of ordinary differential
equations, one equation for the material balance of each compartment. In the type of model I consider in this review, each term
on the right side of these equations represents an individual
flux as a constant fraction of a donor compartment transferred
to a recipient per unit time. As a result, the equations are
linear in the state variables (i.e. quantities of material in
the compartments). It should be noted that the term 'compartment
model' is often applied to a broader class of models which
includes non-linear and variable coefficient formats. The present
review will use the term in the more restrictive sense of
linear, constant coefficient models.

A typical approach to reviewing compartment modeling has
been to contrast linear and nonlinear models, outlining the
advantages and disadvantages of each. I particularly recommend
the presentations of Bledsoe (1976), Wiegert (1975), Patten
(1971, 1974, 1975), and Patten *et al*. (1975a). Such an approach
is useful in helping the reader form an opinion on the
applicability of linear models. However, some authors tend to
adopt an advocate position and argue for one type of model,
almost to the exclusion of the other.

The approach in this review will be to assume that linear
compartment models can be useful in ecological studies. The
argument for this position is simply the large number of
applications which will be documented in the course of this review.
My major goal will be a critique of the literature to specify
the conditions under which compartment models can be legitimately
applied.

The presentation will begin with a brief historical review
of the development of linear compartment models in the context
of the general development of the field of systems ecology.

This historical perspective should help explain the origins of
linear compartment models and why they have been widely applied
in ecology. This will be followed by a discussion of the
analytical advantages of the compartment model and the available
mathematical techniques. The main body of the review will be
concerned with an outline of conditions under which the linear
compartment model can be legitimately applied.

3. HISTORICAL PERSPECTIVE

The modeling of ecological problems is almost as old as the
field of ecology (e.g. Gompertz, 1825; Lotka, 1925; Volterra,
1926). Early models were nonlinear representations of
population processes. Caswell *et al*. (1972) have pointed out
that models resembling compartment models have been applied in
atmospheric sciences since at least 1935 (Koztitzin 1935).
The earliest references to the potential application of compartment models in ecology appears to be Olson (1959, 1960). Thus,
we can date the use of compartment models in ecology, as well
as systems ecology as a field, from about 1960.

The historical roots of systems ecology have been traced
(Shugart and O'Neill, in press) to a number of interacting
factors during the first half of this century. They list as
contributing factors: (1) early developments in population
models predisposing the ecologist to mathematical analysis,
(2) a general trend away from natural history and toward
quantitative ecology, (3) development of the digital computer,
(4) increasing interest in the dynamics of materials in total
ecosystems, and (5) rapid development of the field of systems
analysis during the Second World War. Several of these factors
are particularly relevant to our present discussion.

Another critical ingredient was the development of tracer
experiments in vertebrate physiology and medical research. These
experiments involved the injection of a tracer (initially a dye)
into the body and subsequent measurement of the transport of the
tracer among components of the organism. The analogy between
these experiments and measurement of material transport in
ecosystems can be readily seen.

The tracer experiments encouraged development of a mathematical theory, compartmental analysis, to explain and analyze
the data (Teorell, 1937; Zilversmit *et al*., 1943; Hearon, 1953;
Solomon, 1949; Sheppard and Householder, 1951). The mathematical
theory involved systems of linear differential equations with
constant coefficients and corresponds exactly with the compartment models later applied in ecology. The study of tracer

dynamics was greatly accelerated when radioactive tracers
became available in the 1950s.

Having identified the origins of compartmental analysis in
physiological tracer theory, it remains to be explained how these
techniques came to be applied to ecological problems. The
application resulted from the juxtaposition of several factors.
First, the growing interest in the movement of energy and
material through trophic chains (Lindemann, 1942) and whole
ecosystems (Odum, 1957) presented significant measurement
problems in the field. Radiotracers and compartmental analysis
offered significant advantages for this type of study.
Application of tracer dynamics was precipitated by a second
factor, the post-war problems in radioecology. The problems
involved the effects of radiation and the potential movement
and concentration of fallout radionuclides through foodchains
(e.g. Reichle *et al.*, 1970). In addressing these problems, the
ecologist became familiar with radiotracers and could see their
applicability to the study of ecosystem dynamics.

When these factors are considered along with the increasing
availability of computers and the widespread interest in the
field of systems analysis, it is easy to understand the early
stages in the development of systems ecology. The earliest
discussions (Olson, 1959, 1960; Neel and Olson, 1962) dealt
with models of radionuclide transport in foodchains and
naturally applied the terminology (compartment model) and
techniques (linear differential equations) which had been
developed in tracer dynamics. Although the general applicability
of modeling was pointed out again by Patten (1964, 1966) and
Van Dyne (1966), most models produced until 1969 were compartment
models of radiotracer movement (Bloom *et al.*, 1969; Crossley and
Reichle, 1969; Dahlman and Auerbach, 1968; Eberhardt and Hanson,
1969; Eberhardt and Nakatani, 1969; Gardner, 1967; Kaye and Ball,
1969; Martin, 1966; Martin *et al.*, 1969; Olson, 1963a, 1965;
Patten and Witkamp, 1967; Raines *et al.*, 1969; Witkamp and Frank,
1969). The only exceptions appear to be three papers (Olson,
1963b, 1964; Gore and Olson, 1967) which deal with terrestrial
productivity. It was not until 1968 that compartment models
began to break away from radioecology and be applied to studies
of productivity (Kelly *et al.*, 1969) and element cycling
(Dahlman *et al.*, 1969; McGinnis *et al.*, 1969).

The Biome Programs of the U.S. International Biological
Program became active in the late 1960s and with them came
accelerated development of nonlinear ecosystem models. The
development was strongly influenced by Van Dyne (1966) and his
associates (Bledsoe and Jameson, 1970; Clymer and Bledsoe, 1970;
Jameson, 1970; Swartzman, 1970). As a result of this influence

nonlinear modeling became widespread and compartmental modeling declined in influence. Thus, during the period 1969-1971, I was able to locate 31 studies utilizing compartment models, but only 3 for the period 1976-1978. In spite of this decline in interest, it remains the thesis of this review that compartmental modeling can continue to contribute to ecological studies as long as the proper conditions for applicability are carefully observed (Oak Ridge Systems Ecology Group, 1974).

3. LINEAR ANALYSIS AND COMPARTMENT MODELS

The most important reason for continuing interest in compartment models for ecosystems is mathematical simplicity. The simplicity permits application of a broad spectrum of analytical tools developed for tracer studies (e.g., Atkins, 1969; Jacquez, 1972) and for linear systems analysis (e.g., Wiberg, 1971; Melsa and Schultz, 1969; Shearer et al., 1967; Wymore, 1967). The key factor is that the system of equations can be solved analytically in closed form. I will not attempt to survey the available mathematical techniques, since these are available in the references just given and some will be considered in detail in other papers in this volume. The present section will merely reference applications of these techniques and provide access to discussions of the mathematical methodology in an ecological context.

Solving for transient behavior of a model is relatively simple for compartment models. For small-scale models, the analogue computer provides a virtually instantaneous solution (Patten, 1971; Denmead, 1972) that has considerable advantages in terms of investigator involvement. The advantages encouraged several studies of radioisotope movement using the analog computer (Neel and Olson, 1962; Olson, 1963a; Patten and Witkamp, 1967; Witkamp and Frank, 1969).

Solution of the equations on the digital computer can be done with standard simulation programs, but several codes have been developed specifically for ecological compartment models (Reeves, 1971). These programs include MATEXP (Ball and Adams, 1967), COMSYS1 (Bledsoe and Olson, 1970), MODSOV (Clark et al., 1974) and LEAP (Kercher, 1977). Standard programs also have been developed for frequency response analysis (Kerlin and Lucius, 1966) and for a broad spectrum of linear analyses (Kercher, 1977).

Some excellent general discussions exist on the available analytical techniques and the interpretation of the results for ecosystem models. Funderlic and Heath (1971) provide a useful mathematical background in matrix calculus. Mulholland and

Sims (1976) discuss applications of linear control theory.
Waide *et al.* (1974), Waide and Webster (1976), and Mulholland
and Keener (1974) provide comprehensive discussion and creative
interpretations for the available methods.

Applications and discussions of specific techniques
provide useful examples to guide future work. Astor *et al.*
(1976) discusses sensitivity analysis in considerable detail
while Brylinsky (1972) and Hett and O'Neill (1974) provide
examples of applications. Emanuel and Mulholland (1976)
have applied linear control theory to optimizing fish productivity in a pond. Several workers have considered parameter
estimation techniques for fitting coefficients of the compartment model (Bargmann and Halfon, 1977; Bledsoe and Van Dyne,
1969; Eberhardt *et al.*, 1970; Halfon, 1975a,b; Lewis and Nir,
in press; O'Neill, 1971). Mulholland and Gowdy (in press)
discuss sampling procedures specifically designed with parameter
estimation in mind. Stability analysis has been discussed by
Dudzek *et al.* (1975), Jordan *et al.* (1972), and Webster *et al.*
(1975). Frequency response analysis has been applied to
ecological problems by several workers (Kaye and Ball, 1969;
Child and Shugart, 1972; Shugart *et al.*, 1976). Monte Carlo
analysis of error propagation in compartment models has been
presented by Gardner *et al.* (1976), O'Neill (1973), and
O'Neill and Rust (in press). Particularly creative approaches
have been developed by Astor *et al.* (1976) and Finn (1976)
which are applicable to both linear and nonlinear models.

The number and variety of mathematical techniques available
for analysis of compartment models is indeed impressive. The
techniques are particularly useful because of the efforts of
many workers to adapt and interpret the techniques in an
ecological context. Thus, whenever the use of the compartmental
approach to a problem can be shown to be legitimate, the
available technology argues strongly for its adoption. The
problem, therefore, focuses sharply on the need to specify the
conditions under which compartmental analysis can be utilized.

4. LEGITIMATE APPLICATIONS OF COMPARTMENTAL ANALYSIS IN ECOLOGY

"...the appropriate mathematical modeling technique for
ecosystem analysis depends upon the object to which the model
is to be applied." This quote from Bledsoe (1976) summarizes
the point of view taken in this review. The problem is not to
decide whether compartment models or nonlinear models are
superior in general, but which specific objectives and problem
areas are most appropriate for linear models. Rykiel and
Kuenzel (1971) and Mankin *et al.* (1973) compared nonlinear and
linear models and discovered that the linear models were as

good or even better than nonlinear representations for specific applications. As a result of surveying the ecological literature, I will propose a number of applications in which compartment models appear to be particularly useful.

4.1 The Compartment Model as a Fully Adequate Model of the Dynamics of a Tracer Substance in an Undisturbed Ecological System. It should be obvious from the earlier discussion on historical development, that the compartment model is an adequate descriptor (i.e., simulates actual time series data) of tracer dynamics, and this was among the earliest applications of the approach (Neel and Olson, 1962; Olson, 1963a). Since these early studies, compartment models have been applied to radionuclide movement in marine environments (Bloom and Raines, 1971), grasslands (Dahlman and Auerbach, 1968; Dahlman, 1973), tropical forests (Kaye and Ball, 1969; Raines *et al.*, 1969), deciduous forests (Olson, 1965; Garten *et al.*, 1978), microcosms (Patten and Witkamp, 1967; Witkamp and Frank, 1969, 1970) and a variety of foodchains including arctic (Eberhardt and Hanson, 1969), agricultural (Gardner, 1967) and small mammal (Martin, 1966). A number of general discussions of this application of compartment models are available with particular emphasis on the applied problem of tracing radionuclide movement following chronic or accidental releases from nuclear facilities (e.g., Booth and Kaye, 1971; Booth *et al.*, 1971; O'Neill, 1970; Struxness *et al.*, 1971; Eberhardt and Nakatani, 1969; Martin *et al.*, 1969).

The assumptions required for this type of application are that the tracer be in very small quantity compared to natural pools (e.g., small quantities of ^{32}P introduced into natural phosphorus pools), that the tracer itself does not seriously disrupt the dynamics of the system, and that the ecosystem remains free from unusual major disturbances during the course of the experiment. These conditions seem reasonably satisfied in following the movement of radiotracers through undisturbed ecosystems.

The conditions may also be satisfied for substances other than radionuclides. Thus, the movement of trace substances such as magnesium (Child and Shugart, 1972; Jordan *et al.*, 1973) may also be traced with the compartment model. The movement of DDT and DDE also appears to follow tracer dynamics (Eberhardt *et al.*, 1970; Woodwell *et al.*, 1971; O'Neill *et al.*, 1971).

The basis for these applications can be simply demonstrated from the observation that tracer dynamics in many studies does not need a more complex model in order to explain the data

(Oak Ridge Systems Ecology Group, 1974). This is particularly
clear in studies which directly compare data and model output.
The compartment model appears to explain data from DDT movement
in a freshwater marsh (Eberhardt *et al.*, 1970) and in the human
foodchain (O'Neill *et al.*, 1971). Movement of fallout radionuclides in foodchains (Martin, 1966; Gardner, 1967) appears
to follow simple linear kinetics. Radiocesium dynamics in
microcosms (Patten and Witkamp, 1967; Witkamp and Frank, 1969,
1970) can be closely fitted by linear equations. On a more
ambitious scale, radiocesium kinetics in a forest stand (Olson,
1965) also appears to follow linear tracer dynamics.

A particularly valuable offshoot of compartment models
has been a number of methodologies proposed for field measurement
of ecological parameters. Tracer theory has been used to develop
measures of feeding rates (Kowal and Crossley, 1971; O'Neill, 1971),
turnover times (Crossley and Reichle, 1969; Goldstein and Elwood,
1971) and to deduce the trophic position of natural populations
(McBrayer and Reichle, 1971). Measurements of turnover rates
have been summarized by Reichle *et al.* (1970).

*4.2 The Compartment Model as a Fully Adequate Model of the
Dynamics of a Substance Moving Through an Ecosystem at Steady
State.* The use of the compartment model to describe tracer
dynamics can be expanded to include the movement of non-tracer
substances so long as the ecosystem itself is near steady state.
The system need not be actually at steady state so long as the
changes in the system are slow compared to the rate at which
the substance moves through the system, or subsystem (Witkamp,
1970).

The most common application of this approach is in modeling
the movement of a substance on an annual basis. Although the
ecosystem may show considerable seasonal fluctuations, it can
be argued that behavior can be simulated by linear dynamics
over a longer time frame (Patten, 1972, 1975) such as a year,
or integrated over larger spatial scales (Woodwell *et al.*, 1971;
O'Neill *et al.*, 1971; Baes *et al.*, 1976, 1977; Eriksson and
Welander, 1956; Goldstein and Harris, 1973). In the context
of annual cycles, compartment models have been applied to the
movement of nitrogen (Dahlman *et al.*, 1969), hydrogen (Bloom
and Raines, 1970), energy (Olson, 1963b), calcium (Waide *et al.*,
1974; Shugart *et al.*, 1976) and other elements (Jordan *et al.*,
1972).

*4.3 The Compartment Model as a Minimal (i.e., Less than Fully
Adequate) Model of an Ecosystem When Inadequate Data are
Available to Justify a More Complex Model.* It is not uncommon

in ecological problems to be faced with a need to describe some aspects of dynamic behavior but lack adequate data to describe the complete nonlinear behavior. The data requirements of complex ecosystem models can be enormous. An extreme case, the Grassland ELM model (Innis, 1978), requires thousands of parameter values. The development of a complex model without adequate data is a form of sophistry since model output can only be as good as data input.

The need for a minimal model arises in a number of cases. In applied problems, the need to make some preliminary assessment may require the use of a minimal model, for example, to predict doses to man from a radionuclide contaminated environment (Booth *et al.*, 1971), to assess the potential transport concentration of transuranic elements in food chains (Garten *et al.*, 1978), to evaluate the effectiveness of various control measures on reducing DDT concentrations in human adipose tissue (O'Neill *et al.*,1971). In these cases the immediate need is clearly the most important consideration in choosing the mathematical form of the model. In many of these cases, the alternative to accepting a less-than-adequate compartment model is simple guesswork which does not take into account even the information which is available.

In some applications, regional (Hett, 1971; Hett and O'Neill, 1974), continental (O'Neill *et al.*, 1971) or global spatial scales (Baes *et al.*, 1976, 1977; Woodwell, 1971) are involved. Such scales deal with processes on a simplified, integrated basis and must rely on simple, often coarse, data. Simply stating that a more mechanistic model could be developed is vacuous when time and economic factors are considered.

Another type of scale problem arises when the objective is to compare gross dynamics of a series of elements in the same system, or across systems. In this case, the data requirements are large and the objective can be satisfied by considering only some aspects of total dynamics. The compartment model serves as a convenient and useful tool to summarize data and analyze for general properties (Witkamp and Frank, 1970; Webster *et al.*, 1975; Jordan *et al.*, 1972). As long as the analyses and conclusions are restricted to gross properties of the element cycles, this application appears legitimate and has led to interesting hypotheses about mineral cycling and ecosystem stability (Jordan *et al.*, 1972; Webster *et al.*, 1975).

Preliminary models can also be used as a step toward the development of more complex models (Gore, 1972; Bledsoe, 1976) or as screening tools to determine where problems exist before more ambitious research projects are undertaken. The development and analysis of a compartment model would be a logical and

valuable step toward the evolution of any complex ecosystem
research project.

*4.4 The Compartment Model as a Dynamic Tool for Summarizing
(but Not Explaining) Past Trends and Extrapolating These Trends
into the Future.* The compartment modeling approach can be
legitimately applied when the objective is restricted to analyzing
historical data and extrapolating these trends into the future.
The extrapolation does not attempt to explain the mechanisms
responsible for the trends and offers little insight into
ecological processes. Nevertheless, this simple approach can be
valuable in illustrating the logical consequences of past trends.

There is not a really clear distinction between the use of
compartment models for extrapolation and the previous application
which emphasize inadequate data. The models discussed earlier
also extrapolate in the sense that no mechanistic explanations
are possible and the models merely illustrate the future
consequences of present or past measurements. In both cases,
the model contains the implicit assumption that environmental
variables and the ecological processes responding to these
variables will not undergo significant changes in the future.

The emphasis in the present application is on the analysis of
historical data on compartment values. This approach has been
used for global carbon cycles (Eriksson and Welander, 1956;
Baes *et al.*, 1976, 1977), forest biomass dynamics (Goldstein
and Harris, 1973), land use changes (Hett, 1971) and for
succession (Bledsoe and Van Dyne, 1971; Shugart *et al.*, 1973;
Johnson and Sharpe, 1976). All of these applications depend
either on remeasurement of the system at successive points in
time or on measurement of several plots of known age. Although
framed in the mathematics of Markov processes, many forest
succession models (e.g. Horn, 1975; Leak, 1970; Waggoner and
Stephens, 1970) are closely related to compartment models used
for extrapoloation.

For this application to be legitimate, sufficient historical
data should exist to demonstrate that historical trends can be
simulated by a linear model. In some cases this condition is
satisfied (Baes *et al.*, 1976, 1977). But in most instances,
only a single remeasurement was taken so that only two historical
data points are available (e.g. Johnson and Sharpe, 1976). In
this case, the application of the compartment model is more
tenuous.

4.5 The Compartment Model as a Non-Dynamic Model that Serves as a Convenient Tool for the Summarization of Extensive Ecosystem Data. Large ecosystem studies involve extensive data sets and yet have as an important objective the synthesis of information into a coherent representation of total system dynamics. The format of the compartment model provides a convenient tool which summarizes extensive data into a concise framework, i.e. transfer coefficients and initial (or final state) values for compartments. In this context, the compartment model has been used to synthesize information on biomass and product- ivity (Odum, 1970; Williams and Murdoch, 1972; Kelly *et al.*, 1969; Goldstein and Harris, 1973), plutonium (Garten *et al.*, 1978), sulfur (May *et al.*, 1972), nitrogen (Mitchell *et al.*,1975), and hydrogen (Odum *et al.*, 1970).

Since modeling is often conceived as co-extensive with simulating, this application of compartment models might seem to be of minimal interest and importance. And yet the approach makes it possible to compare biomass or energy dynamics across many different systems (e.g. O'Neill, 1971c; Reichle *et al.*, 1973) or else the dynamics of several different substances with the same system (e.g. Cornaby, 1973; Gist, 1972; Gist and Crossley, 1975; McGinnis *et al.*, 1969; Webster, 1975). The ability to compare dynamics between substances has led to interesting hypotheses about gross features of element cycling, e.g. selective pressure may lead to recycling of essential elements, but not of non-essential elements (Jordan *et al.*, 1972).

Viewed in a slightly different perspective, several authors (Bledsoe, 1976; Swartzman, 1970; Gore, 1972; Dettmann, 1971) have pointed out the utility of the compartment model at early stages in a complex research project. Because of its ability to summarize dynamics, the model can serve as a framework or checklist for the important transfers which must be considered in the study. When developed by a research team, the model serves as a common basis for communication and as a common conceptual structure with each individual clearly understanding his own role within the total project.

4.6 The Compartment Model as a Theoretical Tool for Exploration of Abstract Properties of the Ecosystem. Because of the extensive analytical tools which are available for linear models, the compartment model has been applied in theoretical exploration of ecosystem dynamics (Astor *et al.*, 1976; Brylinsky, 1972; Child and Shugart, 1972; Kaye and Ball, 1969; Patten *et al.*, 1975a; Waide *et al.*, 1974; Webster *et al.*, 1975). In these applications, some aspect of ecosystem behavior is isolated which can be reasonably expected to follow linear dynamics, e.g. elemental cycling on an annual basis in an undisturbed system (Webster *et al.*, 1975; Waide *et al.*, 1974; Child and Shugart, 1972). It has also been argued that many aspects of ecosystem behavior can be simulated with linear models (Patten, 1974, 1975). The most common argument is that a nonlinear model in the neighborhood of an equilibrium will respond to small perturbations as though it were linear. However, Bledsoe (1976) has pointed out that the appropriate linearized model is not identical to the compartment model of the system.

This application of the compartment model would seem to be legitimate as long as the assumptions inherent in the approach are considered carefully. The results must also be considered as hypotheses to be verified by experimentaion. But, despite the limitations, these theoretical explorations have led to interesting hypotheses about the relationship between connectance in a complex system and resultant stability (Gardner and Ashby, 1970; May 1972; DeAngelis, 1975).

5. DISCUSSION

It should be patent from the large number of applications documented in this review that linear compartmental analysis has and should continue to play a useful role in ecosystem science. The region of applicability is undoubtedly more restricted than was envisioned by early developers of the approach but, if the areas outlined above are indeed legitimate, then a broad scope of utility remains.

The greatest limitations of the compartment model are involved with the assumption of linearity. Some aspects of ecosystem behavior appear to follow linear dynamics, but by far the most common problems are distinctively nonlinear. Many of the perturbations imposed on ecosystems, particularly by human society, are not minor and cannot be considered to remain within the neighborhood of a system equilibrium. In its simplest representation, the constant coefficients of the compartment model do not even consider the response of ecosystem dynamics to changes in environmental parameters. Seasonal dynamics which incorporate many of the adaptations and selective pressures operating on natural populations are poorly represented in linear models.

However, considered either as a preliminary step or as a final model within the severe constraints discussed above, the linear compartment model has significant advantages. As a preliminary model, it can be used to help establish logical sampling intervals (Mulholland and Gowdy, in press) and to explore gross properties deserving detailed examination. If the system under investigation is relatively undisturbed, then a compartment model of the movements of a substance through the system has tremendous advantages because of the mathematical tools which can be brought to bear on the problem. The simplicity of the compartment model strongly recommends it, if the necessary restrictions can be met.

Ultimately, the applicability of linear compartmental analysis must be decided on the basis of the individual problem under examination. The objectives of a problem should not be manipulated in order to fit into a linear approach. Neither should the compartment model be rejected offhand. Based on the principle that the simplest possible explanation should be adopted, the compartment model should continue to find useful application in ecosystem science whenever the assumptions of the model fit logically into the objectives of a problem.

ACKNOWLEDGEMENTS

Research sponsored in part by the Division Biomedical and Environmental Research, U. S. Department of Energy, under contract W-7405-eng-26 with Union Carbide Corporation, and in part by the Eastern Deciduous Forest Biome, US-IBP, funded by the National Science Foundation under Interagency Agreement AG-199, DEB76-00761 with the Department of Energy, Oak Ridge National Laboratory. Publication No. 1335, Environmental Sciences Division, Eastern Deciduous Forest Biome Contribution No. 329.

REFERENCES

Astor, P. H., Patten, B. C., and Estberg, G. N. (1976). The sensitivity substructure of ecosystems. In *Systems Analysis and Simulation in Ecology, Vol. 4*, B. C. Patten, ed. Academic Press, New York. 390-430.

Atkins, G. L. (1969). *Multicompartment Models for Biological Systems*. Methuen, London.

Baes, C. F., Goeller, H. F., Olson, J. S., and Rotty, R. M. (1976). *The global carbon dioxide problem.* ORNL-5194, Oak Ridge National Laboratory, Oak Ridge, Tennessee.

Baes, C. F., Goeller, H. F., Olson, J. S., and Rotty, R. M. (1977). Carbon dioxide and climate: the uncontrolled experiment. *American Scientist*, 65, 310-320.

Ball, S. J. and Adams, R. K. (1967). *MATEXP: A general purpose computer program for solving ordinary differential equations by the matrix exponential method.* ORNL/TM-1933, Oak Ridge National Laboratory, Oak Ridge, Tennessee.

Bargmann, R. E. and Halfon, E. (1977). Efficient algorithms for statistical estimation in compartment analysis: Modeling ^{60}Co kinetics in an aquatic microcosm. *Ecological Modelling*, 3, 211-226.

Bledsoe, L. J. (1976). Linear and nonlinear approaches for ecosystem dynamic modeling. In *Systems Analysis and Simulation in Ecology, Volume 4*, B. C. Patten, ed. Academic Press, New York. 283-298.

Bledsoe, L. J. and Jameson, D. A. (1970). Model structure for a grassland ecosystem. In *The Grassland Ecosystem - A Preliminary Synthesis*, R. L. Dix and R. G. Beidleman, eds. Range Science Series Number 2, Range Science Department, Colorado State University, Fort Collins, Colorado. 410-435.

Bledsoe, L. J. and Olson, J. S. (1970). *COMSYS1 - A stepwise compartmental simulation program.* ORNL-2413, Oak Ridge National Laboratory, Oak Ridge, Tennessee.

Bledsoe, L. J. and VanDyne, G. M. (1969). *Evaluation of a digital computer method for analysis of compartmental models of ecological systems.* ORNL/TM-2414, Oak Ridge National Laboratory, Oak Ridge, Tennessee.

Bledsoe, L. J. and VanDyne, G. M. (1971). A compartment model simulation of secondary succession. In *Systems Analysis and Simulation in Ecology, Volume 1*, Academic Press, New York. 479-511.

Bloom, S. G., Levin, A. A., and Raines, G. E. (1969). *Mathematical simulation of ecosystems, a preliminary model applied to a latic freshwater environment.* Battelle Memorial Institute, Columbus, Ohio.

Bloom, S. G. and Raines, G. E. (1970). Kinetic model fo hydrogen flow through the El Verde forest system. In *A Tropical Rain Forest*, H. T. Odum and R. F. Pigeon, eds. TID-24270, Division of Technical Information, USAEC, Washington, D.C.

Bloom, S. G. and Raines, G. E. (1971). Mathematical models for predicting the transport of radionuclides in a massive environment. *Bioscience*, 21, 691-696.

Booth, R. S. and Kaye, S. V. (1971). *A preliminary systems analysis model of radioactivity transfer to man from deposition in a terrestrial environment.* ORNL/TM-3135. Oak Ridge National Laboratory, Oak Ridge, Tennessee.

Booth, R. S., Kaye, S. V., and Rohwer, P. S. (1971). A systems analysis methodology for predicting dose to man from a radioactively contaminated terrestrial environment. In *Proceedings of the Third National Symposium on Radioecology.* Oak Ridge National Laboratory, Oak Ridge, Tennessee.

Brylinsky, M. (1972). Steady state sensitivity analysis of energy flow in a marine ecosystem. In *Systems Analysis and Simulation in Ecology, Vol. 1,* B. C. Patten, ed. Academic Press, New York. 81-101.

Caswell, H., Koenig, H. E., Resh, J. A., and Ross, Q. E. (1972). An introduction to systems science for ecologists. In *Systems Analysis and Simulation in Ecology, Vol. 2,* B. C. Patten, ed. Academic Press, New York. 4-78.

Child, G. I. and Shugart, H. H. (1972). Frequency response analysis of magnesium cycling in a tropical forest ecosystem. In *Systems Analysis and Simulation in Ecology, Vol. 2,* B. C. Patten, ed. Academic Press, New York. 103-135.

Clark, F. H., Booth, R. S., and Vanderploeg, H. A. (1974). *User manual for MODSOV, a terminal operated program solving the linear matrix equation* X = Ax + K. ORNL/TM-4404, Oak Ridge National Laboratory, Oak Ridge, Tennessee.

Clymer, A. G. and Bledsoe, L. J. (1970). A guide to the mathematical modeling of an ecosystem. In *Simulation and Analysis of the Dynamics of a Semi-Desert Grassland,* R. G. Wright and G. M. VanDyne, eds. Range Science Department, Colorado State University, Fort Collins, Colorado. 175-199.

Cornaby, B. W. (1973). *Population parameters and systems models of litter fauna in a white pine ecosystem.* Ph.D. thesis, University of Georgia, Athens, Georgia.

Crossley, D. A. and Reichle, D. E. (1969). Analysis of transient behavior in insect food chains. *Bioscience,* 19, 341-343.

Dahlman, R. C. (1973). Radionuclide concentrations of components of a grassland ecosystem from an acute contamination event. In *Final Report on Postattack Ecology.* ORNL/TM-3837, Oak Ridge National Laboratory, Oak Ridge, Tennessee.

Dahlman, R. C. and Auerbach, S. I. (1968). *Preliminary estimation of erosion and radiocessium distribution in a fescue meadow*. ORNL/TM-2343. Oak Ridge National Laboratory, Oak Ridge, Tennessee.

Dahlman, R. C., Olson, J. S., and Doxtader, K. (1969). The nitrogen economy of grassland and dune soils. In *International Biological Program, Biology and Ecology of Nitrogen*. Conference Proceedings, National Academy of Science, Washington, D.C.

DeAngelis, D. L. (1975). Stability and connectance in food web models. *Ecology*, 56, 238-243.

Denmead, J. K. (1972). Accurace without precision - an introduction to the analogue computer. In *Mathematical Models in Ecology*, J. N. R. Jeffers, ed. Blackwell, London. 215-235.

Dettmann, E. H. (1971). *Use of transfer matrices to organize models of complex systems*. EDFB Memo Report 71-5, Oak Ridge National Laboratory, Oak Ridge, Tennessee.

Dudzek, M., Harte, J., Levy, D., and Sondusky, J. (1975). *Stability indicators for nutrient cycles in ecosystems*. LBL-3264. Lawrence Berkeley Laboratory, Berkeley, California.

Eberhardt, L. L. and Hanson, W. C. (1969). A simulation model for an arctic food chain. *Health Physics*, 17, 793-806.

Eberhardt, L. L., Meeks, R. L., and Peterle, T. J. (1970). *DDT in a freshwater marsh - a simulation study*. BNWL-1297. Battelle Northwest Laboratory, Richland, Washington.

Eberhardt, L. L. and Nakatani, R. E. (1969). Modeling the behavior of radionuclides in some natural systems. In *Proceedings of the Second National Symposium on Radioecology*, D. J. Nelson and F. C. Evans, eds. CONF-670503. Ann Arbor, Michigan. 740-750.

Emanuel, W. R. and Mulholland, R. J. (1976). Linear periodic control with applications to environmental systems. *International Journal of Control*, 24, 807-820.

Eriksson, E. and Welander, P. (1956). On a mathematical model of the carbon cycle in nature. *Tellus*, 8, 155-175.

Finn, J. T. (1976). Measures of ecosystem structure and function derived from analysis of flows. *Journal of Theoretical*

Biology, 56, 363-380.

Funderlic, R. E. and Heath, M. T. (1971). *Linear compartmental analysis of ecosystems*. ORNL/IBP-71/4. Oak Ridge National Laboratory, Oak Ridge, Tennessee.

Gardner, M. R. and Ashby, W. R. (1970). Connectance of large dynamics (cybernetic) systems: Critical values for stability. *Nature*, 228, 784.

Gardner, R. H. Mankin, J. B., and Shugart, H. H. (1976). *The COMEX computer code*. EDFB/IBP-76/4, Oak Ridge National Laboratory, Oak Ridge, Tennessee.

Gardner, R. H. (1967). A mathematical analysis of the transfer of fission products to cow's milk. *Health Physics*, 13, 205-212.

Garten, C. T., Gardner, R. H., and Dahlman, R. C. (1978). A linear compartment model of plutonium dynamics in a deciduous forest ecosystem. *Health Physics* (in press).

Gist, C. S. (1972). *Analysis of mineral pathways in a crypto-zoan foodweb*. Ph.D. thesis, University of Georgia, Athens, Georgia.

Gist, C. S. and Crossley, D. A. (1975). A model of mineral-element cycling for an invertebrate food web in a southeastern hardwood forest litter community. In *Mineral Cycling in Southeastern Ecosystems*, F. G. Howell, J. B. Gentry, and M. H. Smith, eds. USERDA, CONF-740513, Technical Information Center, Springfield, Virginia. 84-106.

Goldstein, R. A. and Elwood, J. W. (1971). A two-compartment, three-parameter model for the absorption and retention of ingested elements by animals. *Ecology*, 52, 935-939.

Goldstein, R. A. and Harris, W. F. (1973). SERENDIPITY - a watershed level simulation model of tree biomass dynamics. In *Proceedings of the 1973 Summer Computer Simulation Conference*. Montreal, Canada.

Gompertz, B. (1825). On the nature of the function expressive of the law of human mortality. *Philosophical Transactions*, 115, 513-585.

Gore, A. J. P. (1972). A field experiment, a small computer, and model simulation. In *Mathematical Models in Ecology*, J. N. R. Jeffers, ed. Blackwell, London. 309-325.

Gore, A. J. P. and Olson, J. S. (1967). Preliminary models for the accumulation of organic matter in an ERIOPHORUM/CALLUNA ecosystem. *Aquilo, Series Botanica*, 6, 297-313.

Halfon, E. (1975a). *A systems identification procedure for large-scale ecosystem models.* Ph.D. thesis, University of Georgia, Athens, Georgia.

Halfon, E. (1975b). The systems identification problem and development of ecosystem models. *Simulation*, 38, 149-152.

Hearon, J. Z. (1953). The kinetics of linear systems with special reference to periodic reactions. *Bulletin of Mathematical Biophysics*, 15, 121-141.

Hett, J. M. (1971). *Land-use changes in East Tennessee and a simulation model that describes these changes for three counties.* ORNL/IBP-71/8, Oak Ridge National Laboratory, Oak Ridge, Tennessee.

Hett, J. M. and O'Neill, R. V. (1974). Systems analysis of the Alent ecosystem. *Arctic Anthropology*, 11, 31-40.

Horn, H. S. (1975). Markovian processes of forest succession. In *Ecology and Evolution of Communities*, M. L. Cody and J. M. Diamond, eds. Harvard University Press, Cambridge, Massachusetts. 196-211.

Innis, G. S., ed. (1978). *Grassland Simulation Model.* Springer Verlag, New York.

Jacquez, J. A. (1972). *Compartmental Analysis in Biology and Medicine.* Elsevier, New York.

Jameson, D. A. (1970). Basic concepts in mathematical modeling of grassland ecosystems. In *Modeling and Systems Analysis in Range Science*, D. A. Jameson, ed. Range Science Series Number 5, Range Science Department, Colorado State University, Fort Collins, Colorado. 1-15.

Johnson, W. C. and Sharpe, D. M. (1976). An analysis of forest dynamics in the northern Georgia Piedmont. *Forest Science*, 22, 307-322.

Jordan, C. F., Kline, J. R., and Sasscer, D. S. (1972). Relative stability of mineral cycles in forest ecosystems. *American Naturalist*, 106, 237-253.

Jordan, C. F., Kline, J. R., and Sasscer, D. S. (1973). A simple model of strontium and manganese dynamics in a tropical rain forest. *Health Physics*, 24, 477–489.

Kaye, S. V. and Ball, S. J. (1969). Systems analysis of a coupled compartment model for radionuclide transfer in a tropical environment. In *Proceedings of the Second National Symposium on Radioecology*, D. J. Nelson and F. Evans, eds. CONF-670503. Ann Arbor, Michigan.

Kelley, J. M., Opstrup, P. A., Olson, J. S., Auerbach, S. I. and VanDyne, G. M. (1969). *Models of seasonal primary productivity in Eastern Tennessee FESTUCA and ANDROPOGON ecosystems.* ORNL-4310, Oak Ridge National Laboratory, Oak Ridge, Tennessee.

Kercher, J. R. (1977). *Mathematical Methods of Analysis of Linear Models: A documentation of the linear ecosystem analysis package.* EDFB/IBP-75/6, Oak Ridge National Laboratory, Oak Ridge, Tennessee.

Kerlin, T. W. and Lucius, J. L. (1966). *The SFR-3 code: a FORTRAN program for calculating the frequency response of a multivariate system and its sensitivity to parameter changes.* ORNL/TM-1575, Oak Ridge National Laboratory, Oak Ridge, Tennessee.

Kowal, N. E. and Crossley, D. A. (1971). The ingestion rates of microarthropods in pine more, estimated with radioactive calcium. *Ecology*, 52, 444–451.

Koztitzin, V. A. (1935). *Evolution de l'atmosphere.* Herman, Paris.

Leak, W. B. (1970). Successional change in northern hardwoods predicted by birth and death simulation. *Ecology*, 51, 794–801.

Lewis, S. and Nir, A. A study of parameter estimation procedures of a model for lake phosphorus dynamics. *Ecological Modelling* (in press).

Lindemann, R. L. (1942). The trophic dynamic aspect of ecology. *Ecology*, 23, 399–418.

Lotka, A. J. (1925). *Elements of physical biology.* Williams and Wilkins, Baltimore.

Mankin, J. B., Shugart, H. H., and Van Hook, R. I. (1973). Comparison of models of an old field arthropod food chain.

In *Proceedings of the 1973 Summer Computer Simulation Conference.* Montreal, Canada.

Martin, W. E. (1966). Transfer of Strontium-90 from plants to rabbits. *Health Physics,* 12, 621-631.

Martin, W. E., Raines, G. E., Bloom, S. G., and Levin, A. A. (1969). Ecological transfer mechanisms - terrestrial. *Proceedings of the Symposium on Public Health Aspects of Peaceful Uses of Nuclear Explosives.* Las Vegas, Nevada.

May, P. F., Till, A. R., and Cummings, M. J. (1972). Systems analysis of Sulfur-35 kinetics in pastures grazed by sheep. *Journal of Applied Ecology,* 9, 25-49.

May, R. (1972). What is the chance that a large complex system will be stable? *Nature,* 238, 413-414.

McBrayer, J. F. and Reichle, D. E. (1971). Trophic structure and feeding rates of forest soil invertebrate populations. *Oikos,* 22, 381-388.

McGinnis, J. T., Golly, F. B., Clements, R. G., Child, G. I., and Dueve, M. J. (1969). Elemental and hydrologic budgets of the Panamanian tropical moist forest. *Bioscience,* 19, 697-700.

Melsa, J. L. and Schultz, D. G. (1969). *Linear control systems.* McGraw Hill, New York.

Mitchell, J. E., Waide, J. B., and Todd, R. L. (1975). A preliminary compartment model of the nitrogen cycle in a deciduous forest ecosystem. In *Mineral Cycling in Southeastern Ecosystems,* F. G. Howell, J. B. Gentry, and M. H. Smith, eds. USERDA, CONF-740513, Technical Information Center, Springfield, Virginia, 41-57.

Mulholland, R. J. and Gowdy, C. M. Theory and application of the measurement of structure and determination of function for laboratory ecosystems. *Journal of Theoretical Biology* (in press).

Mulholland, R. J. and Keener, M. S. (1974). Analysis of linear compartment models for ecosystems. *Journal of Theoretical Biology,* 44, 105-116.

Mulholland, R. J. and Sims, C. S. (1976). Control theory and the regulation of ecosystems. In *Systems Analysis and Simulation in Ecology, Vol. 4,* B. C. Patten, ed. Academic Press, New York. 373-389.

Neel, R. B. and Olson, J. S. (1962). *Use of analog computers for simulating the movement of isotopes in ecological systems.* ORNL-3172, Oak Ridge National Laboratory, Oak Ridge, Tennessee.

Odum, H. T. (1957). Trophic structure and productivity of Silver Spring, Florida. *Ecological Monographs*, 27, 55-112.

Odum, H. T. (1970). Summary, an emerging view of the ecological system at El Verde. In *A Tropical Rain Forest*, H. T. Odum and R. F. Pigeon, eds. TID-24270. Division of Technical Information, USAEC, Washington, DC.

Odum, H. T., Moore, S. M., and Burns, L. A. (1970). Hydrogen budgets and compartments in the rain forest. In *A Tropical Rain Forest*, H. T. Odum and R. F. Pigeon, eds. TID-24270, Division of Technical Information, USAEC, Washington, DC.

Oak Ridge Systems Ecology Group (1974). Dynamic ecosystem models: Progress and challenges. In *Ecosystem Analysis and Prediction*, S. A. Levin, ed. Society for Industrial and Applied Mathematics, Philadelphia.

Olson, J. S. (1959). Health Physics Annual Report, ORNL-4446. Oak Ridge National Laboratory, Oak Ridge, Tennessee.

Olson, J. S. (1960). Health Physics Annual Report, ORNL-4634. Oak Ridge National Laboratory, Oak Ridge, Tennessee.

Olson, J. S. (1963a). Analog computer models for movement of isotopes through ecosystems. In *Proceedings of the First National Symposium on Radioecology*, V. Schultz and A. W. Klements, eds. Reinhold, New York and the American Institute for the Biological Sciences, Washington, DC.

Olson, J. S. (1963b). Energy storage and the balance of producers and decomposers in ecological systems. *Ecology*, 44, 322-332.

Olson, J. S. (1964). Gross and net production of terrestrial vegetation. In *Proceedings of the Jubilee Symposium, British Ecological Society*, A. MacFadyen and P. J. Newbould, eds. (Supplement to the *Journal of Ecology*, 52 and 53).

Olson, J. S. (1965). Equations for cesium transfer in a LIRIODENDRON forest. *Health Physics*, 11, 1385-1392.

O'Neill, R. V. (1970). *Pathway analysis: a preliminary application of systems ecology to nuclear facility safety evaluation.* ORNL-CF-70-3-25, Oak Ridge National Laboratory, Oak Ridge, Tennessee.

O'Neill, R. V. (1971a). Tracer kinetics in total ecosystems: a systems analysis approach. In *Nuclear Techniques in Environmental Pollution*. International Atomic Energy Agency, Vienna, Austria. 693-705.

O'Neill, R. V. (1971b). Systems approaches to the study of forest floor arthropods. In *Systems Analysis and Simulation in Ecology, Vol. 1*, B. C. Patten, ed. Academic Press, New York. 441-477.

O'Neill, R. V. (1971c). *Examples of ecological transfer matrices*. ORNL/IBP-71/3, Oak Ridge National Laboratory, Oak Ridge, Tennessee.

O'Neill, R. V. (1973). Error analysis of ecological models. In *Radionuclides in Ecosystems*, D. J. Nelson, ed. USAEC-CONF-710501, United States Atomic Energy Commission, Washington, DC. 898-908.

O'Neill, R. V., Burke, O. W., and Booth, R. S. (1971). *A simple systems model for DDT and DDE movement in human foodchains*. ORNL/IBP-71/9, Oak Ridge National Laboratory, Oak Ridge, Tennessee.

O'Neill, R. V. and Rust, B. Aggregation error in ecological models. *Ecological Modeling* (in press).

Patten, B. C. (1964). *System approach in radiation ecology*. ORNL/TM-1008, Oak Ridge National Laboratory, Oak Ridge, Tennessee.

Patten, B. C. (1966). Systems ecology - a course sequence in mathematical ecology. *Bioscience*, 16, 593-598.

Patten, B. C. (1971). A primer for ecological modeling and simulation with analog and digital computers. In *Systems Analysis and Simulation in Ecology, Vol. 1*, B. C. Patten, ed. Academic Press, New York. 3-121.

Patten, B. C. (1972). A simulation of the short-grass prairie ecosystem. *Simulation*, 19, 177-186.

Patten, B. C. (1974). The zero state and ecosystem stability. In *Proceedings of the First International Congress of Ecology*. Centre for Agricultural Publications and Documentation, Wageningen, The Netherlands.

Patten, B. C. (1975). Ecosystem linearization: an evolutionary design problem. *American Naturalist*, 109, 529-539.

Patten, B. C., Basserman, R. W., Finn, J. T., and Cole, W. G. (1975a). Propagation of cause in ecosystems. In *Systems Analysis and Simulation in Ecology*, Vol. *4*, B. C. Patten, ed. Academic Press, New York.

Patten, B. C., Egloff, D. A., Rechardson, T. H., *et al.* (1975b). Total ecosystem model for a cove of Lake Texoma. In *Systems Analysis and Simulation in Ecology*, Vol. *3*, B. C. Patten, ed. Academic Press, New York.

Patten, B. C. and Witkamp, M. (1967). Systems analysis of Cesium-134 kinetics in terrestrial microcosms. *Ecology*, 48, 813-825.

Raines, G. E., Bloom, S. G., Levins, A. A. (1969). Ecological models applied to radionuclide transfer in tropical ecosystems. *Bioscience*, 19, 1086-1091.

Reeves, M. (1971). *A code for linear modeling of an ecosystem.* ORNL/IBP-71/2. Oak Ridge National Laboratory, Oak Ridge, Tennessee.

Reichle, D. E., Dunaway, P. B., and Nelson, D. J. (1970). Turnover and concentration of radionuclides in food chains. *Nuclear Safety*, 11, 43-55.

Reichle, D. E., O'Neill, R. V., and Olson, J. S., eds. (1973). *Modeling forest ecosystems.* EDFB/IBP-73/7, Oak Ridge National Laboratory, Oak Ridge, Tennessee.

Rykiel, E. J. and Kuenzel, N. T. (1971). Analog computer models of the waves of Isle Royale. In *Systems Analysis and Simulation in Ecology*, Vol. *1*, B. C. Patten, ed. Academic Press, New York. 513-541.

Shearer, J. L., Murphy, A. T., and Richardson, H. T. (1967). *Introduction to System Dynamics.* Addison-Wesley, Reading, Massachusetts.

Sheppard, C. W. and Householder, A. S. (1951). The mathematical basis of the interpretation of tracer experiments in closed steady-state systems. *Journal of Applied Physics*, 22, 510-520.

Shugart, H. H., Crow, T. R., and Hett, J. M. (1973). Forest succession models: a rationale and methodology for modeling forest succession. *Forest Science*, 19, 203-212.

Shugart, H. H. and O'Neill, R. V., eds. *Benchmark Papers in Systems Ecology.* Dowden, Hutchinson, and Ross, Stroudsburg, Pennsylvania (in press).

Shugart, H. H., Reichle, D. E., Edwards, N. T., and Kercher, J. R. (1976). A model of calcium cycling in an East Tennessee LIRIODENDRON forest: model structure, parameters, and frequency response analysis. *Ecology*, 57, 99-109.

Solomon, A. K. (1949). Equations for tracer experiments. *Journal of Clinical Investigation*, 28, 1297-1307.

Struxness, E. G., O'Neill, R. V., Kaye, S. V., and Booth, R. S. (1971). *Application of environmental systems analysis to the radiological safety analysis of releases to the environment.* ICRP/71/C4-04, International Commission on Radiological Protection.

Swartzman, G. L. (1970). *Some concepts of modeling.* Technical Report 32, Grassland Biome, Natural Resource Ecology Laboratory, Colorado State University, Fort Collins, Colorado.

Teorell, T. (1937). Kinetics of distribution of substances administered to the body. *Archives Internationales de Pharmacodynamic*, 57, 205-240.

Van Dyne, G. M. (1966). *Ecosystems, Systems Ecology, and Systems Ecologists.* ORNL-3957. Oak Ridge National Laboratory, Oak Ridge, Tennessee.

Volterra, V. (1926). Variazioni e fluttuazioni del numero d'individui in specie animali conviventi. *Memoria Academie Lincei*, 2, 31-113.

Waggoner, P. E. and Stephens, G. R. (1970). Transition probabilities for a forest. *Nature*, 225, 1160-1161.

Waide, J. B., Krebs, J. E., Clarkson, S. F., and Setzler, E. M. (1974). A linear systems analysis of the calcium cycle in a forested watershed ecosystem. *Progress in Theoretical Biology*, 3, 261-345.

Waide, J. B. and Webster, J. R. (1976). Engineering systems analysis: applicability to ecosystems. In *Systems analysis and simulation in ecology, Vol. 4,* B. C. Patten, ed. Academic Press, New York. 330-372.

Webster, J. R. (1975). *Analysis of Potassium and calcium dynamics in stream ecosystems on three southern Appalachian watersheds of contrasting vegetation.* Ph.D. thesis, University of Georgia, Athens, Georgia.

Webster, J. R., Waide, J. B., and Patten, B. C. (1975). Nutrient recycling and the stability of ecosystems. In *Mineral Cycling in Southeastern Ecosystems*, F. G. Howell, J. B. Gentry, and M. H. Smith, eds. Technical Information Center, USERDA, CONF-740513. Springfield, Virginia. 1-27.

Wiberg, D. M. (1971). *State Space and Linear Systems*. Schaum's Outline Series, McGraw Hill, New York.

Wiegert, R. G. (1975). Simulation models of ecosystems. *Annual Review of Ecology and Systematics*, 6, 311-338.

Williams, R. B. and Murdoch, M. B. (1972). Compartmental analysis of the production of JUNCUS ROEMERIANUS in a North Carolina salt marsh. *Chesapeake Science*, 13, 69-79.

Witkamp, M. (1970). Mineral retention by epiphyllic organisms. In *A Tropical Rain Forest*, H. T. Odum and R. F. Pigeon, eds. Division of Technical Information, USAEC, TID-24270, Washington, DC.

Witkamp, M. and Frank, M. L. (1969). Cesium-137 kinetics in terrestrial microcosms. In *Proceedings of the Second National Symposium on Radioecology*, D. J. Nelson and F. C. Evans, eds. CONF-670503, Ann Arbor, Michigan. 635-734.

Witkamp, M. and Frank, M. L. (1970). Effects of temperature, rainfall, and fauna on transfer of Cs-137, K, Mg and mass in consumer-decomposer microcosms. *Ecology*, 51, 465-474.

Woodwell, G. M., Craig, P. P., and Johnson, H. A. (1971). DDT in the biosphere: where does it go? *Science*, 174, 1101-1107.

Wymore, A. W. (1967). *A Mathematical Theory of Systems Engineering - the Elements*. Wiley, New York.

Zilversmit, D. B., Entenman, C., and Fishler, M. D. (1943). On the calculation of turnover time and turnover rate from experiments involving the use of labeling agents. *Journal of General Physiology*, 26, 325-331.

[*Received May* 1978. *Revised September* 1978]

J. H. Matis, B. C. Patten, and G. C. White, (eds.),
Compartmental Analysis of Ecosystem Models, pp. 29-42. All rights reserved.
Copyright ©1979 by International Co-operative Publishing House, Fairland, Maryland.

A COMPARTMENTAL MODEL OF A MARINE ECOSYSTEM[1]

G. G. WALTER

Department of Mathematics
The University of Wisconsin
Milwaukee, Wisconsin 53201 USA

SUMMARY. A compartmental model of the interactions between various finfish species on Georges Bank is constructed. The flow rates are estimated by using stomach content data, biomass estimates, and energy consumption estimates for eleven fish species, squid, and other invertebrates. The graph of the model is used to estimate the dimension of the ecological phase space and the trophic status of each species. The Markov chain form of the compartmental model is then used to determine the ultimate energy distribution in the apex predator compartments. The effect of eliminating Mackerel and Herring respectively is compared.

KEY WORDS. compartmental model, fishery, ecosystem, Georges Bank, graph.

1. INTRODUCTION

It has become increasingly evident to fisheries managers that interactions between species and with the environment must be considered in the proper regulation of fisheries (Gulland, 1977). A number of different approaches have been tried from relatively simple competition equations (Walter and Hogman, 1971; Pope, 1976) to more complex simulation models (Anderson, Lassen, and Ursin, 1973). The difficulty seems to be the same in both types; the parameters of the model are difficult to determine. Little use seems to have been made of pure compartmental models in fisheries management; probably because they are not thought of as being

---

[1] Center for Great Lakes Studies Contribution No. 185.

realistic. They often assume that the compartments are homogeneous and that the flow rates are linear functions of the donor compartment. A stock of fish is rarely homogeneous even if fully recruited. It is composed of various age and size classes which may have different feeding habits and may be found in different locations. In some cases it is necessary to combine a number of species into a single compartment thereby increasing its heterogeneity even more. Moreover the non-linearities in, for instance the logistic equation, form the basis for regulation (Schaefer, 1954). Nonetheless, such a model can form a first approximation (Patten, 1975) in the study of an ecosystem. Furthermore, one can often infer certain properties of the ecosystem from the structure of the compartmental model alone without exact knowledge of the flow rates.

In this work we shall attempt to construct a partial model of various finfish species on Georges Bank. This region in the Northwest Atlantic is a traditional fishing ground whose production declined sharply in the 1970's (Clark and Brown, 1977). We shall use available data (Maurer, 1975) on analysis of stomach contents to construct food paths between ten species of finfish, squid, and other invertebrates. We use this information together with biomass estimates and estimates of production and consumption for the same region to obtain rough approximation to the flow rates. This model is then analyzed in a number of ways. First a competition graph is constructed and the dimension of the ecological phase space established. Next, a study of the trophic status of the various species is made. Finally, an important subsystem is identified and the effect of deleting one of the prey species is analyzed.

2. THE FLOW MODEL

The starting point is the predator-prey matrix (Maurer, 1975) derived from a summary of stomach contents of ten predators. These were taken from a region which included, but was not restricted to, Georges Bank during 1969-72. Only those prey of more than 100 g total per 10^6 of predator are included (Table 1).

This matrix is then used to construct a food web, as shown in Figure 1, where the abbreviations are self-explanatory. Each path represents the existence of more than 1 g of the donor compartment per kg of the recipient found in the latter's stomach.

There are a few analyses one can perform without knowing the actual values of the flow rates. One involves using the competition graph to determine the dimension of the ecological phase space. The other involves a determination of the trophic status of each species (see Roberts, 1976, p. 140).

TABLE 1:   *Predator-prey relationships expressed as grams of prey per kilogram of predator   (Maurer, 1975).*

Prey	Predators									
	C	Ha	R	YT	OF_l	He	M	P	SH	OF
Cod	–	–	–	–	–	–	–	–	–	–
Haddock	–	–	–	–	–	–	–	–	–	–
Redfish	.4	–	–	–	–	–	–	–	–	.2
Yellowtail	.3	–	–	–	–	–	–	–	–	–
Other flatfish[a]	.2	–	–	–	–	–	–	–	–	.3
Herring	2.5	–	–	–	–	–	–	–	1.3	–
Mackerel	.4	–	–	–	–	–	–	–	2.4	1.3
Pollock	–	–	–	–	–	–	–	–	–	–
Silver Hake	.1	–	–	–	.2	–	–	.2	.3	.3
Other finfish[b]	7.6	–	–	–	–	–	–	4.4	5.5	6.7
Squid	–	–	–	–	.1	–	–	–	.3	1.2
Other invertebrates	5.3	2.8	4.5	2.0	2.8	2.7	2.2	10.4	3.4	3.9
Total	16.8	2.8	4.5	2.0	3.1	2.7	2.2	15.0	13.2	13.9

[a] This category includes five additional species.

[b] This predator category includes fifteen other finfish, while the prey category includes unidentifiable fish of all species, and fish eggs as well.

The competition graph is obtained from the food web digraph by joining two vertices when they have the same prey. The resulting graph can be studied from the standpoint of inter-section graphs. Each vertex corresponds to a set and two vertices are joined if the sets intersect. These sets may be identified with the ecological niche in some ecological phase space since two niches overlap (i.e., the sets intersect) exactly when two species are in competition.

This ecological phase space is multidimensional in general. Yet not all dimensions may be needed to separate noncompeting species. It may be that the space can be projected onto a two- or three-dimensional subspace which preserves the essential relations between the niches. That is, overlaps and lack of over-laps occur in the subspace as they do in the entire phase space. The problem then is to try to represent the competition graph as the intersection graph of k-dimensional intervals whose dimension is as small as possible.

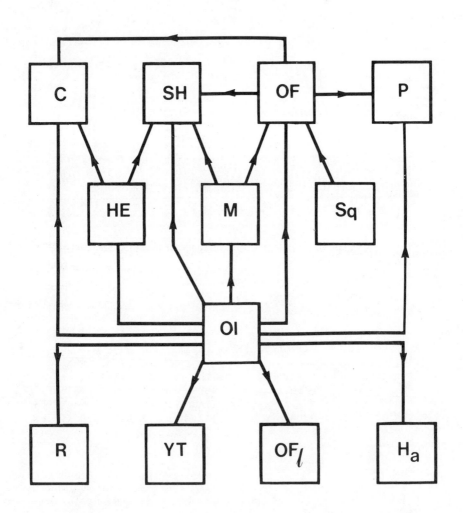

FIG. 1: *Principal food paths for the twelve species on Georges Bank. Only those paths corresponding to more than one gram of food per kilogram of predator are shown.*

In the case of an ecosystem in which all species except for pure food species are in competition with each other, the dimension can be taken to be one. Since then each vertex except the pure food vertices may be represented by the same one-dimensional interval. This is what occurs in the particular food web under study. All species except squid and other invertebrates are in competition with each other.

If we ignore the primary food species and consider only higher level predator-prey relations the same picture emerges. It is shown by the competition graph (see Figure 2) which again can be represented as a one-dimensional interval graph. Indeed let the intervals on the real line be given by (0,3) for Cod, (2,6) for Silver Hake, (1,4) for Pollock, (5,7) for other finfish and any other set of disjoint intervals for the other species. The intersection graph of these intervals is the same as the graph given by Figure 2. If, on the other hand, one dimension may not be sufficient. This could happen if Pollock competed with both other fish and Cod. However, a proper competition analysis will have to await additional data.

In order to determine the trophic status of each species, we again use the digraph and two measures $h_\ell(u)$ and $h_s(u)$ of trophic status. Both measures are of the form $\Sigma\, kn_k$ where n_k = no. of species at level k below a particular compartment u. In the case of h_s the level is determined by the shortest path and in h_ℓ by the longest path. The results are given in Table 2.

Thus Cod appears to be king under both measures of trophic status. These measures are a rough indication of the relative proportion of the total energy residing in each of the various compartments.

3. THE COMPARTMENTAL MODEL

The original flow model is converted into a compartmental model by adding flow rates to the arcs joining the compartments. We shall assume the flows are linear and donor controlled and shall use the numbering given in Table 2. It should be observed that R, YT, OF_ℓ, and Ha are combined into a single compartment (3) since each has input only from compartment 1 and no output flow.

We shall determine the flow rates by using the stomach content data to partition the consumption by each species among its

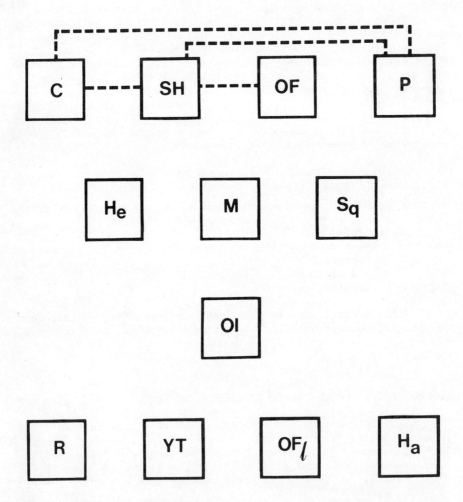

FIG. 2: *Higher level competition graph (shown by dotted lines). Species are connected if they compete for food other than invertebrates.*

TABLE 2: Trophic status of the twelve species as determined by two different measures.

Comp #	u	$h_\ell(u)$	$h_s(u)$
1	OI	0	0
2	Sq	0	0
	R	1	1
	YT	1	1
3	OF_ℓ	1	1
	Ha	1	1
4	He	1	1
5	M	1	1
6	OF	4	3
7	P	8	6
8	SH	8	6
9	C	9	7

prey. The annual consumption in turn is obtained from knowledge of the consumption to biomass ratios and biomass estimates during some reference period. The annual production is obtained by multiplying these by the growth efficiency. The biomass esti- mates are those used in Clark and Brown (1977) for the years 1972-74 on Georges Bank* (see Table 3); the consumption to biomass ratios as well as the growth efficiencies are those calculated by Grosslein, Langton, and Sissenwine (1978).

Since we are assuming a donor controlled model the consumption by the ith compartment of the jth compartment is given by

$$a_{ij} X_j(t)$$

where $X_j(t)$ is the amount of biomass in compartment j at time t. During the reference period this quantity is

$$a_{ij} \bar{X}_j = r_{ij}\, p_i\, \bar{X}_i,$$

where $\bar{X}_j$ is the mean biomass during the reference period, r_{ij} the proportion of food from the jth compartment in the stomach of

*Not given directly, but may be calculated from their Tables 7 and 11.

TABLE 3: Estimates of biomass, growth efficiency, and consumption to biomass ratios for the various species on Georges Bank in 1972-74.

Species	C	SH	P	OF	M	He	Ha	OF$_\ell$	YT	R	Sq
Compartment	9	8	7	6	5	4		3			2
Biom. 10^5MT	2	3	0.2	3	13	5	0.5	0.5	0.2	0.1	6
Growth eff.	0.19	0.12	0.11	0.11	0.10	0.06	0.11	0.11	0.14	0.08	0.21
Cons./biom.	3.2	5.0	4.1	4.1	4.4	4.5	2.8	4.1	4.5	3.0	7.0

the ith compartment, and p_i the consumption to biomass ratio. However, the flow into a compartment is not the consumption, but must be multiplied by the growth efficiency, g_i. The remainder goes primarily for metabolism and thus is lost.

Hence the flow model must be modified by adding arrows out of each compartment. The compartmental model resulting is shown in Figure 3. The numbers on the arrows except for these new arrows refer to the a_{ij}'s where

$$a_{ij} = r_{ij}\, p_i\, \bar{X}_i / \bar{X}_j .$$

The only thing lacking is the estimate for the biomass of compartment 1 which we take to be 10 times the total finfish biomass.

The differential equation corresponding to this model is given by $dX/dt = AX$; where $X^T = [x_1 x_2 x_4 x_5 x_6 x_7 x_8 x_9 x_3]$, and

$$
A = 10^{-3}
\begin{bmatrix}
-329 & 0 & 0 & 0 & 0 & 0 & 0 & 0 & 0 \\
0 & -384 & 0 & 0 & 0 & 0 & 0 & 0 & 0 \\
5 & 0 & -524 & 0 & 0 & 0 & 0 & 0 & 0 \\
19 & 0 & 0 & -417 & 0 & 0 & 0 & 0 & 0 \\
3 & 42 & 0 & 21 & -3369 & 0 & 0 & 0 & 0 \\
0 & 0 & 38 & 0 & 9 & 0 & 0 & 0 & 0 \\
2 & 0 & 39 & 27 & 268 & 0 & 0 & 0 & 0 \\
1 & 0 & 0 & 0 & 200 & 0 & 0 & 0 & 0 \\
2 & 0 & 0 & 0 & 0 & 0 & 0 & 0 & 0
\end{bmatrix} .
$$

Here we have moved the third row and third column to the end.

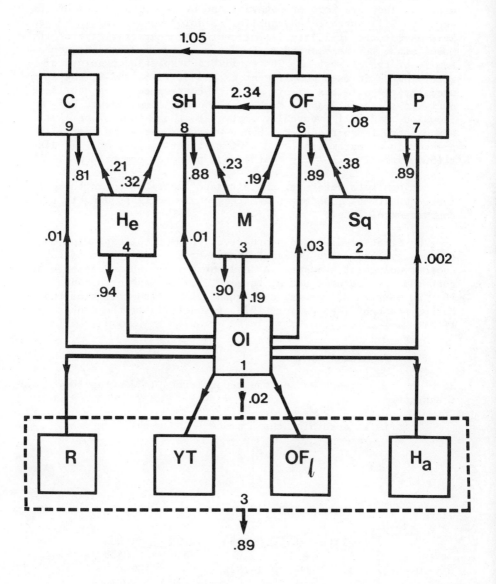

FIG. 3: *Compartmental model for Georges Bank. Weights on arcs correspond to donor controlled flow in years^{-1}. The weights on arrows out of each compartment with no terminal end correspond to that proportion of incoming biomass which is not converted into that species biomass.*

The eigenvalues are the elements on the diagonal of the matrix. They are zero or negative and hence any solution of the equation will approach an equilibrium solution as time increases. This asymptotic stability is, of course, a characteristic of the model since it was assumed to be donor controlled and not necessarily of the ecosystem. The equilibrium solution consists of zero biomass in each compartment except 3, 7, 8, and 9, the apex predator compartments. This solution corresponds to a heuristic analysis of the digraph. Each of the other compartments has flows out and will eventually empty into those four or be lost since there is no feedback. Hence whatever energy is initially present in compartments 1 and 2 will either be lost or will distribute itself ultimately to the apex predator compartments.

In order to determine the ultimate distribution in those four compartments, we use the Markov chain corresponding to this compartmental model.

We first need to add another compartment corresponding to the energy lost. The corresponding matrix results from adding another column of zeros to A and another row whose elements are such that the columns add up to zero. If we denote this new 10×10 matrix B, we may obtain the transition matrix of the Markov chain by (see Walter, 1978, for details of the construction)

$$P = (hB + I)^T.$$

Here h is a time step and the superscript T denotes the transpose.

This is an absorbing chain since $\underset{\sim}{P}$ has the form

$$P = \begin{bmatrix} Q & R \\ 0 & I \end{bmatrix},$$

where

$$Q = \begin{bmatrix} 1-.33h & 0 & .005h & .019h & .003h \\ 0 & 1-.38h & 0 & 0 & .042h \\ 0 & 0 & 1-.52h & 0 & 0 \\ 0 & 0 & 0 & 1-.42h & .021h \\ 0 & 0 & 0 & 0 & 1-.3.37h \end{bmatrix}$$

and

$$R = h10^{-3} \begin{bmatrix} 0 & 2 & 1 & 2 & 297 \\ 0 & 0 & 0 & 0 & 342 \\ 0 & 38 & 32 & 0 & 447 \\ 0 & 27 & 0 & 0 & 369 \\ 9 & 268 & 209 & 0 & 2892 \end{bmatrix} .$$

The matrix N, given by

$$N = (I - Q)^{-1}R,$$

gives the probabilities of being absorbed in state j if the chain starts in state i. The probabilities, as one might expect, are independent of the time step h used

$$(I - Q)^{-1}R = \begin{bmatrix} 3.04 & 0 & .03 & .14 & 0 \\ 0 & 2.60 & 0 & 0 & .03 \\ 0 & 0 & 1.91 & 0 & 0 \\ 0 & 0 & 0 & 2.40 & .02 \\ 0 & 0 & 0 & 0 & .30 \end{bmatrix} R$$

$$= \begin{bmatrix} 0 & .019 & .004 & .006 & .968 \\ 0 & .008 & .006 & 0 & .976 \\ 0 & .072 & .074 & 0 & .854 \\ 0 & .070 & .004 & 0 & .929 \\ .003 & .080 & .060 & 0 & .868 \end{bmatrix} .$$

The first two rows are the important ones. The first gives the distribution of one gram initially in compartment 1 (0.I.) in the last compartment or in the four apex predator compartments P, SH, C, and {Ha,OF$_\ell$,YT,R}. The second row gives the distribution of one gram of Squid in these compartments.

The rows of $(I-Q)^{-1}$ also have a useful interpretation. Their sum is the expected time to absorption. Since our time units are in years, the expected time to absorption of a unit biomass of OI is 3.21 years.

There is another sense in which this model can be analyzed. The Herring and Mackerel compartments form a central link in the food web. They both have the same food source and some of the same predators. If one or the other were eliminated, the distribution of apex predators would certainly change, but it's not clear in which direction. Let us briefly look at this change.

If Herring were eliminated, the matrix (I−Q) would be the
same except for the third row and third column which would be
eliminated and the first element in the first row which would be
.32h. Hence we would have

$$(I-Q)^{-1}R = \begin{bmatrix} 3.09 & 0 & .14 & 0 \\ 0 & 2.60 & 0 & .03 \\ 0 & 0 & 2.40 & .02 \\ 0 & 0 & 0 & .30 \end{bmatrix} R$$

$$= \begin{bmatrix} 0 & .011 & .004 & .006 & .977 \\ x & x & x & x & x \\ x & x & x & x & x \\ x & x & x & x & x \end{bmatrix}.$$

That is, 1.1% of the initial gram of biomass in compartment 1 will
end up in compartment 8, i.e., the SH compartment, while 0.4%
will end up in compartment 9, the C compartment. This compares
with 1.9% and 0.4% previously. This would be a desirable outcome
since Cod are generally considered more valuable than Silver Hake.

If, on the other hand, the Mackerel compartment were eliminated, the outcome would be somewhat different. The matrix I−Q
would then have the fourth row and column eliminated and the
element in the upper left hand corner changed to .31 h. Then we
would have

$$(I-Q)^{-1}R = \begin{bmatrix} 3.20 & 0 & .03 & .004 \\ 0 & 2.60 & 0 & .03 \\ 0 & 0 & 1.91 & 0 \\ 0 & 0 & 0 & .30 \end{bmatrix} R$$

$$= \begin{bmatrix} 0 & .009 & .005 & .006 & .970 \\ x & x & x & x & x \\ x & x & x & x & x \\ x & x & x & x & x \end{bmatrix}.$$

In this case the proportion of energy going to Silver Hake
is decreased even more while that going to Cod is increased
slightly.

The implications for fisheries management if clear: as an intermediate link in the food chain, Herring seem to be more important from an economic standpoint than Mackerel.

4. SHORTCOMINGS

The obvious and most glaring shortcoming in this approach is the calculation of rate coefficients based on the relative biomass and food habits of the various species. Moreover these food habits may not accurately reflect the gut content studies on which they are based. These gut contents are often quite variable, even with large samples, and tend to underrepresent rapidly digested food such as squid. Thus the results should be considered very tentative and the quantities treated as orders of magnitude indication at best.

Another shortcoming is the grouping of all other invertebrates except for squid in one compartment. The grouping includes crustaceans, echinoderms, molluscs, etc., which are preyed on differently by different species. This could have an effect on the competition graph and the trophic status calculations.

Other shortcomings include the lack of testable hypotheses, the over-simplification of a complex natural system, the lack of delays in biomass transfer, etc., which are common to most compartmental models. Nonetheless, certain results such as the relative importance of Herring and Mackerel seem to be rather robust in that changes in parameters seem not to affect them very much. The question of how much awaits further analysis.

REFERENCES

Anderson, K. P., Lassen, H., and Ursin, E. (1973). A multi-species extension of the Beverton and Holt assessment model with an account of primary production. *International Council for the Exploration of the Sea, C. M. Pelagic Fish (Northern) Committee,* H:20.

Clark, S. H. and Brown, B. E. (1977). Changes in biomass of finfish and squids from the Gulf of Maine to Cape Hatteras, 1963-1974, as determined from research vessel survey data. *U. S. Fisheries Bulletin,* 75, 1-21.

Grosslein, M. D., Langton, R. W., and Sissenwine, M. P. (1978). Recent fluctuations in Pelagic fish stocks of the Northwest Atlantic, Georges Bank region in relationship to species interactions. *International Council for the Exploration of the Sea. Symposium on the Biological Basis of the Pelagic Fish Stock Management,* no. 25.

Gulland, J. A. (1977). Goals and objectives of fishery management. *U. N. FAO Fisheries Technical Paper* No. 166.

Maurer, R. (1975). A preliminary description of some important feeding relationships. *International Commission for the Northwest Atlantic Fishery, Research Document* 75/IX/130.

Patten, B. C. (1975). Ecosystem linearization: An evolutionary design problem. In *Ecosystem Analysis and Prediction,* S. A. Levin, ed. SIAM Institute for Mathematics and Society, Philadelphia. 182-201.

Pope, J. G. (1976). The effect of biological interactions on the theory of mixed fisheries. *International Commission for the Northwest Atlantic Fishery, Selected Papers* #1, 157-162.

Roberts, Fred S. (1976). *Discrete Mathematical Models.* Prentice-Hall, Englewood Cliffs, New Jersey.

Schaefer, M. B. (1954). Some aspects of the dynamics of population important to the management of the commerical fisheries. *Bulletin Inter-American Tropic Tuna Commission,* 1, 25-56.

Walter, G. and Hogman, W. (1971). Mathematical models for estimating changes in fish population with applications to Green Bay. *Proceedings 14th Conference on Great Lakes Research,* 170-184.

Walter, G. (1979). Compartmental models, digraphs, and Markov chains. In *Compartmental Analysis of Ecosystem Models,* J. H. Matis, B. C. Patten, and G. C. White, eds. Satellite Program in Statistical Ecology, International Co-operative Publishing House, Fairland, Maryland.

[*Received June* 1978. *Revised November* 1978]

J. H. Matis, B. C. Patten, and G. C. White, (eds.),
Compartmental Analysis of Ecosystem Models, pp. 43-72. All rights reserved.
Copyright ©1979 by International Co-operative Publishing House, Fairland, Maryland.

REVIEW AND EVALUATION OF INPUT-OUTPUT FLOW ANALYSIS FOR ECOLOGICAL APPLICATIONS*

M. CRAIG BARBER, BERNARD C. PATTEN

Department of Zoology and Institute of Ecology
University of Georgia
Athens, Georgia 30602 USA

JOHN T. FINN

Department of Forestry and Wildlife
University of Massachusetts
Amherst, Massachusetts 01003 USA

SUMMARY. Input-output flow analysis, originated in a restricted case by Leontief for economics applications, has recently undergone development for analysis of resource flows in ecological systems. Both static and dynamic methodologies have been formulated in deterministic and stochastic versions. These methodologies are appropriate for analysis of compartment models representing flows of conserved substances. The deterministic methods evolved from the Leontief approach, whereas stochastic extensions treat resource flows as Markov processes.

This paper reviews these flow analysis models. The two basic classes are input oriented and output oriented. The former look backward temporally through the interactive structure of a system to its input boundary. The latter look forward temporally through the system internal structure to its output boundary. In each case, within system input and output flow structures are reconstructed through the analysis. These structures and their char-

---

*University of Georgia, *Contributions in Systems Ecology*, No. 44.

acteristics are represented by various derived measures including
path length, cycling efficiency, cycling index, residence time,
number of intercompartmental transfers since entering or before
exiting the system, and origin of inputs and destination of outputs. Principal results are illustrated with a simple ecosystem
compartment model. Finally, the deterministic and stochastic
approaches are compared with respect to similarities and differences, and evaluated in terms of what they actually measure that
is ecologically significant.

KEY WORDS. ecosystem, compartment model, input-output analysis,
flow analysis, Markov chain.

1.   INTRODUCTION

Compartment models have become a prevalent means of representing resource dynamics in ecological systems. Employing the
principle of mass-energy conservation, such models may be formulated as systems of difference or differential equations, e.g.,
in the latter case,

$$\dot{x}_j(t) = \sum_{i=0}^{n} f_{ij}(t) - \sum_{k=0}^{n} f_{jk}(t) \quad (j=1,2,\cdots,n), \quad (1)$$

where x_j is the instantaneous storage of an energy or matter
resource in compartment j of n compartments, $\dot{x}_j$ is the
first time derivative of x_j, and f_{ij} and f_{jk} are instantaneous flows from compartment i to j and j to k, respectively. The environment is denoted by subscript 0.

Whether generated by such models, or by empirical measurement,
or whatever means, it becomes of interest to know explicitly the
quantitative relationships between various components in a
resource flow system. Recent developments in input-output flow
analysis enable such interactions to be determined. The purpose
of this paper is to review these developments and illustrate kinds
of uses to which they may be put in analysis of ecological compartment models.

Input-output analysis for applications in economics was
introduced by Leontief (1936, 1966). The original motivation was
supply-and-demand. What were the direct and indirect requirements of raw materials, goods and services to produce a unit of
finished product from some sector of the economy? Leontief wrote
equilibrium matrix equations relating each industry's require-

ments to output. Given a demand vector, he was able to solve the
equations for the output of each industry.

Hannon (1973) applied input-output analysis to ecological
systems, introducing a significant conceptual innnovation in the
process. He constructed Leontief type equations, but provided
outputs from components as knowns and formulated a matrix of
production coefficients to implement a matrix production equation.
Solution of this algebraic equation yielded an inverse matrix
whose entries quantified what Hannon referred to as the 'structure'
of the system. This structure consisted of the direct and indirect
flows within a system required to sustain outflow from each
component. It was thus possible to trace backward through static
equilibrium models and identify the origins of outputs. Cale
(1975) applied Hannon's formulation to a large scale simulation
model of carbon flow in a short-grass prairie ecosystem. The
simulation was dynamic, and the static methodology was employed
at selected instants to illustrate how the internal flow structure
of systems changes dynamically.

Patten was interested in the complementary problem, that of
tracing forward through the intrasystem flow network the movement of resources introduced as inputs. His student, Finn (1976a,
b), provided the necessary reorientation of Hannon's formulation.
Finn focused on nutrient cycling and defined several measures of
cycling in flow models. He also formulated (1977) the dynamic
case of input-output flow analysis, making it possible to analyze
changes in flow structure within systems dynamically.

The availability of two complementary methods, one developing
system structure in terms of forward propagation of resources from
inputs, and the other representing structure as traced backward
from outputs, enabled two different modes of intrasystem analysis.
Patten (1975) made a preliminary effort to place these in a general system theory framework, showing that the change in orien-
tation occasioned by considering causality progressing forward
from inputs versus regressing backward from outputs has more than
trivial consequences. Specifically, distinct input and output
environments can be defined and investigated for every system or
system component. This two-environment concept was then developed
comprehensively as an ecological system theory (Patten, Bosserman, Finn, and Cale, 1976). The principal features of the theory
were later reorganized and presented as three propositions
(Patten, 1978a). Finally, a control matrix was defined to examine
intercompartmental control relationships within flow models by
comparison of component input and output environments (Patten,
1978b). The two-environment approach to input-output analysis is
central in the dissertation of Finn (1977).

The Leontief-Hannon-Finn models of input-output flow analysis
are deterministic. However, as noted by Patten, *et al*. (1976, p.

530), similar results obtain from Markov chain theory. Kemeny and Snell (1960, p. 200) pointed out how Leontief's analysis could be represented as a Markov chain. Another student of Patten, Barber (1978a, b), under impetus of the developments outlined above, formulated Markov chain models of both forward flows of resource particles through a system from inputs, and reverse flows corresponding to movements toward outputs. This probabilistic methodology provides an alternative means of analyzing resource flows to the deterministic models. In subsequent sections both approaches will be described, compared and evaluated, and their utility in analysis of ecological flow models illustrated.

In equation (1) it is obvious that any resource flow f_{ij} is simultaneously an outflow from compartment i and an inflow to j. This duality of interpretation motivates two distinct but complementary classes of flow analyses. The first analyzes historical patterns of flows leading to output, and the second analyzes future flows resultant from input. Deterministic formulations of both classes are flux oriented whereas the stochastic methods are particle oriented. These complementary methods are outlined below.

2. DETERMINISTIC FLOW ANALYSIS

In the following, the terms 'compartment model' and 'system' will be used synonymously. The deterministic methods vary in how they have been presented by different authors. Finn's (1977) description, which generalizes previous ones in a unified notation, will be followed here, with some alterations for consistency with stochastic methods to be described later.

The deterministic methodologies are predicated on the assumption that a system's endogeneous flows are determined as the difference between those processes which introduce resources into the system's flow structure and those that remove resources from the flow structure. Thus, both negative compartmental derivatives $\dot{x}_j^-(t)$ and system inflows of $f_{0j}(t) \equiv z_j(t)$ introduce resources into circulation, while both positive compartmental derivatives $\dot{x}_j^+(t)$ and system outflows $f_{j0}(t) \equiv y_j(t)$ remove resources from circulation, $j=1,\cdots,n$. The system's transition function (1) can then be expressed as the difference between processes which introduce and remove resources to and from the circulating flow structure:

$$[\dot{x}_j^+(t) + y_j(t)] - [z_j(t) - \dot{x}_j^-(t)] = \sum_{i=1}^{n} f_{ij}(t) - \sum_{k=1}^{n} f_{jk}(t),$$

$$(2)$$

where $\dot{x}_j^+(t) = \max \{\dot{x}_j(t), 0\}$ and $\dot{x}_j^-(t) = \min \{\dot{x}_j(t), 0\}$. The throughflow $T_j(t)$ of compartment j is either the sum of inflows,

$$T_j(\Sigma in) \triangleq -\dot{x}_j^-(t) + z_j(t) + \sum_{i=1}^{n} f_{ij}(t) \quad (j=1,\cdots,n), \quad (3)$$

or the sum of outflows,

$$T_j(\Sigma out) \triangleq \dot{x}_j^+(t) + y_j(t) + \sum_{j=1}^{n} f_{jk}(t) \quad (j=1,\cdots,n). \quad (4)$$

These concepts are illustrated for a tropical rainforest model in Figure 1 and Example 1 (see following page).

Let the flow $f_{jk}(t)$ from compartment j to k be expressed as a fraction of $T_k(\Sigma in)$:

$$f_{jk}(t) = q'_{jk}(t) T_k(\Sigma in). \quad (5)$$

Substituting (3) and (5) into (2) gives

$$T_j(\Sigma in) = \sum_{k=1}^{n} q'_{jk}(t) T_k(\Sigma in) + y_j(t) + \dot{x}_j^+(t). \quad (6)$$

Equation (6) can be solved for $T_j(\Sigma in)$ by the corresponding matrix equation

$$\underset{\sim}{T}(\Sigma in) = (\underset{\sim}{I} - \underset{\sim}{Q}'(t))^{-1} [\dot{\underset{\sim}{x}}^+(t) + \underset{\sim}{y}(t)], \quad (7)$$

where $\underset{\sim}{T}(\Sigma in)$ is an $n \times 1$ vector of throughflows, $\underset{\sim}{I}$ is an $n \times n$ identity matrix, $\underset{\sim}{Q}'(t)$ is the $n \times n$ matrix $[q'_{ij}(t)]$, $\dot{\underset{\sim}{x}}^+(t)$ is the $n \times 1$ vector $[\dot{x}_i^+(t)]$, and $\underset{\sim}{y}(t)$ is the $n \times 1$ vector $[f_{i0}(t)]$. The matrix $\underset{\sim}{N}'(t) = [\underset{\sim}{I} - \underset{\sim}{Q}'(t)]^{-1} = [n'_{ij}(t)]$ defines the flow structure of the system in the sense of Hannon (1973). Interpretations of the elements $n'_{ij}(t) \in \underset{\sim}{N}'(t)$ are deferred until later.

Example 2. The matrices $\underset{\sim}{Q}'$ and $\underset{\sim}{N}'$ for the tropical rainforest static model of Figure 1 are as follows:

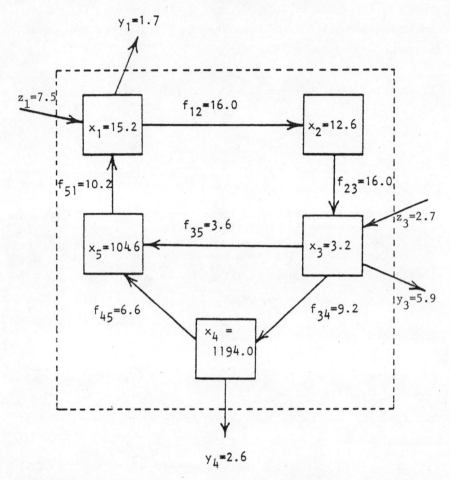

FIG. 1:   *Tropical rainforest static nitrogen model; storages*
(boxes) gN m^{-2}, *and flows (arrows) in* gN m^{-2}y^{-1}.

Example 1:   Figure 1 represents a static model $(\dot{x}_j(t) = 0,$
$j=1,\cdots,5)$ of nitrogen storage and flow in a tropical rainforest,
as described in Patten, *et al.* (1976, p. 574 ff.). Compartments
(boxes) represent storages, and arrows denote flows. The compartments are nitrogen in: x_1, leaves and epiphyllae; x_2, loose
litter; x_3, fibrous roots; x_4, soil; and x_5, wood. Note
that $T_j(\Sigma in) = T_j(\Sigma out)$ $(j=1,\cdots,5)$, since the model is static.
For simplicity, this model rather than a dynamic version will be
utilized for subsequent illustrations.

	$\underset{\sim}{Q}'$					$\underset{\sim}{N}' = (\underset{\sim}{I} - \underset{\sim}{Q}')^{-1}$				
to from	x_1	x_2	x_3	x_4	x_5	x_1	x_2	x_3	x_4	x_5
x_1	0	1	0	0	0	1.97	1.97	1.69	1.69	1.69
x_2	0	0	.86	0	0	0.97	1.97	1.69	1.69	1.69
x_3	0	0	0	1	.35	1.14	1.14	1.97	1.97	1.97
x_4	0	0	0	0	.65	0.74	0.74	0.63	1.63	1.28
x_5	.58	0	0	0	0	1.14	1.14	0.97	0.97	1.97

Deterministic output analysis is performed by letting the flow $f_{ij}(t)$ from compartment i to j be expressed as a fraction of $T_i(\Sigma out)$:

$$f_{ij}(t) = q''_{ij}(t)T_i(\Sigma out). \tag{8}$$

Substituting (4) and (8) into (2), we obtain

$$T_j(\Sigma out) = \sum_{i=1}^{n} q''_{ij}(t)T_i(\Sigma out) + z_j(t) - \dot{x}_j^-(t). \tag{9}$$

Equation (9) can again be solved for $T_j(\Sigma out)$ by the matrix equation

$$\underset{\sim}{T}(\Sigma out) = [-\underset{\sim}{\dot{x}}^-(t) + \underset{\sim}{z}(t)](\underset{\sim}{I} - \underset{\sim}{Q}''(t))^{-1}, \tag{10}$$

where $\underset{\sim}{T}(\Sigma out)$ is a $1 \times n$ vector of throughflows, $\underset{\sim}{I}$ is again the identity matrix of nth order, $\underset{\sim}{Q}''(t)$ is the $n \times n$ matrix $[q_{ij}''(t)]$, $\underset{\sim}{\dot{x}}^-(t)$ is the $1 \times n$ vector $[\dot{x}_i^-(t)]$, and $\underset{\sim}{z}(t)$ is the $1 \times n$ vector $[f_{0i}(t)]$. $\underset{\sim}{N}''(t) = (\underset{\sim}{I} - \underset{\sim}{Q}'')^{-1}$ also represents system structure, but now as propagated forward from inputs rather than backwards from outputs. The elements $n''_{ij}(t) \epsilon \underset{\sim}{N}''(t)$ will be interpreted below.

Example 3. The matrices $\underset{\sim}{Q}''$ and $\underset{\sim}{N}''$ for the tropical rainforest model are as follows:

$$\underset{\sim}{Q}'' \qquad\qquad\qquad \underset{\sim}{N}'' = (\underset{\sim}{I} - \underset{\sim}{Q}'')^{-1}$$

from \ to	x_1	x_2	x_3	x_4	x_5	x_1	x_2	x_3	x_4	x_5
x_1	0	.90	0	0	0	1.97	1.78	1.78	0.80	0.97
x_2	0	0	1	0	0	1.08	1.97	1.97	0.97	1.08
x_3	0	0	0	.49	.19	1.08	0.97	1.97	0.97	1.08
x_4	0	0	0	0	.72	1.42	1.28	1.28	1.63	1.42
x_5	1	0	0	0	0	1.97	1.78	1.78	0.88	1.97

The elements $n'_{ij}(t) \in \underset{\sim}{N}'(t)$ and $n''_{ij}(t) \in \underset{\sim}{N}''(t)$ are now to be interpreted. To motivate these interpretations, let $T_0(t)$ be the 'throughflow' of the system's environment. $T_0(t)$ is defined as either

$$T_0(\Sigma in) \overset{\Delta}{=} \sum_{k=1}^{n} f_{k0}(t) + \sum_{k=1}^{n} \dot{x}^+_k(t), \qquad (11)$$

or

$$T_0(\Sigma out) \overset{\Delta}{=} \sum_{k=1}^{n} f_{0k}(t) - \sum_{k=1}^{n} \dot{x}^-_k(t). \qquad (12)$$

Clearly, if

$$\dot{x}^+_0(t) = - \sum_{k=1}^{n} \dot{x}^-_k(t) \quad \text{and} \quad \dot{x}^-_0(t) = - \sum_{k=1}^{n} \dot{x}^+_k(t), \qquad (13,14)$$

then (11) is consistent with equation (3) and (12) with equation (4). These consistency relations are confirmed by the following consideration. Since the total amount of resource in the union of a system and its environment is constant, the sum

$$\sum_{i=0}^{n} \dot{x}_i(t) = \sum_{i=0}^{n} [\dot{x}^+_i(t) + \dot{x}^-_i(t)]$$

must equal zero. However, since equations (13) and (14) ensure this, definitions of T_0 given by (11) and (12) are consistent with the definitions of $T_j(t)$, $j \neq 0$, given by equations (3) and (4), respectively. Therefore, equations (7) and (10) can be expressed as

$$\underset{\sim}{T}(\Sigma in) = \underset{\sim}{N}'(t) \; \underset{\sim}{q}'_0(t) \; T_0(\Sigma in) \tag{15}$$

and

$$\underset{\sim}{T}(\Sigma out) = T_0(\Sigma out) \; \underset{\sim}{q}''_0(t) \underset{\sim}{N}''(t), \tag{16}$$

respectively, where $\underset{\sim}{q}'_0(t)$ is the column vector $\underset{\sim}{q}'_0(t) = [\dot{x}_i^+(t) + f_{i0}(t)]/[\Sigma_k \dot{x}_k^+(t) + f_{k0}(t)]$, and $\underset{\sim}{q}''_0(t)$ is the row vector $\underset{\sim}{q}''_0(t) = [-\dot{x}_i^-(t) + f_{0i}(t)]/[\Sigma_k f_{0k}(t) - \dot{x}_k^-(t)]$.

Since equation (15) expresses each throughflow $T_j(\Sigma in)$ defined by inflows as a function of $T_0(t)$, which is the flux of resources to the system's environment, (15) embodies the system input history. Specifically, $\underset{\sim}{q}'_0(t)$ transforms $T_0(t)$ into $T_i(\Sigma in)$, and $\underset{\sim}{N}'(t)$ transforms $T_i(\Sigma in)$ into $T_j(\Sigma in)$. Each entry $n'_{ij}(t) \; \varepsilon \; \underset{\sim}{N}'(t)$ can thus be interpreted in the following ways:

1) throughflow $T_i(\Sigma in)$ in compartment i required to generate one unit of output $T_j(\Sigma in)$ in compartment j;

2) total flow $\phi_{ij}(t)$, direct $f_{ij}(t)$ plus indirect over all paths of all lengths from compartment i to j, to sustain one unit of outflow $T_j(\Sigma in)$ from compartment j; and

3) number of times a unit of flow in compartment j has passed through i since entering the system.

Example 4. In the $\underset{\sim}{N}'$ matrix of Example 2, consider element $n'_{25} = 1.69$. This is the throughflow out of compartment 2 required to generate one unit of output from compartment 5. It is also the total flow propagated from 2 to 5 to sustain each unit of outflow from 5. And, a unit of flow at 5 has passed through 2 this number of times since entering the system.

For the output case, since equation (16) expressed each throughflow $T_i(\Sigma out)$ defined by outflows as a function of $T_0(t)$, the resource flux from the system's environment, (16) specifies the system's output future. That is, $\underset{\sim}{N}''(t)$ transforms $T_i(\Sigma out)$ to $T_j(\Sigma out)$ and $\underset{\sim}{q}''_0(t)$ transforms $T_j(\Sigma out)$ into $T_0(t)$. Thus each entry $n''_{ij}(t) \in \underset{\sim}{N}''(t)$ can be interpreted as:

1) throughflow $T_j(\Sigma out)$ in compartment j produced by one unit of $T_i(\Sigma out)$ in compartment i;

2) total flow $\phi_{ij}(t)$, including both direct $f_{ij}(t)$ and indirect over all paths of all lengths from i to j, generated by one unit of inflow $T_i(\Sigma out)$ into compartment i; and

3) number of times a unit of flow in compartment i will pass through j before exiting the system.

Example 5. In the $\underset{\sim}{N}''$ matrix of Example 3, consider element $n''_{25} = 1.08$. This is the throughflow into 5 generated by one unit of input to 2. It is the flow passing from 2 to 5 due to one unit of input into 2. Also, it is the number of times a unit of flow at 2 will pass through 5 before leaving the system.

By using the matrices $\underset{\sim}{N}'(t)$ and $\underset{\sim}{N}''(t)$, several measures of system structure have been defined (Finn, 1976a, b; 1977).

1) *Path length.* This is the number of compartments that an inflow to compartment i will pass through before leaving, or that an outflow from j has passed through since entering the system, counting multiple visits to the same compartment. Inflow path length is

$$ZPL_i(t) = \sum_{j=1}^{n} n''_{ij}(t) \quad (i=1,\cdots,n), \tag{17}$$

and outflow path length

$$YPL_j(t) = \sum_{i=1}^{n} n'_{ij}(t) \quad (j=1,\cdots,n). \tag{18}$$

System path lengths are weighted sums of compartment path lengths:

$$ZPL(t) = \sum_{i=1}^{n} \frac{z_i(t)}{\Sigma z(t)} ZPL_i(t), \tag{19}$$

and

$$YPL(t) = \sum_{j=1}^{n} \frac{y_j(t)}{\Sigma y(t)} YPL_j(t). \tag{20}$$

Example 6. Compartment and system inflow and outflow path lengths for the tropical rainforest nitrogen model are:

Inflow		Outflow	
$ZPL_1 = 7.4$	$ZPL_4 = 7.0$	$YPL_1 = 6.0$	$YPL_4 = 8.0$
$ZPL_2 = 7.1$	$ZPL_5 = 8.4$	$YPL_2 = 7.0$	$YPL_5 = 8.6$
$ZPL_3 = 6.1$	$\overline{ZPL} = 7.0$	$YPL_3 = 7.0$	$\overline{YPL} = 7.0$

2) Cycling efficiency. This is the ratio of cycled flow to total flow entering or leaving a compartment or system. It is given by the diagonal elements of the $\underset{\sim}{N'}$ and $\underset{\sim}{N''}$ matrices. As these elements are always equal, $n'_{ii}(t) = n''_{ii}(t)$, compartment cycling efficiency is

$$CE_i(t) = (n'_{ii} - 1)/n'_{ii} = (n''_{ii} - 1)/n''_{ii} \quad (i=1,\cdots,n),$$

$$\tag{21}$$

and system cycling efficiency is

$$CE(t) = \sum_{i=1}^{n} \frac{z_i(t)}{\Sigma z(t)} CE_i(t) = \sum_{j=1}^{n} \frac{y_j(t)}{\Sigma y(t)} CE_j(t). \tag{22}$$

The numerator terms in (21) represent cycled flow, the denominator terms represent cycled flow plus input or output flow at i.

Example 7. Cycling efficiencies for the rainforest nitrogen model are:

$$CE_1 = .49 \qquad CE_3 = .49 \qquad CE_5 = .49$$
$$CE_2 = .49 \qquad CE_4 = .38 \qquad \overline{CE = 0.49} \; .$$

3) *Cycling index.* This is the ratio of cycled to uncycled flow in a system, or of cycled to total flow. Define total system throughflow as

$$TST(t) = \sum_{i=1}^{n} T_i(\Sigma in) = \sum_{j=1}^{n} T_i(\Sigma out). \tag{23}$$

Cycled throughflow is

$$TST_c(t) = \sum_{i=1}^{n} CE_i(t)T_i(\Sigma in) = \sum_{j=1}^{n} CE_j(t)T_j(\Sigma out), \tag{24}$$

and uncycled throughflow is

$$TST_s(t) = TST(t) - TST_c(t). \tag{25}$$

The cycling indexes are then

$$CI_1 = TST_c/TST_s \quad \text{and} \quad CI_2 = TST_c/TST. \tag{26,27}$$

Example 8. For the tropical rainforest nitrogen model:

$$CI_1 = 46/25.8 = 1.78 \quad \text{and} \quad CI_2 = 46/71.8 = 0.64.$$

The structure matrices $\underset{\sim}{N}'(t)$ and $\underset{\sim}{N}''(t)$ in effect define intrasystem input and output environments for each compartment. Patten (1978) has termed these input and output *environs*, E'_i and E''_i respectively, and has shown that the sets $\{E'_i: i=1,\cdots,n\}$ and $\{E''_i: i=1,\cdots,n\}$ each form a system partition.

Example 9. Figure 2 depicts normalized input environs E'_1, E'_3, and E'_4 defined by one unit of outflow in each case from compartments 1, 3, and 4 of the tropical rainforest nitrogen model. Figure 3 shows the output environs E''_1 and E''_3 generated by one unit of inflow, respectively, into compartments 1 and 3.

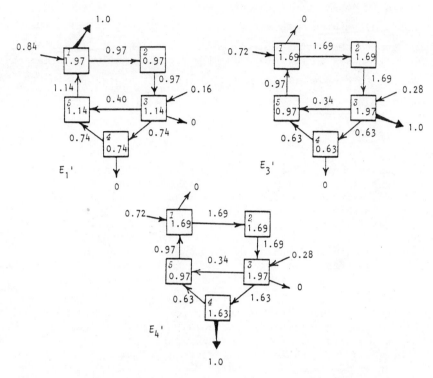

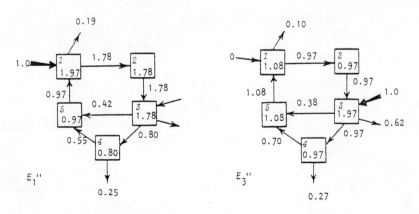

FIG. 2: *Input environs* E'_1, E'_3, *and* E'_4 *for the tropical rainforest nitrogen model.*

FIG. 3: *Output environs* E''_1 *and* E''_3 *for the tropical rainforest nitrogen model.*

Numbers in boxes denote throughflows T_i (i=1,$\cdots$,5), and numbers associated with arrows denote the flows f_{ij} (i,j=0,$\cdots$,5) that sum to the throughflows. Note the columns 1, 3, and 4 of the $\underset{\sim}{N'}$ matrix in Example 2 provide the throughflow information required to construct Figure 2, and rows 1 and 3 of the $\underset{\sim}{N''}$ matrix in Example 3 do the same for Figure 3.

---

3. STOCHASTIC FLOW ANALYSIS

Markov chains have been the principal model for construction of stochastic flow analysis methodologies. Finn (1977) reformulated the deterministic analyses discussed above in accordance with the Markovian interpretation of Leontief economic input-output analysis provided by Kemeny and Snell (1960). This was motivated by prior work of Barber (1978a, b) on Markov methodologies which are independent of the Leontief and above deterministic approaches. The latter methods analyze two random processes, $\{\chi(h\tau)\varepsilon I_\chi\colon$ h=0,-1,-2,$\cdots\}$ for input analysis, and $\{\chi(h\tau)\varepsilon I_\chi\colon$ h=0,1,2,$\cdots\}$ for output analysis, where $\chi(h\tau)$ designates the position in a system of a resource particle at time $\pm$ hτ. It is these methods that are described below. Both input and output analyses concern the statistical properties of Markov models of resource distribution in systems whose environment is either a taboo or an absorbing state (Kemeny and Snell, 1960; Chung, 1967). The models lead to measures of cycling efficiency, intrasystem residence time, and number of intercompartmental transfers.

Other workers have also used Markov chains as models for ecological resource flow. Horn (1975) modelled the climax frequency of tree species as the stationary distribution of a Markov chain. Komota, *et al*. (1976) investigated ergodic properties of resource flow within energetically open but biochemically closed systems with Markov chains.

To trace historic and future movements of a resource particle through a system, assume that flow of the resource can be represented by the discrete Markov chain $\{\chi(h\tau)\}$. Given that a particle is in compartment j at time t, then for an infinitely small $\tau > 0$, $\{\chi(h\tau)\}$ models that particle's immediate history either as

$$\{\cdots,\chi(t) = X_j, \chi(t-\tau) = X_j, \chi(t-2\tau) = X_j,\cdots\} \qquad (28)$$

if it has remained in X_j, or as

$$\{\cdots, \chi(t) = X_j, \; \chi(t-\tau) = F_{ij}, \; \chi(t-2\tau) = X_i, \cdots\} \tag{29}$$

if it has been transferred from X_i to X_j via flow F_{ij}. Similarly, that resource's immediate future is either

$$\{\cdots, \chi(t) = X_j, \; \chi(t+\tau) = X_j, \; \chi(t+2\tau) = X_j, \cdots\} \tag{30}$$

if it remains in X_j, or

$$\{\cdots, \chi(t) = X_j, \; \chi(t+\tau) = F_{jk}, \; \chi(t+2\tau) = X_k, \cdots\} \tag{31}$$

if it moves from X_j to X_k via flow F_{jk}.

For input flow analysis (Barber, 1978b) the assumption that τ be infinitely small is relaxed and the following state space established:

$$I_\chi = \{X_0, X_1, \cdots, X_n, \; Z_1, \cdots, Z_n, \; T_1, \cdots, T_n\}. \tag{32}$$

Here, X_0 denotes the system's environment; X_i is compartment i ($i=1, \cdots, n$); Z_i is inflow F_{0i}; and T_i is throughflow computed as summed outputs:

$$T_i(\Sigma\text{out}) = \sum_{j=1}^{n} F_{ij}. \tag{33}$$

The Markov chain of interest is then $\{\chi(h\tau)\varepsilon I_\chi\}$. The following one-step transition matrix $\underset{\sim}{P}'(h\tau)$ applies to this Markov chain for each time $h\tau$, $h=0,-1,-2,\cdots$:

from \ to	$Z_1 \cdots Z_i \cdots Z_n$			X_0	$X_1 \cdots X_i \cdots X_n$				$T_1 \cdots T_j \cdots T_n$		
$Z_{.1}$				1							
$Z_{.i}$				1							
Z_n				1							
X_0				1							
$X_{.1}$	b_{10}				a_{1}				$b_{.11} \cdots b_{.1j} \cdots b_{.1n}$		
$X_{.i}$		b_{i0}				a_{i}			$b_{.i1} \cdots b_{.ij} \cdots b_{.in}$		
X_n			b_{n0}				a_{n}		$b_{n1} \cdots b_{nj} \cdots b_{nn}$		
$T_{.1}$					1						
$T_{.i}$						1					
T_n							1				

$$(34)$$

where blank entries are zero and where

$$
b_{ij} = \frac{\int_{(h-1)\tau}^{h\tau} F_{ji}(t)\,dt}{X_i((h-1)\tau) + \sum_{k=0}^{n} \int_{(h-1)\tau}^{h\tau} F_{ki}(t)\,dt}
\tag{35}
$$

is the probability that a particle was transitioned from compartment j to i by entering flow F_{ji} (which will ultimately contribute to throughflow $T_j(\Sigma out)$), and

$$
a_i = 1 - \sum_{j=0}^{n} b_{ij}
\tag{36}
$$

is the probability that the particle has remained in compartment i.

Example 10. The one-step transition matrix $\underset{\sim}{P}'$ for input analysis
of the tropical rainforest nitrogen model, calculated according
to equations (35) and (36) with $\tau=1$ year, is

from \ to	z_1	z_3	X_0	X_1	X_2	X_3	X_4	X_5	T_1	T_2	T_3	T_4	T_5
z_1	$\underset{\sim}{0}$		1	0	0	0	0	0	$\underset{\sim}{0}$				
z_3			1	0	0	0	0	0					
X_0	0	0	1	0	0	0	0	0	0	0	0	0	0
X_1	.23	0	0	.46	0	0	0	0	0	0	0	0	.31
X_2	0	0	0	0	.44	0	0	0	.56	0	0	0	0
X_3	0	.12	0	0	0	.16	0	0	0	.71	0	0	0
X_4	0	0	0	0	0	0	.99	0	0	0	.01	0	0
X_5	0	0	0	0	0	0	0	.91	0	0	.03	.06	0
T_1			0	1	0	0	0	0					
T_2			0	0	1	0	0	0					
T_3	$\underset{\sim}{0}$		0	0	0	1	0	0	$\underset{\sim}{0}$				
T_4			0	0	0	0	1	0					
T_5			0	0	0	0	0	1					

For output flow analysis (Barber, 1978a) the Markov chain
$\{\chi(h\tau)\varepsilon I_\chi: h=0,1,2,\cdots\}$ could be used by simply defining a forward transition matrix. However, some calculations are facili-
tated by the Markov chain $\{\xi(h\tau)\varepsilon I_\xi\}$, where:

$$I_\xi = \{X_0,X_1,\cdots,X_n, \ T_0,T_1,\cdots,T_n\}. \tag{37}$$

Here, X_0 is the system's environment; X_j is compartment j
$(j=1,\cdots,n)$; and T_j is throughflow computed as summed inputs:

$$T_j(\Sigma in) = \sum_{i=0}^{n} F_{ij}. \tag{38}$$

Thus, the only difference between $\{\xi(h\tau)\}$ and $\{\chi(h\tau):$
$h=0,1,2,\cdots\}$ is the way they account for intercompartmental
transfers. That is, although $\chi(h\tau) = \xi(h\tau) = X_j$ for all j,
if $\chi(h\tau) = T_j \ \varepsilon \ I_\chi$, then $\xi(h\tau) = T_k \ \varepsilon \ I_\xi$. The following one-
step transformation matrix $\underset{\sim}{P}''(h\tau)$ is appropriate for $\{\xi(h\tau)\}$
at each time $h\tau$, $h=0,1,2,\cdots$:

to \ from	X_0	$X_1 \cdots X_i \cdots X_n$	$T_0 \cdots T_j \cdots T_n$
X_0	1		
$X_{.1}$ ⋮ $X_{.i}$ ⋮ X_n		$a_1 . \\ . \\ a_i . \\ . \\ a_n$	$b_{.10} \cdots b_{.1j} \cdots b_{.1n} \\ b_{.i0} \cdots b_{.ij} \cdots b_{.in} \\ b_{n0} \cdots b_{nj} \cdots b_{nn}$
$T_{.0}$ ⋮ $T_{.i}$ ⋮ T_n		$\underset{\sim}{I}$	

$$(39)$$

where $\underset{\sim}{I}$ is an identity matrix, blanks denote zero, and where

$$b_{ij} = \frac{\int_{h\tau}^{(h+1)\tau} F_{ij}(t)\,dt}{X_i(h\tau) + \sum_{k=0}^{n} \int_{h\tau}^{(h+1)\tau} F_{ki}(t)\,dt} \qquad (40)$$

represents the probability of a particle transitioning from compartment i to j by entering flow F_{ij} (which is derived from throughflow $T_j(\Sigma\text{in})$), and a_i is as defined in equation (36), denoting the probability that the particle will remain in compartment i.

Example 11. The one-step transition matrix $\underset{\sim}{P}''$ for output analysis of the rainforest nitrogen model, computed according to equations (36) and (40) with $\tau=1$ year, is:

from \ to	X_0	X_1	X_2	X_3	X_4	X_5	T_0	T_1	T_2	T_3	T_4	T_5
X_0	1	0	0	0	0	0	0	0	0	0	0	0
X_1	0	.46	0	0	0	0	.05	0	.49	0	0	0
X_2	0	0	.44	0	0	0	0	0	0	.56	0	0
X_3	0	0	0	.16	0	0	.27	0	0	0	.41	.16
X_4	0	0	0	0	.99	0	.002	0	0	0	0	.006
X_5	0	0	0	0	0	.91	0	.09	0	0	0	0
T_0	1	0	0	0	0	0						
T_1	0	1	0	0	0	0						
T_2	0	0	1	0	0	0				$\underset{\sim}{0}$		
T_3	0	0	0	1	0	0						
T_4	0	0	0	0	1	0						
T_5	0	0	0	0	0	1						

Input analysis is concerned with determining historical features of resource flows in systems. Let $V'_{ij}(h\tau)$ be a random variable denoting the number of times a resource particle in $i\varepsilon S$, where $S = I_X - X_0$, has frequented $j\varepsilon S$, since it most recently entered the system from the environment X_0. Let $e'_{ij}(h\tau)$ be the expectation of $V'_{ij}(h\tau)$. Computational formulas for this and other moments are given in Barber (1978b). The following quantities may now be defined:

1) *Residence time.* The time, in units of τ, that resource particles in X_i at time $h\tau$ have spent in X_j since their most recent entry into the system is:

$$RT'_{X_i X_j}(h\tau) = V'_{X_i X_j}(h\tau). \tag{41}$$

The time particles in X_i at time $h\tau$ have remained within the system since their last entry is:

$$RT'_{X_i S}(h\tau) = \sum_{j=1}^{n} V'_{X_i X_j}(h\tau). \tag{42}$$

The expectation of (42) is:

$$ERT'_i(h\tau) = \sum_{j=1}^{n} e'_{X_i X_j}(h\tau). \tag{43}$$

The variance of (42) can also be calculated.

2) *Number of intercompartmental transfers.* The number of times resource particles in X_i at time $h\tau$ have passed through X_j since their most recent entry into the system is:

$$IT'_{X_iX_j}(h\tau) = V'_{X_iT_j}(h\tau).\tag{44}$$

The total number of compartments through which a particle in X_i at time $h\tau$ has been transferred since its latest entry into the system is:

$$IT'_{X_iS}(h\tau) = \sum_{j=1}^{n} V'_{X_iT_j}(h\tau).\tag{45}$$

The expectation of (45) is:

$$EIT'_i(h\tau) = \sum_{j=1}^{n} e'_{x_iT_j}(h\tau).\tag{46}$$

Variance of (45) can also be calculated.

3) *Input origin.* The probability that a resource particle in X_i at time $h\tau$ entered the system in inflow F_{0j} in its most recent entry is given by:

$$IO'_{X_iZ_j}(h\tau) = e'_{X_iZ_j}(h\tau).\tag{47}$$

4) *Cycling efficiency.* The efficiency with which resource particles in X_i at time $h\tau$ are cycled can be defined as the probability that a particle in X_i at time $h\tau$ was already in X_i at least once since its last entry into the system. This probability for steady state systems is

$$CE'_i(h\tau) = Pr[V'_{X_iT_i}(h\tau) \geq 1]$$

$$= e'_{X_iT_i}(h\tau)/[e'_{X_iT_i}(h\tau) + 1].\tag{48}$$

Total system cycling efficiency can be calculated according to the expression given by Barber (1978b, eq. (22)).

A $Q'(h\tau)$ matrix analogous to the $Q'(t)$ matrix of deterministic flow analysis is formed from $P'(h\tau)$ [equation (34)] as described by Barber (1978b). Then the matrix $N'(h\tau)$ analogous to $N'(t)$ in the deterministic method is determined for steady state systems as

$$N'(h\tau) = [I - Q'(h\tau)]^{-1} = \sum_{k=0}^{\infty} \prod_{j=0}^{k} Q'((h-j)\tau). \qquad (49)$$

The elements of $N'(h\tau)$ are the expectations $e'_{ij}(h\tau)$ of $V'_{ij}(h\tau)$. A variance matrix can also be computed, given $N'(h\tau)$ (Barber, 1978b, eq. (24)).

Example 12. A combined matrix of expectations and variances for the tropical rainforest nitrogen model is given in Display 1.

Consider, for example, row X_2. *Residence times*: The expected time that nitrogen in compartment 2 at time $h\tau$ has spent in compartment 5 is 12.8 years with an associated standard deviation of 19.8 years. The expected time that nitrogen in compartment 2 has already resided in the system since entering is 113.9 years, with standard deviation 187.4 years. *Intercompartmental transfers*: The expected number of times that nitrogen in 2 has been in 5 since entering the system is 1.14, with a variance of 2.06. The expected value and variance of the total number of compartments through which nitrogen in compartment 2 has already passed since in the system are 5.96 and 42.75, respectively. *Input origins*: The system has inputs Z_1 and Z_3. For compartment 2, about five times the nitrogen in compartment 2 originated in Z_1 (expectation 0.84, variance 0.14) as originated in Z_3 (expected value 0.16, variance 0.14). This is also true for compartment 1. For compartments 3, 4, and 5, however, only about three times as much nitrogen originated in Z_1 (mean 0.72, variance 0.20) as in Z_3 (mean 0.28, variance 0.20). Note, in Figure 1, that Z_1 is three times greater than Z_3 (7.5 vs 2.7 gN m^{-2}y^{-1}). *Cycling efficiencies:* For compartments 1, 2, 3, and 5, expected cycling efficiency is $0.97/1.97 = 0.49$. For compartment 4, the expected cycling efficiency is $0.63/1.63 = 0.39$. Total system cycling efficiency expected value is 0.40, indicating that 40% of the nitrogen in the system's compartments has been cycled at least one time.

DISPLAY 1: *Input analysis expectations and variances (in parentheses) for the tropical rainforest nitrogen model (see Example 12).*

	Z_1	Z_3	X_1	X_2	X_3	X_4	X_5	ΣX
X_1	.84(.14)	.16(.14)	3.67(9.78)	1.74(7.52)	1.36(3.18)	92.6(29160)	12.8(390.3)	112.2(35120)
X_2	.84(.14)	.16(.14)	3.67(9.78)	3.53(8.93)	1.36(3.18)	92.6(29160)	12.8(390.3)	113.9(35120)
X_3	.72(.20)	.28(.20)	3.14(10.0)	3.02(9.18)	2.35(3.18)	79.2(26010)	10.9(354.2)	98.6(31650)
X_4	.72(.20)	.28(.20)	3.14(10.0)	3.02(9.18)	2.35(3.18)	204.2(41510)	10.9(354.2)	223.6(47150)
X_5	.72(.20)	.28(.20)	3.14(10.0)	3.02(9.18)	2.35(3.18)	160.7(39660)	22.2(469.2)	191.4(45420)

	T_1	T_2	T_3	T_4	T_5	ΣT
X_1	0.97(1.92)	0.97(1.92)	1.14(2.06)	0.74(1.13)	1.14(2.06)	4.96(42.75)
X_2	1.97(1.92)	0.97(1.92)	1.14(2.06)	0.74(1.13)	1.14(2.06)	5.96(42.75)
X_3	1.69(2.12)	1.69(2.12)	0.97(1.92)	0.63(1.04)	0.97(1.92)	5.95(42.56)
X_4	1.69(2.12)	1.69(2.12)	1.97(1.92)	0.63(1.04)	0.97(1.92)	6.95(42.56)
X_5	1.69(2.12)	1.69(2.12)	1.97(1.92)	1.29(1.26)	0.97(1.92)	7.61(42.79)

Stochastic output analysis follows analogously the input analysis methodology, and concerns future aspects of resource flows. Let $V''_{ij}(h\tau)$ denote the number of times a particle in $i\varepsilon S$ will frequent $j\varepsilon S$ before it exits the system to the environment X_0. Then, as before, moments can be computed (Barber, 1978a). A $\underset{\sim}{Q}''(h\tau)$ matrix is formed from $\underset{\sim}{P}''(h\tau)$ [equation (39)], and then the structure matrix $\underset{\sim}{N}''(h\tau)$ is for steady state systems

$$\underset{\sim}{N}''(h\tau) = [\underset{\sim}{I} - \underset{\sim}{Q}''(h\tau)]^{-1} = \sum_{k=0}^{\infty} \sum_{j=0}^{k} \underset{\sim}{Q}''((j+h)\tau). \qquad (50)$$

The elements of $\underset{\sim}{N}''(h\tau)$ are expectations $e''_{ij}(h\tau)$ of $V''_{ij}(h\tau)$, and a variance matrix is also computed (Barber, 1978a, eq. (23)). Then, similar quantities as before (residence time, number of transfers, output destination, and cycling efficiency), only pertaining to the future, can be derived by substituting $V''_{ij}(h\tau)$, and $e''_{ij}(h\tau)$ into equations (41)-(48). The modified output notations for these quantities are $RT''_{ij}(h\tau)$, $IT''_{ij}(h\tau)$, $IO''_{ij}(h\tau)$, and $CE''_i(h\tau)$.

Example 13. A combined matrix of expectations and variances for the rainforest nitrogen model is given in Display 2.

For illustration, consider row X_2 again. *Residence times:* the expected time nitrogen in compartment 2 will spend in 5 before it leaves the system is 12.1 years, with a standard deviation of 19.4 years. The mean time nitrogen in 2 will spend within the system before exiting is 141.2 years, standard deviation 200.6 years. *Intercompartmental transfers:* The expected number of times nitrogen in 2 will be in 5 before leaving the system is 1.07, variance 2.00. The mean and variance of the total number of compartments through which nitrogen in 2 will pass before exiting are 6.05 and 42.24 respectively. *Output destinations:* Since the state set I_ξ excludes system outputs Y_1, Y_3, and Y_4 [equation (37)], the contributions of compartments of each of these cannot be computed with the present version of stochastic output analysis. *Cycling efficiencies:* These are as given by the input analysis in Example 12 since the principal diagonals are identical for both throughflow matrices.

DISPLAY 2: Output analysis expectations and variances (in parentheses) for the tropical rainforest nitrogen model (see Example 13).

	X_1	X_2	X_3	X_4	X_5	ΣX
X_1	3.66(9.75)	3.19(9.12)	2.12(3.34)	109.5(32480)	10.9(353.0)	129.4(38110)
X_2	2.00(8.64)	3.53(8.90)	2.35(3.17)	121.2(34520)	12.1(376.5)	141.2(40240)
X_3	2.00(8.64)	1.74(7.49)	2.35(3.17)	121.2(34520)	12.1(376.5)	139.4(40240)
X_4	2.63(9.71)	2.29(8.60)	1.52(3.31)	203.2(41230)	15.9(435.0)	225.5(46920)
X_5	3.66(9.75)	3.19(9.12)	2.12(3.34)	109.5(32480)	22.1(468.0)	140.6(38220)

	T_1	T_2	T_3	T_4	T_5	ΣT
X_1	0.97(1.91)	1.78(2.07)	1.78(2.07)	0.88(1.21)	0.97(1.91)	6.38(42.50)
X_2	1.07(2.00)	0.97(1.91)	1.97(1.91)	0.97(1.25)	1.07(2.00)	6.05(42.24)
X_3	1.07(2.00)	0.97(1.91)	0.97(1.91)	0.97(1.25)	1.07(2.00)	5.05(42.24)
X_4	1.41(2.16)	1.28(2.12)	1.28(2.12)	0.63(1.02)	1.41(2.16)	6.01(44.72)
X_5	1.97(1.91)	1.78(2.07)	1.78(2.07)	0.88(1.21)	0.97(1.91)	7.38(42.50)

Example 14. Input analysis gives historical information during the time interval since particle entry into the system until $h\tau$, and output analysis provides corresponding data for the future from $h\tau$ until exit. Residence times and number of intercompartmental transfers for the whole period when the resource unit is within the system are obtained by summing the input and output analysis results. *Residence times:* The time nitrogen in compartment 2 is expected to reside in 5 while it is in the system is 24.9 years, with a standard deviation of 27.7 years. The total residence time within the system of nitrogen in 2 is 255.1 years, standard deviation 274.5 years. *Intercompartmental transfers:* The expected number of times nitrogen in compartment 2 will contact 5 while it is in the system is 2.21, with a variance of 4.06. The total number of compartments that nitrogen in 2 will contact during its traverse through the system is 11.0, and the associated variance is 84.99.

---

4. COMPARISON AND EVALUATION

4.1 Convergent Aspects. If a compartment model is in static steady state, $\dot{\underset{\sim}{x}}(t) = 0$, then the deterministic and stochastic analyses yield several essentially identical results. At steady state the relation between elements of the $\underset{\sim}{N}'$ and $\underset{\sim}{N}''$ matrices is:

$$n'_{ji}(t) = e'_{X_i T_j}(h\tau) - \delta_{ij} \tag{51}$$

$$n''_{ij}(t) = e''_{X_i T_j}(h\tau) - \delta_{ij}, \tag{52}$$

where δ_{ij} is the Kronecker delta. Note if $T_i \varepsilon I_\chi$ had been defined as $T_i \varepsilon I_\xi$ and conversely, $n'_{ji}(t)$ and $n''_{ij}(t)$ would have been identical to $e'_{X_i T_j}(h\tau)$ and $e''_{X_i T_j}(h\tau)$, respectively. Two immediate consequences of equations (51) and (52) are:

$$ZPL_i(t) = EIT'_i(h\tau) - 1 \tag{53}$$

$$YPL''_i(t) = EIT''_i(h\tau) - 1. \tag{54}$$

The difference between $ZPL_i(t)$ and $EIT'_i(h\tau)$ is that the former represents the number of compartments contacted since entry into the system by resources in compartment i, including present occupancy of X_i; $EIT_i(h\tau)$ is the mean number of compartments through which resources in i have cycled prior to latest entrance into X_i. Similarly, the difference between $YPL_i(t)$ and $EIT''_i(h\tau)$ is that $PL''_i(t)$ is the number of compartments through which resources in i will cycle inclusive of starting in i, while $EIT''_i(h\tau)$ is the mean number of compartments through which resources in i will cycle after their next exit from i.

Since $n'_{ii}(t) = n''_{ii}(t)$ for all t in the deterministic method, only one cycling efficiency measure, $CE_i(t)$, can be defined for each compartment [equation (21)]. In the stochastic approach both historical and future cycling measures, $CE'_i(h\tau)$ and $CE''_i(h\tau)$, are definable for each compartment i [equation (48) and second sentence following equation (50)]. These measures are, in the general nonsteady state condition, unequal. In static systems, however, all three of these measures are identically equal since:

$$[n'_{ii}(t) - 1]/n'_{ii}(t) = e'_{X_i T_i}(h\tau)/[e'_{X_i T_i}(h\tau) + 1], \quad (55)$$

$$[n''_{ii}(t) - 1]/n''_{ii}(t) = e''_{X_i T_i}(h\tau)/[e''_{X_i T_i}(h\tau) + 1]. \quad (56)$$

For the stochastic methods the weighted means

$$\sum_{i=1}^{n} \frac{X_i(h\tau)}{\Sigma X_j(h\tau)} CE'_i(h\tau) \quad \text{and} \quad \sum_{i=1}^{n} \frac{X_i(t)}{\Sigma X_j(t)} CE''_i(h\tau) \quad (57,58)$$

evaluate, respectively, the historical and future resource cycling of an entire system. These measures are analogous to one of the cycling indexes [equation (27)] of the deterministic methodology except for the method by which the individual compartment cycling measures are weighted.

4.2 *Unique Aspects.* The deterministic input-output analyses provide two unique interpretive advantages. First is the propagative interpretation that can be given matrix entries $n'_{ij}(t)$ and $n''_{ij}(t)$. Specifically, these entries express direct and indirect relationships between compartment throughflows. These relationships are discussed in great detail by Patten, *et al.* (1976). Since the Markovian methodologies merely statistically analyze paths along which a resource unit may flow, $e'_{ij}(h\tau)$ and $e''_{ij}(h\tau)$ lack such a propagative interpretation.

A second unique aspect of the deterministic methods is the use of $\underset{\sim}{N}'(t)$ and $\underset{\sim}{N}''(t)$ matrices to construct digraphs (environs) of system causality (e.g., Figures 2 and 3). A procedure for doing this is described by Finn (1977). Such a procedure is not available with the Markovian analyses.

The stochastic methods do, however, possess several unique features of their own. The measurement of mean compartment and system residence times, numbers of intercompartmental and intrasystem transfers, and origins of inputs and (potentially) destinations of outputs cannot be duplicated by the deterministic approaches. The computation of variances associated with each random variable is an obviously unique aspect of the Markovian approaches. Consequently, for example, an expected bound on the time a resource in compartment i at time t will reside in the system before exiting can be delineated as

$$ERT''_i(h\tau) \pm \sigma \; [\; \sum_{j=1}^{n} V''_{X_i X_j}(h\tau)].$$

4.3 Evaluation. To evaluate the flow analysis methodologies it is necessary to compare how each represents and subsequently analyzes system flow dynamics. Both approaches give essentially convergent results in the static steady state case. Suppose, however, it was desired to analyze the output flow structure associated with compartment i at time $t=0$ if the throughflow $T_i(\Sigma \text{out})$ was some continuous function $T_i(t)$ on the interval $t \geq 0$. How do the deterministic and stochastic methods depict the function $T_i(t)$?

The deterministic approach provides an analysis at each time t, using only instantaneous flow values. Consequently, it represents $T_i(t)$ as a constant function $c_i(t)$ on $t \geq 0$. However, the Markov chain approach represents $T_i(t)$ as a discrete function $s_i(h\tau)$ whose step length equals τ and whose integral is such that

$$\int_{h\tau}^{(h+1)\tau} T_i(t) \; dt = \int_{\tau}^{(h+1)\tau} s_i(t) \; dt.$$

Thus, the stochastic model provides a more realistic estimate of dynamic flow relationships than the deterministic one.

It should be noted, however, that under certain conditions only small errors would result from approximating $T_i(t)$ by $c_i(t)$ rather than $s_i(t)$. For example, if $|T_i(t) - c_i(t)| < \delta$ and $|T_i(t) - s_i(t)| < \varepsilon$, for $t\varepsilon[0,T)$ such that δ and ε are small, and a large fraction of the resources in i at time $t=0$ has left the system by time $t=T$, then $T_i(t)$ can reasonably be approximated by either $c_i(t)$ or $s_i(t)$. There are at least two classes of ecosystem models that fulfill this criterion. The first is those which are in a small amplitude oscillatory steady state. The second is rapid turnover systems which are near steady state only in the time frame of their turnover times. Certain carbon or energy flow systems could be examples of this latter type of ecosystem model.

Since resources exist physically as compartmental standing crops and only abstractly as flows, a second criterion to consider in evaluating flow analysis methods is how storages are incorporated in the different models. Standing crops are explicitly formulated in the Markov methods, but they are only indirectly implicated in the deterministic approaches by compartment derivatives. In fact, compartment storages are conceptualized as outside the system of circulating flows. The stochastic methodologies are more realistic in this respect.

In general, the deterministic methods require less data on flows and standing crops than the stochastic approaches. If field data comprise the data set for a flow analysis, this may be the major criterion in selecting one approach or the other. However, if a computer simulation model is to be analyzed, the problem of abundance or paucity of data exists only at the level of requirements to construct and validate the model.

The method chosen to perform input-output flow analyses in ecological applications is predicated on three considerations: (1) the questions to be answered in relation to the unique aspects of each methodology; (2) the data available to perform the analyses directly or on computer generated flows; and (3) the nature of resource flow dynamics and how closely the different methodologies represent these dynamics.

REFERENCES

Barber, M. C. (1978a). A Markovian model for ecosystem flow analysis. *Ecological Modelling*, 5, 193-206.

Barber, M. C. (1978b). A retrospective Markovian model for ecosystem resource flow. *Ecological Modelling*, 5, 125-135.

Cale, W. G. (1975). *Simulation and systems analysis of a short-grass prairie ecosystem*. Ph.D. thesis, University of Georgia, Athens, Georgia.

Chung, K. L. (1967). *Markov Chains with Stationary Transition Probabilities*. Springer-Verlag, Berlin.

Finn, J. T. (1976a). Measures of ecosystem structure and function derived from analysis of flows. *Journal of Theoretical Biology*, 56, 363-380.

Finn, J. T. (1976b). Cycling index: a general measure of cycling in compartment models. In *Environmental Chemistry and Cycling Processes Symposium*, D. C. Adriano and I. L. Brisbin, eds. U. S. Energy Research and Development Administration, Washington.

Finn, J. T. (1977). *Flow analysis: A method for analyzing flows in ecosystems*. Ph.D. thesis, University of Georgia, Athens, Georgia.

Hannon, B. (1973). The structure of ecosystems. *Journal of Theoretical Biology*, 41, 535-546.

Horn, H. S. (1975). Markovian properties of forest succession. In *Ecology and Evolution of Communities*, M. L. Cody and J. M. Diamond, eds. Harvard University Press, Cambridge, Massachusetts. 196-211.

Kemeny, J. G. and Snell, J. L. (1960). *Finite Markov Chains*. Van Nostrand-Reinhold, Princeton.

Komota, Y., Futatsugi, K., and Kimura, M. (1976). On Markov chains generated by Markovian control systems. 1. Ergodic properties. *Mathematical Biosciences*, 32, 81-106.

Leontief, W. W. (1936). Quantitative input-output relations in the economic system of the United States. *Reviews of Economic Statistics*, 18, 105-125.

Leontief, W. W. (1966). *Input-Output Economics*. Oxford, London.

Patten, B. C. (1975). Ecosystem as a coevolutionary unit: A theme for teaching systems ecology. In *New Directions in the Analysis of Ecological Systems, Part 1*, G. S. Innis, ed. Society for Computer Simulation, La Jolla. 1-8.

Patten, B. C. (1978a). Systems approach to the concept of
    environment. *Ohio Journal of Science*, 78, 206-222.

Patten, B. C. (1978b). Energy environments in Ecosystems. In
    Energy Use Management, Vol. *3*, R. A. Fazzolare and C. B.
    Smith, eds. Pergamon, New York. 853-857.

Patten, B. C., Bosserman, R. W., Finn, J. T., and Cale, W. G.
    (1976). Propagation of cause in ecosystems. In *Systems
    Analysis and Simulation in Ecology*, Vol. *4*, B. C. Patten ed.
    Academic Press, New York. 457-579.

[*Received July* 1978. *Revised November* 1978]

J. H. Matis, B. C. Patten, and G. C. White, (eds.),
Compartmental Analysis of Ecosystem Models, pp. 73-97. All rights reserved.
Copyright ©1979 by International Co-operative Publishing House, Fairland, Maryland.

AN APPLICATION OF COMPARTMENTAL MODELS TO MESO-SCALE MARINE ECOSYSTEMS

IVAN T. SHOW, JR.

Science Applications, Inc.
1200 Prospect Street
La Jolla, California 92038 USA

SUMMARY. The background of a project designed to investigate the effects of chronic low-level hydrocarbon contamination in an active oil field is presented. The structure and role of the modeling effort is outlined. Some of the premise and philosophy of the use of compartmental models in marine ecosystems is also presented.

KEY WORDS. marine, ecosystems, compartments, time-varying, stochastic.

1. INTRODUCTION

This paper is presented as a background to modeling meso-scale ecosystems in the oceans. Since the modeling effort is not complete, the primary thrust is toward the conceptual and logical framework of such an effort. A complete conceptualization is presented. Therefore, it is possible to present only very preliminary results based on individual components of the overall model. Some of these results are derived from early work in the same area while others are presented in defense of the approach used in this effort.

Special applications of compartmental models are often used to model marine ecosystems. These applications have their own special concepts and assumptions. Also, the various properties of compartmental models take on special meaning within our concept of the structure and functioning of marine ecosystems.

Marine ecosystems are basically identical in structure and function to the more familiar terrestrial systems. Energy flows through a complex food-web consisting of procedures, consumers, and decomposers. The special problems which arise in dealing with marine systems result from a lack of 'visibility' and from spatial heterogeneity which is both dynamic and complex. Marine systems are less 'visible' because it is not possible to take a census of marine biota as one might do in a forest or grassland ecosystem. Spatial problems are more active because the biota as well as environmental variables are continually moved about by physical advection and diffusion.

The approach is taken in marine ecosystem modeling that a model is merely an abstraction of nature. As in systems theory, it is held that any theory of system structure and function is based on an abstract in the form of a model (Mesarovic, 1968). Also the more fine detail a model attempts to incorporate, the less chance the model has of elucidating the basic structure and function of the system (Patten, 1968).

If an ecosystem model is regarded as a set of related hypotheses, then only the detail directly bearing on the hypotheses need be included. The model is then accepted or rejected according to the comparison of model results to observations. Hence, the approach in no way violates the experimental philosophy of modern science (Kowal, 1971).

Most interest in marine ecosystems has been with the trophodynamic nature of the system. Therefore, most marine models have dealt with energy flow between components of a food web. Few attempts have been made to incorporate dynamic spatial patterns and most of these have been based on non-motile organisms, primarily phytoplankton. In models dealing with more active organisms, the problem of spatial variation has been largely ignored or the models have been formulated in such a way as to minimize the spatial effects. Clearly, the reason for this trend is that the incorporation of spatial variability not only increases the complexity of the model, but also increases the time and expense involved in developing and using the model.

The recent models of Powell, *et al.* (1975), Kremer and Nixon (1975), Wroblewski (1976), and Show (1977) investigate the physical processes that lead to dynamic spatial patterns. Based on the physical concepts of momentum and continuity, Powell, *et al.* (1975) were able to conclude that treating phytoplankton populations as conservative properties fails because of other factors such as variable nutrient availability and zooplankton grazing. Kremer and Nixon (1975) developed a model of Narragansett Bay in which the biota and chemical variables are advected by currents. Compartmentalization is on two levels: the bay is divided into spatially defined compartments within each of which is a trophic model. Wroblewski (1976) modeled phytoplankton in the upwelling

plume off Oregon. The model incorporates physical advection and
diffusion and the dependence of phytoplankton productivity on
certain chemical and physical variables. Show (1977) investigated
the dependence of zooplankton spatial patterns on physical advection
and diffusion as well as behavior movements of the organisms. Show
also demonstrated the magnitude of errors which can occur as the
results of ignoring spatial patterns.

Due to the nature of marine ecosystems, marine modeling has
evolved in certain directions. There are two basic forms of marine
models: point models and spatial models. Point models are the
most basic. They ignore or attempt to compensate for the fact that
most biotic and non-biotic components of the system are constantly
in motion and that these components rarely are confined within
recognizable boundaries. The models are typically linear, donor-
controlled systems. They are often forced by such factors as
solar insolation and temperature. The coefficients are periodic
due to seasonality and irreversible due to the character of energy
flow through a food web. Feedback is almost universally utilized,
again based on the structure of marine food webs.

Spatial models attempt to account for the physical dynamic
processes which determine the ever-changing spatial patterns of
both biotic and abiotic components of marine ecosystems. The
models usually have two levels of compartmentalization. The first
level divides the area to be modeled into geometrically defined
compartments. The components of the total system are then moved
from compartment to compartment by a simulation of physical advection and diffusion. In a compartmental sense the first level is
linear or non-linear, donor-controlled and forced. The coefficients
are time-dependent, often periodic, and reversible. Feedback is
not explicitly included. The second level of the model is the
imposition of a point model in each first level spatial compartment.

Mathematically, point and spatial models can involve both
systems of ordinary differential equations and systems of partial
differential equations. Point models are usually formulated as
systems of ordinary differential equations of the form

dx/dt = time rate of gain - time rate of loss.

Spatial models, however, are formulated in one of two manners. The
first technique involves a mix of ordinary and partial differential
equations. The first level (spatial) is formulated as a system of
partial differential equations (Section 4.1). The results from
this level drive the second (point) level. The second level is
formulated as a system of ordinary differential equations. The
second technique involves only a system of partial differential
equations of the form

$$\frac{\partial A}{\partial t} = - u \frac{\partial A}{\partial x} - v \frac{\partial A}{\partial y} - w \frac{\partial A}{\partial z} + \frac{\partial}{\partial x} \left(D_x \frac{\partial A}{\partial x} \right)$$

$$+ \frac{\partial}{\partial y} \left(D_y \frac{\partial A}{\partial y} \right) + \frac{\partial}{\partial z} \left(K \frac{\partial A}{\partial z} \right) + R,$$

where all terms are defined as in Section 4.1 and R represents non-conservative biological terms such as birth and death processes.

There are some enormous problems involved in the use of marine ecosystems models. Most of the problems evolve from the fact that it is usually not possible to match the model results closely to field sampling results. In general, this problem can be traced to the nature of the field data. Not only is the variability normally very high, but variability attributable to sampling and to the processes of interest is inseparably confounded. In addition, there are no tests of statistical significance available. Another basic problem is that we usually lack independent data sets for parameter estimation and for testing. In general, it is too costly to collect the second data set.

Despite the problems mentioned above, many models have been found to be functionally similar to real-world processes. The philosophy which has evolved, therefore, has been two-fold: first, to judge the success of a model on its ability to simulate processes and not on its ability to numerically match field data; second, to use the models to answer questions not heavily dependent on the ability of the model to numerically match field data. A more basic philosophy is that, given the infinite number of models which can constructed, field data can always be matched. Of what use is such a model, however, if no understanding of underlying structure and function is gained?

What appropriate questions can be asked, given that a model is an adequate abstraction of certain ecosystem process? The first basic question might concern the stability of the system. Linear, donor-controlled models are guaranteed to be mathematically stable due to the nature of the ordinary differential equation formulation of the model. The stability is, therefore, a characteristic of the model and not the real ecosystem. If the transfer coefficients of such a model are treated as functions of external forcing variables, donor and recipient compartments, etc., however, numerical linearity is no longer guaranteed. Further, if the transfer coefficients are formulated according to our best knowledge of the mechanisms involved, it is possible to make inferences about the stability of the real ecosystem. Overall stability could be considered or periodic stability. Based on the periodic behavior of eigenvalues, Yakubovich and Starzhinsky (1975) give techniques for investigating periodic stability.

Another appropriate question involves the external or internal variables to which the model is most sensitive. Here, classical sensitivity analysis is used. Sensitivity analysis involves varying model parameters and observing the effect on model outputs. Again the technique depends primarily on the model adequately simulating the structure and function of the real ecosystem. Patten (1968) and Steinhorst (1979) treat this subject.

Other appropriate questions might involve the nature of the flow of material or energy through the system. Finn (1976) gives techniques for estimating the number of times a particle (or parcel) of matter might pass through a given compartment before leaving the system. The technique could be used to determine where bioaccumulation of a toxin is most likely to occur. The input-output flow analysis given by Barber, Patten, and Finn (1979) could be used to determine the relative amount of control on the entire system exercised by each component.

Finally, Show (1977) has used the technique of Thakur and Rescigno (1978) to estimate the within spatial compartment variance at each time t for a marine copepod. The time dependent behavior of the variability was used to make inferences about the patch formation behavior of the copepod.

2.   THE BUCCANEER OIL FIELD PROGRAM

The present program is indicative of the type of applied modeling being carried out on marine ecosystems. The objective is to construct a model of the sources, fates, and effects of petroleum hydrocarbon contaminants produced by an operating oil and gas field. The modeling effort is part of the Buccaneer Oil and Gas Field Environmental Assessment program. A look at the background of this program, its objectives and implementation, and the role played by the modeling effort is very instructuve before continuing to a discussion of the model itself.

Oil and gas fields are common in the northern Gulf of Mexico. In spite of the fact that these fields have been present for several decades, there is relatively little information on the impact of drilling and production related platforms and their associated activities on marine ecosystems. Recently, federally sponsored research programs have been initiated to assess the impacts of oil and gas development. Most of these programs are designed to collect baseline data in predevelopment areas. The Buccaneer Oil Field program, however, is designed to assess impacts in an area that has been fully developed and is near the end of its production lifetime.

In spite of its uniqueness, the approach in the Buccaneer program is a valid one. Most management decisions in offshore oil developments are presently being made on the basis of hypotheses put forward by scientists as to what may happen. The data provided by the Buccaneer study will provide decision-makers and scientists alike with much needed hindsight information.

Any program which attempts to assess the impact of environmental perturbations might satisfy certain requirements. First, the program must be designed with well-defined objectives. Second, the program must be scientifically defensible. Third, the results should be applicable to a wide range of contingencies in order to facilitate management decision. The last requirement can only be realized through a thorough understanding of the ecosystem. Ecosystem modeling is the means chosen to study the structure and function of the ecosystems. It is also the means chosen to predict the effects of various environmental perturbations.

The program is now in its third year. Data from the oil field is available on physical oceanography, plankton, nekton, benthos, fouling organisms, sediments, trace metals, hydrocarbons, and the microbiota. In addition, a great deal of general background information is available on the general geographic area.

3. HYDROCARBON FATES AND EFFECTS

The primary objective of the modeling effort is to describe and predict the fates and effects of petroleum hydrocarbons on the Buccaneer Oil Field ecosystem. To complete the background for the discussion of the modeling effort, an overview of our understanding of hydrocarbon fates and effects in the marine environment is presented here.

Fates and effects research is a relatively new area. Most of the significant advances have come during the last decade as a result of laboratory studies, but comparatively few field studies. Three field studies, however, have been performed in the northwest Gulf of Mexico. The Guld Universities Research Consortium (GURC) Offshore Ecology Investigation has been the most comprehensive study to date. No significant biological effects were documented in the GURC study (Oppenheimer, 1977), although the influence of the Mississippi River may have been sufficient to mask any possible ecological damage (Fucik, 1974). Mackin (1971) investigated the effects of bleedwater brine effluent on estuarine organisms in several Texas bays. His results varied from no effects in Lavaca and Matagorda Bays to a localized decrease in benthic populations around oil separator platforms in Trinity Bay. In a similar study, Armstrong, *et al.* (1977) correlated hydrocarbons in sediments with decreases in benthic populations in Trinity Bay.

Since there have been no major oil spills in the Buccaneer Oil Field, hydrocarbons associated with brine discharged from wells probably account for the major input. The amount can be significant since it is believed that venting of waste gases and discharging brine are the major sources of nonmethane light hydrocarbons in upper Gulf of Mexico coastal waters (Brooks, *et al.*, 1977). Armstrong (1977) found that brine water discharges resulted in a large increase of petroleum hydrocarbons in the sediments of Trinity Bay.

When oil enters the marine environment, it becomes associated with the water by emulsion, dispersion, accommodation, and dissolution (Rice, *et al.*, 1977). Oil on the surface undergoes chemical and photo-oxidation. Some of the lighter components evaporate (Smith and MacIntyre, 1971; Parker, *et al.*, 1971). Oil in the water column is subject to microbial degradation, biological uptake, and absorption onto suspended particulates (Fesly and Cline, 1977). Some of the hydrocarbons in the water column will be advected or diffused horizontally (Burwood and Speers, 1974). while some will reach the bottom by sedimentation, sinking, and transport in fecal material. Once in the sediments, oil may persist for a long period under certain conditions.

Biological uptake can be either an active or passive process. Effects range from none to acute toxicity (Cradock, 1977). Sublethal concentration can influence physiology, growth, reproduction, development, and behavior (Johnson, 1977; Patten, 1977).

Karrick (1977) has stated that a model which treats the fates and effects of petroleum hydrocarbons must consider physical, chemical, and biological processes. Most previous models have been concerned with predicting the trajectory of spilled oil. Some work has been done on the microbial degradation of petroleum. Of particular importance to our application is the comprehensive petroleum sources and fates model developed by Kolpack (1978). This model is a three-dimensional treatment of petroleum advection, diffusion, chemical changes, and biological changes. The model is driven by ambient environmental conditions. It will be dealt with in more detail later.

4. MODEL COMPONENTS

The model used for this application consists of three components: a three-dimensional physical dynamic model, a hydrocarbon sources and fates model, and a spatial biological model. Figure 1 shows how the model components are coupled to one another and to the external ambient environment. The arrows indicate the direction in which information flows between components. For instance, the physical dynamic model influences the biological model but not vice versa.

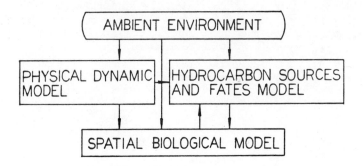

FIG. 1: *Interrelationships of model components.*

4.1 *Physical Dynamic Model.* The physical model is a generalized three-dimensional model of any large-scale turbulent flow. It is based on finite difference analogues of the vertically integrated form of the equations of motion, continuity, and conservation in an incompressible fluid.

The primitive equations for incompressible flow used in the model are:

$$\frac{\partial u}{\partial t} + u\frac{\partial u}{\partial x} + v\frac{\partial u}{\partial y} + w\frac{\partial u}{\partial z} - fv - \frac{1}{\rho}\left(\frac{\partial \tau xx}{\partial x} + \frac{\partial \tau xy}{\partial y} + \frac{\partial \tau xz}{\partial z}\right) + \frac{1}{\rho}\frac{\partial p}{\partial x} = 0$$

$$\frac{\partial v}{\partial t} + u\frac{\partial v}{\partial x} + v\frac{\partial v}{\partial y} + w\frac{\partial v}{\partial z} + fu - \frac{1}{\rho}\left(\frac{\partial \tau yx}{\partial x} + \frac{\partial \tau yy}{\partial y} + \frac{\partial \tau yz}{\partial z}\right) + \frac{1}{\rho}\frac{\partial p}{\partial x} = 0$$

$$\frac{\partial p}{\partial z} + \rho g = 0$$

$$\frac{\partial u}{\partial x} + \frac{\partial v}{\partial y} + \frac{\partial w}{\partial z} = 0$$

$$\frac{\partial s}{\partial t} + u\frac{\partial s}{\partial x} + v\frac{\partial s}{\partial y} + w\frac{\partial s}{\partial z} - \frac{\partial}{\partial x}\left(D_x \frac{\partial s}{\partial x}\right) - \frac{\partial}{\partial y}\left(D_y \frac{\partial s}{\partial y}\right) - \frac{\partial}{\partial z}\left(K \frac{\partial s}{\partial z}\right) = 0$$

$$\frac{\partial T}{\partial t} + u\frac{\partial T}{\partial x} + v\frac{\partial T}{\partial y} + w\frac{\partial T}{\partial z} - \frac{\partial}{\partial x}\left(D_x \frac{\partial T}{\partial x}\right) - \frac{\partial}{\partial y}\left(D_y \frac{\partial T}{\partial y}\right) - \frac{\partial}{\partial z}\left(K \frac{\partial T}{\partial z}\right) = 0$$

where x,y,z are Cartesian coordinates; u,v,w are velocity components; t is time; f is coriolis parameter; p is pressure; S is salinity ^{o}/oo; T is temperature ^{o}C; ρ is density; K is the vertical eddy diffusion coefficient; τ's are the components of stress tensor; and D_x, D_y are the horizontal eddy diffusion coefficients.

The model is arranged in levels of fixed thickness. Integration of a level yields the vertically integrated equations of motion.

Figure 2 shows the grid system used in this model. The staggered grid cuts down on the number of computations and also simplifies the boundary conditions. The latter is particularly important in dealing with an open ocean area with extensive open boundaries. The boundary conditions are such that only a velocity term, u, v, or w, is found on the boundary. Therefore, for a closed boundary the variables on that boundary are set to zero. For open boundaries, the variable on the boundary is specified at each step. In addition, the variables centered in the grid (T, S, ρ, P) are specified at the points outside the computational grid which are immediately adjacent to the boundaries. One of the major advantages of this grid and boundary system is that the boundary specifications need only be made on the upstream boundaries.

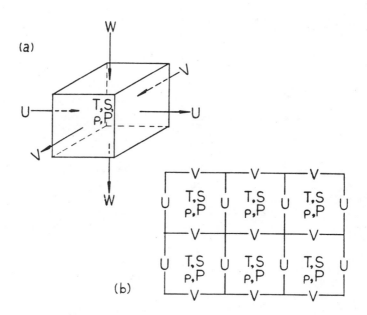

FIG. 2: *Physical model grid system.*   (a) *a single grid box,* (b) *a top view of a number of grid boxes.*

An explicit solution called the leap-frog method is used. In general, this method uses the computed values at time n-1 and n to compute changes which are then applied to time n-1 to yield the state of the system at time n+1. The order of computation is critical because at no point in the computations are any variables determined simultaneously.

The stability criteria is defined as:

$$\Delta t \leq \min \left[\frac{\Delta X}{\sqrt{2gh}} , \frac{\Delta Y}{\sqrt{\cdot 2gh}} \right] \quad ,$$

where h is the maximum depth of the water body being modeled and g is the acceleration due to gravity.

The model has been subjected to considerable theoretical and applied testing. It has been found to be very accurate in all cases.

4.2 Hydrocarbon Sources and Fates Model. This model is based on the work of Kolpack and Plutchak (1978). Kolpack has developed a mass balance model based on rate-of-change data derived from the literature. Certain portions can be extracted for use in the present study.

Three components of the Kolpack model are considered here: composition of the oil, advection and diffusion of the oil, and biological changes in the oil. All of these processes are inter-related to one another and are effected by the external environment. Each model component is discussed separately below.

Oil is a complex mixture of literally thousands of organic compounds. There is, therefore, a need to simplify this structure for the purpose of the model. The simplification is accomplished by constructing a two-dimensional matrix of petroleum components. One dimension is based on Kovat numbers which reflect the molecular weight. Eleven ranges of Kovat numbers are assigned: 1-5, 6-10, ...,46-50, and >50. The other dimension is based on chemical species: paraffins, naphthenes, aromatics, and asphaltics. A 44 element matrix which is updated at regular time intervals is there-fore created. The justification for this amount of elaboration is that the kinds of chemical changes which occur and the rate at which they occur is dependent on both molecular weight of the component and its chemical category.

Petroleum concentration at a given point will depend on the advection and diffusion of all of the components as well as the

chemical changes which occur. Besides the usual physical processes
associated with mass and energy balance, evaporation into the
atmosphere and sedimentation by absorption onto particulates is
considered.

Petroleum components which have absorbed onto particulate
matter are ingested by certain animals. These ingested components
can be either accumulated in the organism's body or incorporated
into fecal material. This is one manner in which the oil is changed
biologically. In addition, many of the components undergo bacterial
degradation. Both of the above processes are considered.

4.3 Spatial Biological Model. The Buccaneer Oil Field is divided
into a three-dimensional set of geographically fixed compartments.
These compartments are presently set at 1.0 km horizontally with
two layers vertically 20 m thick. Figure 3 gives a top view of
this scheme. Within each spatial component will be a trophodynamic
submodel. The compartments of the submodel are determined by our
ability to conceptualize the ecosystem and by the data available.

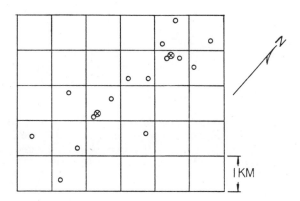

*FIG. 3: A top view of the spatial grid for the biological model.
0 - position of satellite well jackets (548 m^2 of hard substrate
habitat). ⊠ - position of production platforms (3800 m^2 of hard
substrate habitat).*

Not all of the submodel compartments will be included in every
spatial compartment. For instance, there will be no fouling flora,
fouling fauna, or grazers on fouling communities in spatial compartments with no platforms. This scheme is consistent with the actual
ecosystem structure.

The trophodynamic submodels are coupled through the mass
transport results of the physical model. This coupling is conceptually simple. The mass transport of water is determined at
the interface of two spatial boxes. The mass transport of the
substance of interest is then directly proportional to the amount
of water transported.

There are a number of justifications for the elaboration
involved in the spatial approach: (1) the oil platforms create a
heterogeneous environment as concerns platform associated fauna
and flora; (2) the petroleum contamination sources are best treated
as spatially fixed point sources; (3) contaminants are advected
'downstream' from the sources and not dispersed evenly; (4) both
near-field and far-field effects can be considered; and (5) the
physical, chemical, and biological processes involved are more
closely approximated.

5. MODELING APPROACH

Since this model is a simulation of an open ocean environment,
time-varying boundary conditions must be specified. These boundary
conditions are used to force all three of the basic model components.

The physical model requires time series of wind velocity and
and direction, current velocities, temperature, and salinity. The
data are available to specify these variables at the appropriate
points on the boundaries. Show (1977) was successful in using
sine-cosine Fourier series fit to the observed data.

The hydrocarbon sources and fates model requires the appropriate point source inputs from within the oil field. In addition,
the background levels of hydrocarbons advected through the field
must be specified. The data are again available.

Biological inputs must result from the consideration of two
distinctly different processes. First, certain organisms (prin-
cipally plankton) and the nutrients or food supply on which they
depend are advected into the oil field. This process can be handled
by specifying time series of input concentrations at the boundaries.
Second, certain other organisms such as grazers on fouling communities and large predators move into the oil field as a behavioral
response to the presence of the platforms. This process can be
handled by specifying time series of mean concentrations of the
attached organisms at the platforms themselves.

Table 1 gives the basic structure of the coupling between
model components. For the most part, the values from one model
act on the transport coefficients in the other model. In some

TABLE 1: *Model coupling. The terms through which the various model components are coupled are indicated. The direction of influence is denoted by the arrows.*

I.   FORCING FUNCTIONS ————————————> Physical model

    Temperature fields ⎫
    Salinity fields   ⎬ ————————> Density and pressure terms
                                   Computed temperature and salinity fields
    Mass transport ————————————> Current velocity terms

II. FORCING FUNCTIONS ————————————> Hydrocarbon model

    Point source inputs ——————————> Initialize composition matrix
    Background inputs ————————————> Modify composition matrix

III. FORCING FUNCTIONS ————————————> Biological model

    Bacterial input ————————————> Bacterial compartments
    Phytoplankton input ——————————> Phytoplankton compartments
    Nutrient input ————————————> ⎧ Phytoplankton compartments
                                      ⎩ Fouling flora compartments
    Zooplankton input ————————————> Zooplankton compartments
                                      ⎧ Grazers on fouling communities
    Behavioral migration ————————> ⎨ Benthic feeders
                                        ⎩ Large predators

IV. PHYSICAL MODEL ————————————> Hydrocarbon model

                                        Transports between spatial compartments
                                             Nutrients
                                             Contaminants
    Current velocities ——————————>   Phytoplankton
                                             Zooplankton
                                             Bacteria
                                     ⎧ Primary production
                                          ⎪   Phytoplankton
                                          ⎪   Fouling flora
    Temperature fields ——————————> ⎨ Feeding rates
                                          ⎪   (all others)
                                          ⎩ Bacterial growth rate

V.   PHYSICAL MODEL ————————————> Hydrocarbon model

    Current velocities ——————————> Advection and diffusion of hydrocarbons
    Temperature fields ——————————> ⎧ Rate of chemical reactions
                                        ⎩ Rate of bacterial degradation

VI. BIOLOGICAL MODEL ————————————> Hydrocarbon model

    Bacterial concentrations ——————> Rate of bacterial degradation
    Zooplankton concentrations ————> Rate of sedimentation

VII. HYDROCARBON MODEL ————————————> Biological model

    Hydrocarbon concentrations ————> ⎧ Bacteria compartment
                                        ⎩ Benthos compartment
    Toxic component concentrations —> All compartments

cases, the coefficient effected is a birth or death coefficient internal to the compartment.

The basic modeling approach is therefore to establish the physical environment first and then to impose on this the biological portion of the system. Last, the man-made perturbations are included.

6. PRELIMINARY RESULTS AND FOUNDATIONS

At this point in time, the modeling effort is not complete. However, preliminary results can be presented as well as some of the unique formulations in the model. These results are based primarily on Show (1977), Kolpack and Plutchak (1978), Gallaway and Margraf (1978), the results of the Buccaneer Oil and Gas Field Environmental Assessment Program (NMFS, 1976; NMFS, 1977; NMFS, 1978), and the results of the BLM South Texas Outer Continental Shelf Study (BLM, 1978). Concentration will be primarily on the biological aspects of the model. Complete details on the development and testing of the physical model can be found in Show (1977) and complete details of the hydrocarbon model in Kolpack and Plutchak (1978).

Gallaway and Margraf (1978) developed a non-spatial biological model of a single production platform in the Buccaneer Oil Field. The model covered a radius of influence of 45.5 m from the center of the platform; the model has 10 trophically defined compartments. The present form of the model given above is a modification of Gallaway and Margraf's model. They include compartments for Fouling Flora, Fouling Fauna, Meroplankton, Holoplankton, Fouling Gragers, Plankton Feeders, Benthos, Benthic Feeders, Large Predators, and Particulate matter in their model. Annual steady-state eigenvalues are given in Table 2. These values indicate an asymptotic stability of their form of the model. Table 3 gives throughflow through each compartment (kg wet wt/year), cycling efficiency

TABLE 2: *Annual steady-state eigenvalues (from Gallaway and Margraf, 1978).*

Fouling flora	−0.23	Plankton feeders	−9.60
Fouling fauna	−0.27	Benthos	−9.86
Meroplankton	−5.07	Benthic feeders	−30.05
Holoplankton	−6.96	Large predators	−547.60
Fouling gragers	−7.51	particulate matter	−1238.58

TABLE 3: *System throughflow parameters (from Gallaway and Margraf, 1978).*

	Throughflow	Cycling Efficiency	Paths Lengths	
			Inflow	Outflow
Fouling flora	313	0.0	36	1
Fouling fauna	107,403	0.87	37	37
Menoplankton	5,084	0.25	37	37
Holoplankton	1,779	0.10	38	37
Fouling feeders	88,868	0.84	35	38
Plankton feeders	100,099	0.86	37	37
Benthos	2,903	0.16	38	37
Benthic feeders	12,950	0.26	24	25
Large predators	50,310	0.61	20	26
Particulate matter	206,776	0.93	37	36

(cycled throughflow/total throughflow), and inflow and outflow path lengths (average number of times an inflow or outflow passes through a compartment). Total inflow for the system was 57.6×10^4 g wet wt yr^{-1} while total cycled matter for all compartments was 483.1×10^3 g wet wt yr^{-1}. These figures with a high cycling efficiency of 0.838 indicate a system in which inputs are cycled effectively through the system many times before being lost to the outside environment. In addition, particulate matter is shown to be a critical intermediary; thus, the central position of particulates and bacteria in the structure shown in Figure 4.

The innovative application to the Buccaneer Oil Field which is being developed is the spatial formulation of the biological model. The development is based on Show (1977), a model of the spatial distribution of a single planktonic species of copepod.

The model is best understood if we consider a hypothetical water body. The compartments are defined as rectangular parallelopipeds. Figure 5 gives a three-dimensional view of such a compartment. The U's represent X-directed transport of plankton which results from both advection and behavioral response to environmental gradients. The V's and W's represent the same in the y and z directions respectively. The variables, Z and P, at the center are species density and production or interaction terms respectively. Where more than one species or trophic level is involved, Z is a species or trophic level density vector and P is an interaction or trophodynamic submodel.

The single species model yields both expectations and variances of the variable Z. The variance computations are based on the

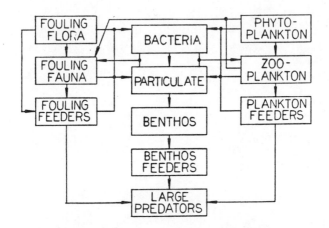

FIG. 4: *Trophodynamic model. A conceptualization of the general-ized model. Arrows indicate energy or biomass flow paths.*

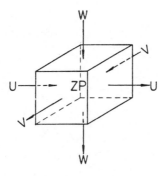

FIG. 5: *A single biological spatial compartment. U's, V's, and W's are mass transports. P and Z are production and density terms.*

work of Thakur and Rescigno (1977). They showed that variances depend approximately on the amount of material entering a compart-ment only through the expectations of the precursor compartments. Their approximation is given by:

$$\mathrm{Var}[Z_n(t)] = Z_{n0}\, e^{-h(t)}[1-e^{-h(t)}] + E[Z_n'(t)],$$

where Z_{n0} is the initial value of Z in compartment n at time $t=0$, $h(t) = \int_0^t u(s)ds$, $Z_n'(t)$ is the amount of Z in compartment n at time t, which was not in compartment n at time $t=0$, and $u(s)$ is the fractional rate of transport out of compartment n.

Figure 6 shows results from Show (1977) for a vertical set of three compartments. The correspondence of computed and observed values clearly justifies pursuing the concept further as the more complex Buccaneer Oil Field model develops.

In view of the fact that a complete development in the above view cannot be given for all aspects of the model, the preceding are given as examples of the background development involved. Some of the conceptual components are shown to be accurate. It remains to be seen whether or not these components can be coupled into a complete, multi-faceted model which is both ecologically sound and provides adequate information to aid in management decisions.

Table 1 shows the direction in which the biological and hydro-carbon models are coupled. There are two processes operating: the change or removal of hydrocarbons and the accumulation by organisms of hydrocarbons. Change or removal comes about primarily by the action of microorganisms (fungi, yeast, and bacteria). Accumulation is primarily by macrophytes (in this case, fouling flora) and attached filter feeders (fouling fauna).

Zobell, *et al.* (1943) and Polyakova (1963) have shown that many marine bacteria are well-adapted to oil degradation; there are many of these organisms in areas of chronic oil release. Oil degradation rates vary from 100 to 960 mg/m^3-day (Zobell, 1969) with rates of 120 mg/m^3-day (Prokop, 1950), 500 mg/m^3-day (Johnston, 1970), and 750 mg/m^3-day (Zobell, 1969) also reported. Several figures for percentage degradation rate have been reported: 40 - 57% in 12 days (Atlas and Bartha, 1921), 65% in three days (Ahearn, 1972), and 35 - 55% in five days (Miget, *et al.*, 1969). All of the above degradation rates are temperature dependent; Kolpack, *et al.* (1973) report the values shown in Table 4.

Hydrocarbon change by the biota is not dealt with in the model. The removal of hydrocarbons and subsequent biotic growth is handled in the bacteria compartment. The bacteria compartment is

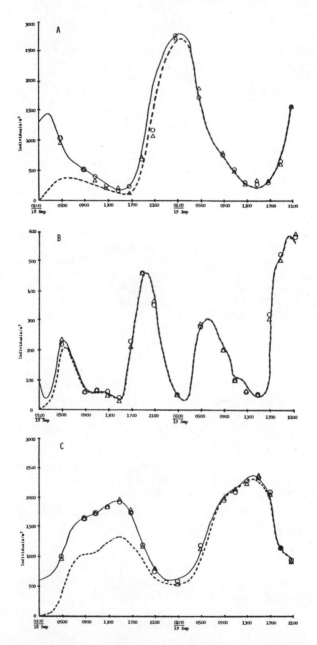

FIG. 6: *Comparison of observed and computed means and variances of (a) the surface compartment, (b) the mid-level compartment, and (c) the bottom compartment.*

TABLE 4:   Percent degradation of fuel oil at the end of six weeks.

Temperature °C	% Oil degraded
4	0
10	20-30
20	30-50
25	50-80

treated much like a primary producer compartment with Michaelis-Menton kinetics. Change in the bacterial compartment is expressed as:

$$\frac{dx(t)}{dt} = \frac{R[X_{max} - X(t-1)] \, e^R t - k_1 e^{(T-T_o)}}{X_{max}} \quad , \tag{1}$$

where

$x(t)$ = population size at time t;

X_{max} = upper bacteria population limit, 10^9 organisms/m^3;

T = concentration of toxic hydrocarbons, ppm;

T_o = lower toxicity threshold; 0.0001%;

k_1 = proportionality constant, 0.01;

Δt = length of model Tinn step;

R = dB/dt - dD/dt;

B = births; D = deaths;

and

$$\frac{dB}{dt} = \frac{B(t-1)}{X(t-1)\Delta t} \, k_1 e^{k_2 (TO - 270)FC}$$

$$\frac{dD}{dt} = \frac{D(t-1)}{S(t-1)\Delta t} \, k_1 e^{T-T_o} + |280 - TO| - \frac{FC + ONPH}{k_3 \, X(t-1)} \quad ,$$

where TO = oil temperature, °K;   F = oil concentration available, ppm;   C = sum of oil preference coefficient for each element of the hydrocarbon matrix (values from Kolpack and Plutchak, 1978);

O = oxygen concentration, ml/l;   N = nitrate level, millimoles;
P = phosphate level, millimoles;   H = pH of the water; k_2 = constant, 2.5; k_3 = constant, 9×10^{-5}.

The first term in the numerator of equation (1) represents growth rate using petroleum hydrocarbons as a food base. The second term is a catastrophic death term which is operative in the case of a massive spill containing high concentrations of toxic hydrocarbons.

The total amount removed from the hydrocarbon matrix is equal to the growth in bacteria expressed in equation (1). The amount removed from each element of the hydrocarbon matrix depends on the oil preference coefficient for that element (the coefficients for all elements of the matrix sum to one).

The above computation is in addition to transfers involving the bacteria compartment within the biotic system given in Figure 4. Additional flows are similar to those in the rest of the model. Losses to the bacteria from toxicity are transferred in their entirety to the particulate compartment.

Accumulations of hydrocarbons is handled by input directly into the fouling flora and particulate compartments. The hydrocarbons then flow through the system by the normal transfer coefficients; however, the values are contained in a separate state variable. Hydrocarbons are allowed to accumulate in the second state variable. Possible toxicity is handled by modifying the coefficients directing flow into the compartment which has reached a toxic level of accumulation. If k_{ij} is a transfer coefficient in question, then it is expressed as

$$k_{ij} = f(\text{natural environmental variables}) - C_{ij}e^{T-T_o},$$

where C_{ij} = proportionality constant specific to k_{ij} and T and T_o are defined above.

ACKNOWLEDGEMENTS

This work is a result of research sponsored by the Environmental Protection Agency, National Oceanic and Atmospheric Administration and the Department of Commerce, National Marine Fisheries Service, Southeast Fisheries Center, Galveston Laboratory, under Contract No. 03-78-800-0036. The United States Government is authorized to produce and distribute reprints.

REFERENCES

Ahearn, D. G., ed. (1972). *The Microbial Degradation of Oil Pollutants*. Proceedings volume. Georgia State University, Atlanta, Georgia.

Armstrong, H. W., Fucik, K. W., Anderson, J. W., and Neff, J. M. (1977). *Effects of Oil Field brine effluent on benthic organisms in Trinity Bay, Texas*. API Publication No. 4291. American Petroleum Institute, Washington, D. C.

Atlas, R. M. and Bartha, R. (1971). *Degradation and mineralization of petroleum by two bacteria isolated from coastal waters. B. Degradation and mineralization of petroleum in sea water: Limitation of nitrogen and phosphorus*. Technical Report No. 2. ONR project no. 137-843.

Barber, M. C., Patten, B. C., and Finn, J. T. (1979). Review and evaluation of input-output flow analysis for ecological applications. In *Compartmental Analysis of Ecosystem Models*, J. H. Matis, B. C. Patten, and G. C. White, eds. Satellite Program in Statistical Ecology, International Co-operative Publishing House, Fairland, Maryland.

BLM. (1978). *Environmental studies of the South Texas outer-continental shelf*. NOAA Report to the BLM. Southeast Fisheries Center, Galveston Laboratory, Galveston, Texas.

Brooks, J. M., Bernard, B. B., and Sackett, W. M. (1977). Input of low molecular weight hydrocarbons from petroleum operations into the Gulf of Mexico. In *Fate and Effects of Petroleum Hydrocarbons in Marine Organisms and Ecosystems*, D. A. Wolfe, ed. Pergamon Press, Oxford. 373-384.

Burwood, R. and Speers, G. C. (1974). Photo-oxidation as a factor in the environmental dispersal of oil. *Estuarine Coastal Marine Science*, 3, 117-135.

Craddock, D. R. (1977). Acute toxic effluents of petroleum on arctic and subarctic marine organisms. In *Effects of Petroleum on Arctic and Subarctic Marine Environments and Organisms, Vol. II*, D. C. Malins, ed. Academic Press, New York. 1-93.

Feely, R. A. and Cline, J. D. (1977). *The distribution, composition, transport, and hydrocarbon adsorption characteristics of suspended matter in the Gulf of Alaska, Lower Cook Inlet, and Shelikof Shelf*. NOAA/OSCEAP Report No. 152.

Finn, J. T. (1976). Measures of ecosystem structure and function derived from analysis of data. *Journal of Theoretical Biology*, 56, 363-380.

Fucik, K. W. (1974). *The effects of petroleum operation on the phytoplankton ecology of the Louisiana coastal waters.* M.S. thesis, Texas A&M University.

Galloway, B. J. and Margraf, F. J. (1978). *Simulation modeling of biological communities associated with a production platform in the Buccaneer Oil and Gas Field.* Final report to NOAA/NMFS. Galveston Laboratory, Galveston, Texas.

Johnson, F. G. (1977). Sublethal effects of petroleum hydrocarbon exposures: bacteria, algae, and invertebrates. In *Effects of Petroleum on Arctic and Subarctic Marine Environments and Organisms, Vol. I,* D. C. Malins, ed. Academic Press, New York. 271-318.

Johnston, R. (1970). The decomposition of crude oil residues in sand columns. *Journal of the Marine Biological Association of the United Kingdom,* 50, 925-937.

Karick, N. L. (1977). Alterations in petroleum resulting from physio-chemical and microbiological factors. In *Effects of Petroleum on Arctic and Subarctic Marine Environments and Organisms, Vol. I,* D. C. Malins, ed. Academic Press, New York. 225-300.

Kolpack, R. L., Mechalas, B. J., Meyers, T. J., Plutchak, N. B., and Eaton, E. (1973). *Fate of Oil in a Water Environment, Vol. I.* American Petroleum Institute, Washington, D. C.

Kolpack, R. L. and Plutchak, N. B. (1978). *Fate of Oil in a Water Environment, Vol. II.*   American Petroleum Institute, Washington, D. C.

Kowal, N. E. (1971). A rationale for modeling dynamic ecological systems. In *Systems Analysis and Simulation in Ecology, Vol. I,* B. C. Patten, ed. Academic Press, New York. 123-134.

Kremer, J. N. and Nixon, S. W. (1975). An ecological simulation model of Narragansett Bay - the plankton community. In *Estuarine Research, Vol. I,* L. E. Cronin, ed. Academic Press, New York. 672-690.

Mackin, R. (1971). *A study of the effect of oil field brine effluents on biotic communities in Texas estuaries.* Texas A&M Research Foundation Project 735.

Mesarovic, M. P. (1968). *Systems Theory and Ecology.* Springer-Verlag, New York.

Miget, R. J., Oppenheimer, C. H., Kator, H. I., and LaRock, P. A. (1969). Microbial degradation of normal paraffin hydrocarbons in crude oil. *Proceedings of the Joint Conference on Prevention and Control of Oil Spills.* American Petroleum Institute, Washington, D. C. 327–331.

NOAA. (1977). *Environmental assessment of an active oil field in the northwestern Gulf of Mexico.* Southeast Fisheries Center, Galveston Laboratory, Galveston, Texas.

NOAA. (1976). *Pilot study of the Buccaneer Oil Field.* Southeast Fisheries Center, Galveston Laboratory, Galveston, Texas.

NOAA. (1978). *Environmental assessment of an active oil field in the Northwestern Gulf of Mexico: 1977-1978.* Southeast Fisheries, Center, Galveston Laboratory, Galveston, Texas.

Oppenheimer, C. H. (1977). The offshore ecology investigation 1972-1974. International Council for the Exploration of the Sea. *Rapports et Proces-vervaux des Reunions,* 171, 147–154.

Parker, C. A., Frugarde, M., and Hatchard, C. G. (1971). The effect of some chemical and biological factors on the degradation of crude oil at sea. In *Water Pollution by Oil,* P. Hepple, ed. Institute of Petroleum, London. 237–244.

Patten, B. C. (1968). Mathematical models of plankton production. *Internationale Revue der Gesamten Hydrobiologie,* 53, 357–408.

Patten, B. C. (1977). Sublethal biological effects of petroleum hydrocarbon exposures: Fish. In *Effects of Petroleum on Arctic and Subarctic Marine Environments and Organisms, Vol. II,* D. C. Malins, ed. Academic Press, New York. 319–335.

Polyakova, I. N. (1963). Distribution of hydrocarbon - oxidizing microorganisms in water of Neva Bay. *Mikrobiologiya* (trans.), 31, 872–876.

Powell, T. M., Richerson, P. J., Dillon, T. M., Agee, B. A., Dizier, B. J., Godden, D. A., and Myrup, L. O. (1975). Spatial scales of current speed and phytoplankton biomass fluctuations in Lake Tahoe. *Science,* 189, 1088–1089.

Prokop, J. R. (1950). *Report on a study of the microbial decomposition of crude oil.* Texas A&M University Research Foundation, Project 9.

Rice, S., Korn, S., and Karinen, J. (1977). *Lethal and sublethal effects on selected Alaskan marine species after acute and long-term exposure to oil and oil components.* NOAA/OCSEAP RU #73, Annual Report.

Show, I. T.   (1977).   *A spatial modeling approach to pelagic ecosystems*.   Ph.D. dissertation, Texas A&M University.

Smith, C. L. and MacIntyre, W. G.   (1971).   Initial aging of fuel oil films of sea water. *Proceedings of 1971 Joint Conference on Prevention and Control of Oil Spills*.   American Petroleum Institute, Washington, D. C.

Soli, G. and Bens, H. M.   (1971).   Hydrocarbon-oxidizing bacteria and their possible use as controlling agents of oil pollution in the ocean.   Manuscript.   Naval Weapons Center, China Lake, California.

Steinhorst, R. K.   (1979).   Parameter identifiability, validation, and sensitivity analysis.   In *Systems Analysis of Ecosystems*, G. S. Innis and R. V. O'Neill, eds.   Satellite Program in Statistical Ecology, International Co-operative Publishing House, Fairland, Maryland.

Thakur, A. K. and Rescigno, A.   (1978).   On the stochastic theory of compartments.   III. General, time-dependent reversible systems.   *Bulletin of Mathematical Biology*, 40, 237-246.

Wroblewskii, J. S.   (1976).   *A model of the spatial structure and production of phytoplankton populations during variable upwelling off the coast of Oregon*.   Ph.D. dissertation, Florida State University.

Yakubovich, V. A. and Starzhinskii, V. M.   (1975).   *Linear Differential Equations with Periodic Coefficients*.   Wiley, New York.

Zobell, C. E., Grant, C. W., and Haas, H. F.   (1943).   Marine microorganisms which oxidize petroleum hydrocarbons. *Bulletin of the American Association of Petroleum Geologists*, 27, 1175-1193.

[*Received October* 1978. *Revised March* 1979]

IDENTIFIABILITY AND
STATISTICAL ESTIMATION
OF PARAMETERS IN
COMPARTMENTAL MODELS

INTRODUCTION TO SECTION II

The four papers of this section provide tools necessary to build models from data. C. Cobelli, A. Lepschy, and G. R. Jacur discuss identification experiments. Before data are collected, they determine whether the compartments will be observed in such a way that parameters are identifiable. Next, M. Berman provides background on the SAAM compartment model simulator. SAAM can be used to estimate parameter values from observed data, and to simulate various system manipulations. G. C. White and G. M. Clark then present a new approach to handling the natural variability of observed data. Rather than assume that the data vary due to sampling error, they suggest models where the variability is incorporated into the process being studied. In the final paper R. E. Bargmann considers how long data collection should continue. Given the expense of field experiments, the researcher wants to collect adequate data, but not waste money by collecting too much data.

These four papers provide a good overview of model building. The problems of identifiability, estimation, and design are all covered.

J. H. Matis, B. C. Patten, and G. C. White, (eds.),
Compartmental Analysis of Ecosystem Models, pp. 99-124. All rights reserved.
Copyright ©1979 by International Co-operative Publishing House, Fairland, Maryland.

IDENTIFICATION EXPERIMENTS AND IDENTIFIABILITY CRITERIA FOR
COMPARTMENTAL SYSTEMS

C. COBELLI, A. LEPSCHY, AND G. ROMANIN JACUR

Laboratorio per Ricerche di Dinamica dei Sistemi
e di Bioingegneria, C.N.R., Padova, Italy and
Istituto di Elettrotecnica e di Elettronica
Universita di Padova, Padova, Italy

SUMMARY. In this paper *a priori* identifiability criteria for
compartmental models are presented. *A priori* identifiability
deals with the problem of stating whether a planned identification
experiment allows the estimation of all unknown parameters of the
system. Compartmental models and identification experiments are
briefly reviewed. A formal definition of *a priori* identifiability
is given. Structural identifiability tests are presented and
necessary conditions are given which can be applied directly to
the compartmental diagram. Examples of application of the presented identifiability tests are given.

KEY WORDS. identification, input-output identification experiments, compartmental systems, structural properties, structural
(*a priori*) identifiability.

1. INTRODUCTION

The identification problem is usually concerned with the
determination of both structure and parameter values of a mathematical model adopted for representing a physical system in such
a way that the model describes the system behavior in accordance
with predetermined criteria. *Compartmental analysis,* which is
often employed for a quantitative description of bio- and ecosystems, directly provides the structure (compartments and
material flows among them) of the model to be considered, so that
the identification problem often reduces to a parameter estimation
problem. For estimating the parameters of a given model, an
identification experiment has to be planned, i.e., some suitable

---

Address correspondence to: Claudio Cobelli, LADSEB-CNR, P. O. Box
1075, 35100 Padova, Italy.

perturbations are applied to the system and the consequent time behavior of relevant variables is measured.

A necessary condition related to identification of compartmental models from input-output data is the *a priori* identifiability. *A priori* identifiability deals with the problem of determining whether a planned input-output identification experiment is able to supply the desired information about the unknown parameters of the compartmental system; such a problem is to be faced only on the basis of the chosen compartmental structure and of the planned input-output experiment. The literature on the matter presents many recent results, particularly on the class of linear time invariant compartmental systems (e.g., Bellman and Åström, 1970; Cobelli and Romanin Jacur, 1975, 1976a, b; Cobelli, *et al.*, 1978a, b, c, d; DiStefano, 1976; DiStefano and Mori, 1977; Grewal and Glover, 1976).

This paper deals with the problem of *a priori* identifiability of linear time invariant compartmental models. In particular Section 2 briefly reviews some fundamentals on compartmental systems and formalizes the equations of the input-output identification experiment. In Section 3 the *a priori* identifiability problem is rigorously stated on the basis of system theory concepts. In Section 4 two identifiability tests are presented and their use for testing structural identifiability of compartmental systems are discussed. Necessary conditions for structural identifiability which can be checked directly on the compartmental diagram are presented in Section 5. Section 6 provides some examples which show how to practically employ the presented identifiability tests. Finally in Section 7 some conclusive remarks are made.

2.   COMPARTMENTAL MODELS AND IDENTIFICATION EXPERIMENTS

As is well known, compartmental analysis (Jacquez, 1972) is a phenomenological and macroscopic approach for modeling physico-chemical processes. The essential ingredients are compartments and intercompartmental flows. A compartment may be considered as a homogeneous and well mixed amount of material which kinetically behaves in a distinctive way; the intercompartmental flows correspond to material exchanges between compartments, due to transport phenomena or chemical reactions.

A compartmental model is based on *mass balance equations:*

$$\dot{Q}_i = \sum_{j=0}^{n} R_{ij} - \sum_{j=0}^{n} R_{ji} \quad (i=1,2,\cdots,n), \tag{1}$$

where Q_i is a compartmental quantity, n is the number of considered compartments, R_{ij} is the flow from compartment j to compartment i, R_{0i} and R_{i0} are the flows from compartment i to the environment and from the environment to compartment i.

Usually the flows are functions of the compartmental quantities; in particular the flow R_{ij} is often assumed to be dependent only on the quantity Q_j from where it originates:

$$R_{ij} = R_{ij}(Q_j). \tag{2}$$

Biological systems described by equation (1) can exhibit two peculiar patterns: a cycling pattern (i.e., each Q_i is a periodic function of time) and a constant pattern or constant steady state (i.e., each Q_i and each R_{ij} assume constant values Q_{is} and R_{ijs}). In the following, compartmental systems in a constant steady state will be considered:

$$0 = \sum_{j=0}^{n} R_{ijs} - \sum_{j=0}^{n} R_{jis}. \tag{3}$$

In order to give a quantitative description of the system, i.e., for writing explicitly equation (3), it is necessary to evaluate parameters which characterize relation (2). If the system can be considered linear, then

$$R_{ij}(Q_j) = k_{ij}Q_j,$$

where the only parameter to be taken into account is k_{ij}, and equation (3) becomes:

$$0 = \sum_{j=1}^{n} k_{ij}Q_{js} + R_{i0s} - \sum_{j=0}^{n} k_{ji}Q_{is} \quad (i=1,2,\cdots,n).$$

If relation (2) exhibits a saturation it may be approximated for instance by:

$$R_{ij}(Q_j) = \frac{\alpha_{ij}Q_j}{1+\beta_{ij}Q_j},$$

where α_{ij} and β_{ij} are constant positive parameters. In many cases the whole behavior of (2) is not of interest and only an approximation of it in the neighborhood of the working point (Q_{js}, R_{ijs}) is needed. In such a case a truncated Taylor expansion may be considered:

$$R_{ij}(Q_j) = R_{ijs} + \sum_{h=1}^{H} \gamma_{ijh}(Q_j - Q_{js})^h.$$

In general, equation (2) can be written in the form:

$$R_{ij}(Q_j) = f_{ij}(\alpha_{ij}, \beta_{ij} \ldots Q_j) \tag{4}$$

where α_{ij}, β_{ij} ... are constant parameters.

The unknown parameters α_{ij}, β_{ij}, etc. cannot be determined experimentally by referring to equation (2) as the intercompartmental flows R_{ij} cannot, in general, be directly measured. Therefore it is necessary to resort to equation (3) by substituting the flows R_{ij} by means of their expressions (4). In such a way the following equation is obtained:

$$0 = R_{i0s} + \sum_{j=1}^{n} f_{ij}(\alpha_{ij}, \beta_{ij} \ldots Q_{js})$$

$$- \sum_{j=0}^{n} f_{ji}(\alpha_{ji}, \beta_{ji} \ldots Q_{is}). \tag{5}$$

Usually, however, equation (5) cannot supply the desired information about the unknown parameters α_{ij}, β_{ij}, etc. as several quantities Q_{is} and inflows R_{i0s} cannot be directly measured and therefore the number of unknown parameters greatly exceeds the number of independent equations.

The only way to get the desired information is then to resort to equation (1) by forcing an artificial dynamics via a planned input-output experiment, i.e., by applying test inputs and by measuring the consequent time behavior of some compartmental quantities Q_i.

An appealing class of input-output experiments are the so-called first order perturbation experiments, e.g., tracer experiments. Let us consider a compartmental system in a constant steady state described by equation (5). If n_B test inputs u_r are applied, the consequent deviation x_i of the compartmental quantity Q_i from its steady state value Q_{is} exhibits a dynamics which is described by the equation:

$$\dot{x}_i = \sum_{j=1}^{n} k_{ij}x_j - \sum_{j=0}^{n} k_{ji}x_i + \sum_{r=1}^{n_B} b_{ir}u_r \quad (i=1,2,\cdots,n), \quad (6)$$

where the generic coefficient k_{ij}, which depends on the intrinsic system parameters α_{ij}, β_{ij}, etc., is given by:

$$k_{ij} = \left.\frac{\partial f_{ij}}{\partial Q_j}\right|_{Q_j = Q_{js}}$$

and where the coefficients b_{ir} are introduced to account for a possible fractioning of some test inputs. The coefficients k_{ij} are usually called transport rate constants.

For describing the experiment on the system, the measurement equations are to be considered, which relate to the n_C actually measured quantities y_m to the deviations x_i. Usually such relations are linear in the $x_i's$ and may be written in the form:

$$y_m = \sum_{i=1}^{n} c_{mi}x_i \quad (m=1,2,\cdots,n_C). \quad (7)$$

The system and the experiment configuration are usually represented by means of a compartmental diagram, that is by a set of blocks representing the compartments and of directed arcs representing the transport rate constants, where test inputs and measured outputs are suitably evidenced (Figure 1).

The set of equations (6) and (7) defines the model for the system and the experiment and may be written, in the usual system theory context, in matrix form:

$$\dot{x} = A x + B u, \quad y = C x \quad (8)$$

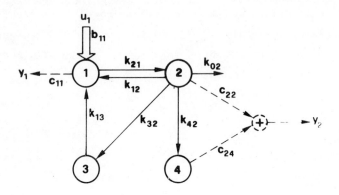

FIG. 1. *The compartmental diagram for a one-input two-output experiment on a 4-compartment system.*

where $\underset{\sim}{x} = [x_1 \ x_2 \ldots x_n]^T$ is the state vector (T means transpose); $\underset{\sim}{u} = [u_1 \ u_2 \ldots u_{n_B}]^T$ is the input vector; $\underset{\sim}{y} = [y_1 \ y_2 \ldots y_{n_C}]^T$ is the output vector; $\underset{\sim}{A} = (a_{ij})$ is an $n \times n$ matrix with

$$a_{ij} = k_{ij} \quad (i \neq j), \qquad a_{ii} = - \sum_{\substack{0 \\ i \neq j}}^{n} k_{ji}; \tag{9}$$

$\underset{\sim}{B} = (b_{ir})$ is an $n \times n_B$ matrix; $\underset{\sim}{C} = (c_{mi})$ is an $n_C \times n$ matrix.

Fundamental properties of the class of linear time invariant compartmental models may be found in Hearon (1963).

It may be noted that equation set (8) also holds in a more general context of first order perturbation experiments (DiStefano, 1976), i.e., if the flows depend not only on the compartmental quantities from where they originate but also on other quantities. In general, if equation (1) is written in the form:

$$\dot{Q}_i = F_i(Q_1, \ Q_2 \ \cdots \ Q_n, \ u_1, \ u_2 \ \cdots \ u_{n_B}),$$

a first order perturbation experiment on the system in a steady state may be approximately described by the equation set (8) where

$$a_{ij} = \frac{\partial F_i}{\partial Q_j} \quad \text{and} \quad b_{ir} = \frac{\partial F_i}{\partial u_r}$$

and the derivatives are evaluated in the steady state.

It may be remarked that the 'gentle' perturbation on the system assures the linearity of the dynamical model while the constant steady state assures time invariance of the parameters of the linear model. If, on the contrary, a steady state cycling pattern is present (e.g., circadian rhythms) or in general if the relevant quantities vary independently of the perturbation experiment, the small signal dynamics of the system is still linear but the parameters are time varying.

3. STRUCTURAL IDENTIFIABILITY: PROBLEM STATEMENT AND DEFINITIONS

The identification of a physical system may be considered as the process of setting up a mathematical model of the system and of estimating numerical values of its parameters from experimental data.

In the identification of compartmental systems via input-output experiments the following steps are to be performed:

i) *compartmentalization:* to identify meaningful compartments and flows among them;

ii) *design of the input output experiment:* to select the compartments for perturbation inputs and for measurements;

iii) *modeling:* according to i) and ii), to write the model equations (8);

iv) *test of a priori identifiability:* to check whether the planned experiment (see step ii) allows the estimation of the unknown parameters of the model (see step iii);

v) *performance of the experiment;*

vi) *estimation of model parameters:* to evaluate the numerical values of the unknown parameters from input-output data by using suitable estimation procedures;

vii) *test of a posteriori identifiability:* to evaluate the accuracy of the estimates and the consistency of the model response to experimental data;

viii) *model validation*.

This paper is concerned with step iv, i.e., with *a priori* identifiability. *A priori* identifiability is strictly linked to the experiment design (step ii) and its importance follows from the fact that the possibly expensive steps v and vi are meaningful from a theoretical point of view and may be successful only if the identifiability test is satisfied. If the test reveals nonidentifiability, it is necessary either to modify the compartmentalization (step i) and/or to modify the experiment (step ii) and/or to look for some external constraints. If the test is positive, identifiability analysis may also suggest the possibility of designing simpler experiments, of adopting a more complex model structure and of obtaining uniqueness of parameter values.

The analysis of *a priori* identifiability must be based only on the knowledge of the assumed model structure and of the chosen experimental configuration. In order to formalize the problem of *a priori* identifiability, it is useful to resort to some preliminary notions and in particular to the one of fixed structure system.

In the classical approach to system theory, a system is considered as a set of input-output pairs and the state vector $\underset{\sim}{x}$ is an intermediate variable between input $\underset{\sim}{u}$ and output $\underset{\sim}{y}$. State variables do not necessarily have a precise physical meaning and instead of them another state vector $\underset{\sim}{\hat{x}}$ may be considered; more precisely, instead of the set of equations (8) the new set:

$$\underset{\sim}{\dot{\hat{x}}} = \hat{\underset{\sim}{A}} \, \underset{\sim}{\hat{x}} + \hat{\underset{\sim}{B}} \, \underset{\sim}{u}, \quad \underset{\sim}{y} = \hat{\underset{\sim}{C}} \, \underset{\sim}{\hat{x}},$$

where:

$$\underset{\sim}{\hat{x}} = \underset{\sim}{T} \, \underset{\sim}{x}; \quad \hat{\underset{\sim}{A}} = \underset{\sim}{T} \, \underset{\sim}{A} \, \underset{\sim}{T}^{-1}; \quad \hat{\underset{\sim}{B}} = \underset{\sim}{T} \, \underset{\sim}{B}; \quad \hat{\underset{\sim}{C}} = \underset{\sim}{C} \, \underset{\sim}{T}^{-1}$$

may be considered, which exhibits the same input-output behavior by connecting $\underset{\sim}{u}$ and $\underset{\sim}{y}$ by means of the new state vector $\underset{\sim}{\hat{x}}$.

It may be realized the convenience of this transformation in dealing with several problems, e.g., for transforming the original matrix $\underset{\sim}{A}$ into a matrix $\hat{\underset{\sim}{A}}$ of a simple form (a companion or diagonal matrix).

In many cases, however, a 'natural' state representation may be considered, in which the state variables and the non null elements of matrix $\underset{\sim}{A}$ (and possibly of matrices $\underset{\sim}{B}$ and $\underset{\sim}{C}$) have a physical meaning. In particular compartmental analysis suggests a natural choice for the state variables: i.e., the compartmental quantities.

In the conventional system theory approach, special interest is devoted to the properties which are common to any different realization $(\underset{\sim}{\hat{A}}, \underset{\sim}{\hat{B}}, \underset{\sim}{\hat{C}})$ of the same system. On the contrary, in dealing with physically based models with the natural state representation, emphasis is given to the properties which are common to any system having the same structure, i.e., the same null elements in original matrices $\underset{\sim}{A}$, $\underset{\sim}{B}$, and $\underset{\sim}{C}$.

The last approach is particularly important in compartmental analysis, where the estimation of the non *a priori* zero transport rate constants k_{ij} is the main objective of the identification and an experiment has been designed by stating the coefficients b_{ir} and c_{mi} which are different from zero.

The above considerations lead to the following definitions which have been already presented by Cobelli, *et al.* (1978b, c, d):

Definition 1. A system is said to be a *fixed structure system* if the identically null parameters of a suitable analytical description of it are *a priori* stated, while other parameters are free and mutually independent.

For a fixed structure system the free parameters may be ordered in a *parameterization vector* $\underset{\sim}{p}$, which corresponds to a point in the parameter space P and characterizes a particular model among all other ones having the same structure.

Definition 2. A property of a fixed structure system is said to be a *structural property* if it holds almost everywhere in the parameter space P.

All the desired information on structural properties may be derived from the topology of a suitable graphical representation of the system, in particular from the compartmental diagram.

Clearly *a priori* identifiability is a structural property and therefore it is often called *structural identifiability*. In order to rigorously define this property we will now formalize some preliminary notions (Cobelli, *et al.*, 1978b, c, d):

Definition 3. Given a fixed structure system, two parameterization vectors $\underset{\sim}{p}'$ and $\underset{\sim}{p}''$ in P are said to be (output) *indistinguishable* if the outputs of the corresponding systems $\underset{\sim}{y}(\underset{\sim}{p}')$ and $\underset{\sim}{y}(\underset{\sim}{p}'')$ are identical for every input $\underset{\sim}{u}$ and every initial state $\underset{\sim}{x}_o = \underset{\sim}{x}(0)$. Otherwise the two parameterization vectors are said to be (output) *distinguishable*.

Definition 4. A fixed structure system is said to be *locally identifiable* in $\underset{\sim}{p}'$ if there exists a neighborhood $\varepsilon(\underset{\sim}{p}')$ such that, for every $\underset{\sim}{p}'' \neq \underset{\sim}{p}'$ in it, the pair $(\underset{\sim}{p}', \underset{\sim}{p}'')$ is distinguishable, while it is said to be *globally identifiable* in $\underset{\sim}{p}'$ if, for every $\underset{\sim}{p}'' \neq \underset{\sim}{p}'$ in the whole space P, the pair $(\underset{\sim}{p}', \underset{\sim}{p}'')$ is distinguishable.

Definition 5. A fixed structure system is said to be *almost everywhere locally (globally) identifiable* if it is locally (globally) identifiable in every $\underset{\sim}{p}'$ of the space P, except at most a subspace of zero measure (that is except at most a line in a bidimensional parameter space, at most a surface in a tridimensional space, etc.).

With reference to the problem of structural identifiability of compartmental systems, we are now able to state the following definitions:

Definition 6. A compartmental system is said to be *structurally identifiable* if it is almost everywhere locally identifiable.

Definition 7. A compartmental system is said to be *uniquely structurally identifiable* if it is almost everywhere globally identifiable.

4. STRUCTURAL IDENTIFIABILITY TESTS

The identifiability problem has been analyzed by Grewal and Glover (1976) for the general case of linear dynamical systems. We will now use the above results for the analysis of structural identifiability of linear time invariant compartmental systems.

Let us refer to equations (8) and to the set of definitions given in Section 3. The following two theorems have been proved by Grewal and Glover (1976).

Theorem 1. A pair of parameterization vectors $(\underset{\sim}{p}', \underset{\sim}{p}'')$ in P is indistinguishable if and only if the Markov parameters

$$\underset{\sim}{M}_r(\underset{\sim}{p}') = \underset{\sim}{M}_r(\underset{\sim}{p}'') \tag{10}$$

for all r, where

$$\underset{\sim}{M}_r(\cdot) = \underset{\sim}{C}(\cdot)\underset{\sim}{A}^r(\cdot)\underset{\sim}{B}(\cdot). \tag{11}$$

Theorem 2. The system described by (8) is globally identifiable in $\underset{\sim}{p}'$ if and only if $\underset{\sim}{p}'' \in P$ and (10) together imply $\underset{\sim}{p}' = \underset{\sim}{p}''$.

Therefore one approach for checking global identifiability in $\underset{\sim}{p}'$ is to consider the Markov parameter matrix

$$\underset{\sim}{M}^T(\underset{\sim}{p}') = [\underset{\sim}{M}_0^T(\underset{\sim}{p}'), \underset{\sim}{M}_1^T(\underset{\sim}{p}'), \cdots, \underset{\sim}{M}_{(2n-1)}^T(\underset{\sim}{p}')] \tag{12}$$

and to check whether, as a function of $\underset{\sim}{p}$, it is one to one in $\underset{\sim}{p}'$.

For testing local identifiability in $\underset{\sim}{p}'$ one has to check whether the mapping from the parameter space P into the Markov parameters is locally one to one in $\underset{\sim}{p}'$. This is equivalent to check whether the rank of the Jacobian $\underset{\sim}{J}_M$ of $\underset{\sim}{M}^T(\underset{\sim}{p}')$ is equal to the number q of the unknown parameters:

$$\text{rank } \underset{\sim}{J}_M(\underset{\sim}{p}') = q. \tag{13}$$

As we are interested in identifiability as a *structural property* (see Definition 2 of Section 3), i.e., with a property which holds almost everywhere in the parameter space, the above conditions have to hold for almost all $\underset{\sim}{p}' \in P$. If conditions (12) or (13) hold almost everywhere in the parameter space P, unique structural identifiability (see Definition 7) and, respectively, structural identifiability (see Definition 6) are assured.

It may be remarked that condition (13) may be used as an operative rule for checking structural identifiability, while condition (12) does not present the same characteristics for checking unique structural identifiability.

A condition similar to (13) may be obtained by considering $n_C \times n_B$ transfer matrix $\underset{\sim}{G}(s)$ which can be obtained by finding the Laplace transform of equations (8)

$$\underset{\sim}{G}(s) = \underset{\sim}{C}(s\underset{\sim}{I}-\underset{\sim}{A})^{-1}\underset{\sim}{B} = \frac{\underset{\sim}{C} \, adj(s\underset{\sim}{I}-\underset{\sim}{A})\underset{\sim}{B}}{det(s\underset{\sim}{I}-\underset{\sim}{A})} . \qquad (14)$$

More precisely the symbolic expressions of the coefficients of the $G_{hk}(s)$ in reduced form are to be considered and their Jacobian $\underset{\sim}{J}_G$ (with respect to the q unknown parameters) is to be computed. The condition for structural identifiability is:

$$rank \, \underset{\sim}{J}_G = q \qquad (15)$$

almost everywhere in the parameter space.

Note that both above criteria allow the incorporation of possible additional knowledge on the system (numerical value of some parameters and/or additional relations among them, not deriving from the system structure).

5. NECESSARY CONDITIONS FOR STRUCTURAL IDENTIFIABILITY

A structural identifiability test based on the use of (13) or (15) may often be cumbersome, as will be shown in the examples of Section 6. Therefore it is useful to refer to necessary conditions for structural identifiability which can be easily tested on the compartmental diagram without symbolically writing the analytical expressions of Markov parameters or transfer matrix coefficients. In fact the test of such conditions allow us to quickly rule out all those system structures and experiment configurations not satisfying the above considered conditions.

Necessary conditions of this type have been presented by Cobelli, *et al.* (1978b, c). In order to present such rules the following definitions (exemplified in Figure 2 of the next section) are given with reference to the compartmental diagram of the considered system (Cobelli, *et al.*, 1978b, c, d):

Path: a succession of arcs such that the compartment entered by an arc is the same from which the next arc is starting and every compartment is crossed only once; the *length* of the path is the number of its intercompartmental arcs.

Input and output connectability: a compartment is input- and/or output-connectable if there exists a path from a compartment directly entered by an input to the considered one and/or a path from it to a measured compartment; a system is input-output connectable if all its compartments are input and output connectable.

hk *subsystem:* the subsystem formed by all compartments connectable to input k and to output h; the hk subsystems are arbitrarily ordered from 1 to $n_B \cdot n_C$.

Closed subsystem: a set of compartments such that no arcs connect a compartment of the set either to compartments outside it or to the environment; closed subsystems may reduce to one compartment (closed compartment) and may be contained in a larger closed subsystem.

Common cascade part (between two hk and ℓm subsystems): a set of compartments such that each of its compartments is influenced by compartments outside it only through one compartment f and influences compartments outside the common cascade part only through one compartment g. Compartment f may coincide with input compartments k and m of both subsystems; similarly, g may coincide with the output compartments h and ℓ of both subsystems.

The first of the necessary conditions we will present is based on a topological property of the compartmental graph. The second one is based on the comparison between the number n_u of unknown model parameters and a certain number n_e of equations among them, which can be obtained from the experiment. The n_u unknown parameters are the non-zero transport rate constants k_{ij} and often also the non-zero output coefficients c_{mi}, while the input coefficients b_{ir} are usually known.

The necessary conditions are: (1) the system is input and output connectable and (2) the number n_u of unknown parameters shall not exceed the number n_e:

$$n_e = n - n' + \sum_{\substack{h=1,n_C \\ k=1,n_B}} w_{hk} - \sum_{\substack{h=1,n_C \\ k=1,n_B}} z_{hk}, \quad \text{where:} \qquad (16)$$

n   is the number of compartments of the system;

n'   is the number of closed subsystems (if a closed subsystem is
      contained in a larger closed subsystem, the latter is not
      taken into account);

w_{hk} refers to the hk subsystem and corresponds to the number
      n_{hk} of its compartments minus the length of the shortest hk
      path; if the length of the shortest path is zero (input and
      output in the same jth compartment) and if the output
      coefficient c_{hj} is known, then w_{hk} is computed as above
      minus one;

z_{hk} refers to the possible cascade parts, common between the hk
      subsystem and any of the previously considered ones (following
      the chosen order); for each common cascade part the corres-
      ponding addendum of z_{hk} is the difference between the number
      of compartments of the common cascade part minus the length of
      the shortest path between f and g, minus one.

     Note that if the system is formed by two or more independent
parts (no connection arcs are present among them at all) identifiability analysis is to be performed separately on each independent part.

     If the system is not input-output connectable, then the
compartments which are not input connectable or not output connectable may be neglected together with their respective outgoing
arcs (the arcs which enter the neglected compartments become arcs
towards the external environment). Then identifiability analysis
may be performed for the remaining part of the system (which is
obviously input and output connectable) and refers clearly only
to the transport rate constants of the remaining part.

     The formal proof of the above identifiability conditions has
been given by Cobelli, *et al.* (1978b, c).

     As far as input and output connectability is concerned, the
following remarks hold. The input-output relation remains
unchanged if we eliminate the compartments which are not input
and output connectable together with the transport rate constants
outgoing from them. In fact, taking into account equations (8),
it is easy to realize that any variation in the measured outputs
is determined only by a variation of output connectable compartments, while a variation on any compartmental quantity may be
induced by perturbation inputs only if the corresponding compartment is input connectable. As a consequence, the transport rate
parameters we have thus eliminated cannot appear in the input-

output relations and therefore they cannot be estimated via the designed experiment.

The second identifiability condition is based on the transfer function matrix $G(s)$.

The generic element G_{hk} of G is the transfer function of the experiment on the system with input u_k and output y_h. It may be written in the non-reduced form as:

$$G_{hk}(s) = \frac{\sum\limits_0^{n-1} \beta_{hkt} s^t}{\sum\limits_0^n \alpha_t s^t} \quad , \quad \alpha_n = 1. \tag{17}$$

The experiment allows the determination of the coefficients β_{hkt} and α_t which are functions of the unknown parameters k_{ij} and c_{mi}. In such a way, a certain number of relations among the unknown parameters are available and the problem is to be faced of ruling out a certain number of dependencies among them.

The denominator $\det(sI-A)$ which is common to all G_{hk} is a polynomial of degree n, where n is the number of system compartments. It has at most n independent coefficients, as the coefficient of s^n is equal to 1. If n' closed subsystems are present, the coefficients $\alpha_{n-1}, \alpha_{n-2} \cdots \alpha_{n'}$ are greater than zero, while the coefficients $\alpha_{n'-1}, \alpha_{n'-2} \cdots \alpha_o$ are equal to zero (Hearon, 1963). In general $n-n'$ equations can be written from the denominator coefficients.

The numerator of G_{hk} is a polynomial the degree of which is at most $n-1$ and in such a case it exhibits n coefficients. In order to reduce the number of equations to be considered, the coefficients equal to zero are to be excluded and the factors, if any, common to the denominator or common to previously considered numerators are also to be excluded.

More precisely, for evaluating the number of equations obtainable from a single G_{hk} numerator, the reduced form of G_{hk} must be considered, i.e., the form where the common factor between numerator and denominator has been eliminated. The degree of the reduced form numerator is at most $n_{hk} - 1$, where n_{hk} is the

number of compartments connectable to input u_k and to output
y_h. The actual degree of such numerator is $n_{hk} - 1$ if input
u_k and output y_h take place in the same compartment; it is
equal to $n_{hk} - 2$ if the shortest path between input and output
is formed by one intercompartmental arc and so on. In general
the number of equations obtainable from the numerator of the hk
subsystem transfer function is given by the number w_{hk} of
equation (16).

The factor, which is common between the numerator of G_{hk}
and the numerator of a previously considered transfer function
$G_{\ell m}$, corresponds clearly to the numerator of the transfer
function of the cascade part, if any, which is common between hk
and rm subsystems. The degree of such factor is given by the
number z_{hk} of equation (16).

Note that the evaluation of n_e, beside its use for the
second necessary condition, is also useful for reducing the
amount of symbolic computation relative to the second criterion
of Section 4. In fact only n_e equations are to be considered;
in particular the coefficients are to be taken into account only
for the numerator polynomials of the transfer functions concerning: (1) the hk subsystem formed by the compartments
connected only to input k and output h and (2) the ℓm
subsystem obtained by neglecting the cascade part which is common
to previously considered hk subsystems.

6. EXAMPLES

In this section the identifiability criteria previously discussed will be applied to some compartmental models and the
modalities for their checking will be presented.

Let us first consider the example of Figure 2 and the set of
necessary conditions of Section 5. The first condition is not
satisfied as compartment 7 is not input connectable and compartment 8 is not output connectable; therefore the system is not
structurally identifiable. After neglecting such compartments
and their outgoing arcs the new model is obviously input-output
connectable and the second condition can be tested. Assuming the
output coefficients as known, n_u equals 9; n_e is evaluated
as follows: $n=6$, $n'=2$, $w_{11}=2$, $w_{21}=3$, $z_{11}\equiv0$, and $z_{21}=1$.
Therefore, $n_e=8<n_u$; the condition is not satisfied and the system

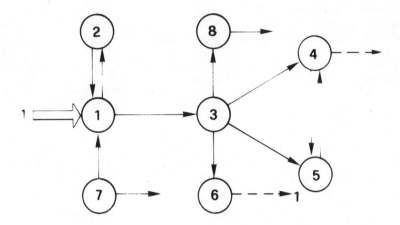

Input 1 in compartment 1

Outputs 1 and 2 in compartments 6 and 4, resp.

Path from 1 to 6: k_{31}, k_{63} (length = 2)

Paths from 1 to 4: k_{31}, k_{43} (length = 2)

$\qquad\qquad\qquad\qquad$ k_{31}, k_{53}, k_{45} (length = 3)

Input connectable compartments: 1,2,3,4,5,6,8

Output connectable compartments: 1,2,3,4,5,6,7

Input-output connectable subsystem: {1,2,3,4,5,6}

hk subsystems: h = 1, k = 1 : {1,2,3,6}

$\qquad\qquad\qquad$ h = 2, k = 1 : {1,2,3,4,5}

Closed subsystems: {6}, {4,5}

Common cascade parts between 1,1 and 2,1

$\quad$ subsystems: {1,2,3}; f ≡ 1; g ≡ 3

FIG. 2: Compartmental diagram for a one-input two-output experiment on an 8-compartment system.

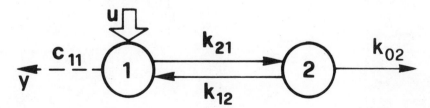

FIG. 3: Compartmental diagram for a one-input one-output experiment on a two-compartment system.

is not identifiable. The example shows the usefulness of the
necessary conditions in detecting the situations of non identifiability without resorting to the symbolic expressions of Markov
parameters or transfer matrix coefficients.

The second example refers to a very simple two compartment
model, with the only purpose of illustrating how the details of
the computation are to be worked out. The compartmental diagram
for a one-input one-output experiment on the two compartment
system is represented in Figure 3.

The conditions presented in Section 5 will be considered
first. The system is input and output connectable. The unknown
parameters are k_{21}, k_{12}, k_{02}, and c_{11}; then $n_u=4$. The number
n_e may be computed as follows: $n=2$, $n'=0$ as no closed subsystems are present, $w_{11}=2-0=2$ as the length of the shortest
path between input and output in zero, and $z_{11}=0$. Thus $n_e=$
$n-n'+w_{11}-z_{11}=4$. Therefore the set of necessary conditions is
satisfied.

In order to apply the first identifiability criterion of
Section 4, i.e., rank $J_{\underset{\sim}{M}}=q$ almost everywhere in the parameter
space, the matrix $\underset{\sim}{M}$ has to be computed. Matrices $\underset{\sim}{A}$, $\underset{\sim}{B}$, and $\underset{\sim}{C}$
are given by

$$\underset{\sim}{A} = \begin{bmatrix} -k_{21} & k_{12} \\ k_{21} & -k_{12}-k_{02} \end{bmatrix} ; \quad \underset{\sim}{B} = \begin{bmatrix} 1 \\ 0 \end{bmatrix} ; \quad \underset{\sim}{C}^T = \begin{bmatrix} c_{11} \\ 0 \end{bmatrix} .$$

Matrix $\underset{\sim}{M}$ is therefore

$$\underset{\sim}{M} \overset{\Delta}{=} \begin{bmatrix} \underset{\sim}{C}\,\underset{\sim}{B} \\ \underset{\sim}{C}\,\underset{\sim}{A}\,\underset{\sim}{B} \\ \underset{\sim}{C}\,\underset{\sim}{A}^2\,\underset{\sim}{B} \\ \underset{\sim}{C}\,\underset{\sim}{A}^3\,\underset{\sim}{B} \end{bmatrix} = \begin{bmatrix} c_{11} \\ -c_{11}k_{21} \\ c_{11}(k_{21}^2+k_{12}k_{21}) \\ -c_{11}(k_{21}^3+2k_{12}k_{21}^2+k_{12}^2k_{21}+k_{12}k_{21}k_{02}) \end{bmatrix} .$$

Then

$$
\underset{\sim}{J}_M \triangleq
\begin{bmatrix}
\dfrac{\partial M_{\sim 11}}{\partial c_{11}} & \dfrac{\partial M_{\sim 11}}{\partial k_{21}} & \dfrac{\partial M_{\sim 11}}{\partial k_{12}} & \dfrac{\partial M_{\sim 11}}{\partial k_{02}} \\[2ex]
\dfrac{\partial M_{\sim 21}}{\partial c_{11}} & \dfrac{\partial M_{\sim 21}}{\partial k_{21}} & \dfrac{\partial M_{\sim 21}}{\partial k_{12}} & \dfrac{\partial M_{\sim 21}}{\partial k_{02}} \\[2ex]
\dfrac{\partial M_{\sim 31}}{\partial c_{11}} & \dfrac{\partial M_{\sim 31}}{\partial k_{21}} & \dfrac{\partial M_{\sim 31}}{\partial k_{12}} & \dfrac{\partial M_{\sim 31}}{\partial k_{02}} \\[2ex]
\dfrac{\partial M_{\sim 41}}{\partial c_{11}} & \dfrac{\partial M_{\sim 41}}{\partial k_{21}} & \dfrac{\partial M_{\sim 41}}{\partial k_{12}} & \dfrac{\partial M_{\sim 41}}{\partial k_{02}}
\end{bmatrix}
=
$$

$$
=
\begin{bmatrix}
1 & 0 & 0 & 0 \\[1ex]
-k_{21} & -c_{11} & 0 & 0 \\[1ex]
k_{21}^2 + k_{12}k_{21} & c_{11}(2k_{21}+k_{12}) & c_{11}k_{21} & 0 \\[1ex]
-(k_{21}^3 + 2k_{12}k_{21}^2 + k_{12}^2 k_{21} + k_{12}k_{21}k_{02}) & -c_{11}\left(3k_{21}^2 + 4k_{12}k_{21} + k_{12}^2 + k_{12}k_{02} \right) & -c_{11}\left(2k_{21}^2 + 2k_{12}k_{21} + k_{21}k_{02} \right) & -c_{11}k_{12}k_{21}
\end{bmatrix}
$$

and det $J_{\sim M} = c_{11}^3 k_{21}^2 k_{12}$ which is always different from zero as c_{11}, k_{21}, and k_{12} are greater than zero.

For the application of the second criterion of section 4, i.e., rank $J_{\sim G} = q$ almost everywhere in the parameter space, the symbolic expression of the transfer function has to be computed:

$$G(s) = \frac{\beta_1 s + \beta_0}{s^2 + \alpha_1 s + \alpha_0} = \frac{c_{11}s + c_{11}(k_{12}+k_{02})}{s^2 + (k_{12}+k_{21}+k_{02})s + k_{21}k_{02}} \quad .$$

Then

$$J_{\sim G} \triangleq \begin{bmatrix} \dfrac{\partial\beta_1}{\partial c_{11}} & \dfrac{\partial\beta_1}{\partial k_{21}} & \dfrac{\partial\beta_1}{\partial k_{12}} & \dfrac{\partial\beta_1}{\partial k_{02}} \\[2em] \dfrac{\partial\beta_0}{\partial c_{11}} & \dfrac{\partial\beta_0}{\partial k_{21}} & \dfrac{\partial\beta_0}{\partial k_{12}} & \dfrac{\partial\beta_0}{\partial k_{02}} \\[2em] \dfrac{\partial\alpha_1}{\partial c_{11}} & \dfrac{\partial\alpha_1}{\partial k_{21}} & \dfrac{\partial\alpha_1}{\partial k_{12}} & \dfrac{\partial\alpha_1}{\partial k_{02}} \\[2em] \dfrac{\partial\alpha_0}{\partial_{11}} & \dfrac{\partial\alpha_0}{\partial k_{21}} & \dfrac{\partial\alpha_0}{\partial k_{12}} & \dfrac{\partial\alpha_0}{\partial k_{02}} \end{bmatrix} =$$

$$\begin{bmatrix} 1 & 0 & 0 & 0 \\[1em] k_{12}+k_{02} & 0 & c_{11} & c_{11} \\[1em] 0 & 1 & 1 & 1 \\[1em] 0 & k_{02} & 0 & k_{21} \end{bmatrix}$$

and det $J_{\sim G} = - c_{11}k_{21} \neq 0.$

Let us now discuss structural identifiability of a four compartment model of calcium cycle in a forest watershed ecosystem (Cobelli, *et al.*, 1978b). The model is shown in Figure 4a. A perturbation input can be applied to compartment 3 only, therefore compartment 4 is not input connectable. In such condition the system is not identifiable and only the subsystem formed by compartments 1, 2, and 3 can be taken into account. For the choice of the output to be actually measured, it may be remarked that measurement of 3 is less complex than the one of 2, which is less complex than the one of 1. The minimal experiment that assures input and output connectability is therefore the one shown in Figure 4b (where the constant inflows into 1, 2, and 3 have been neglected as the system is in steady state). By referring to the compartmental diagram of Figure 4b we find that the system is input and output connectable and the number n_u of unknown parameters is 6. The number n_e can be computed as follows: $n=3$, $n'=0$, $w_{11}=3$, and $z_{11}\equiv 0$. Then $n_e=6=n_u$.

The necessary conditions are satisfied and the identifiability of the proposed system-experiment configuration is to be analyzed. By applying the criteria of Section 4, matrix $\underset{\sim}{M}$ is given by equations (11) and (12) where

$$
\underset{\sim}{A} = \begin{bmatrix} -k_{21} & 0 & k_{13} \\ k_{21} & -k_{32}-k_{02} & 0 \\ 0 & k_{32} & -k_{13}-k_{03} \end{bmatrix} ; \quad \underset{\sim}{B} = \begin{bmatrix} 0 \\ 0 \\ 1 \end{bmatrix} ; \quad \underset{\sim}{c}^T = \begin{bmatrix} 0 \\ 0 \\ c_{13} \end{bmatrix} .
$$

It is evident that to test whether rank $\underset{\sim M}{J} = 6$ is a very hard task even for this relatively simple 3-compartment structure.

Alternatively, the transfer function $G(s)$ and the Jacobian $\underset{\sim G}{J}$ of its coefficients may be considered. In the case of Figure 4b we have:

$$
G(s) = \frac{\beta_2 s^2 + \beta_1 s + \beta_o}{s^2 + \alpha_2 s^2 + \alpha_1 s + \alpha_o}
$$

with $\beta_2 = c_{13}$, $\beta_1 = c_{13}(k_{21}+k_{32}+k_{02})$, $\beta_0 = c_{13}(k_{21}k_{32}+k_{21}k_{02})$,

$$\alpha_2 = k_{13}+k_{21}+k_{32}+k_{02}+k_{03},$$

$$\alpha_1 = k_{13}k_{21}+k_{13}k_{32}+k_{13}k_{02}+k_{21}k_{32}$$

$$+k_{21}k_{02}+k_{21}k_{03}+k_{32}k_{03}+k_{02}k_{03},$$

$$\alpha_0 = k_{13}k_{21}k_{02}+k_{21}k_{32}k_{03}+k_{21}k_{02}k_{03},$$

and the rank of the Jacobian matrix (see Display 1) results to be less than 6; therefore the system is not identifiable. Other experiment configurations on the same system structure which lead to structural and unique structural identifiability have been analyzed by Cobelli, *et al.* (1978b). In particular, with a measurement also of compartment 1 and assuming c_{21} and c_{13} as known unique structural identifiability is achieved.

7. CONCLUSIONS

The problem we have considered in this paper concerns the structural identifiability of a compartmental system, according to Definition 6 (Section 3). In other words, given a compartmental system and an input-output experiment, the problem has to be faced of evaluating whether the planned experiment allows the determination of all unknown parameters only on the basis of the knowledge of the topological structure of the compartmental diagram (i.e., all parameters k_{ij} and possibly c_{mi} which are different from zero are assumed to be unknown and mutually independent).

Compartmental models and identification experiments have been briefly reviewed. The role of structural identifiability in the overall identification problem has been analyzed. Formal definitions for structural identifiability and unique structural identifiability have been provided, by resorting to system theory notions.

Two identifiability criteria have been presented and their use for testing structural identifiability of compartmental systems has been discussed. Such criteria require, in general, a

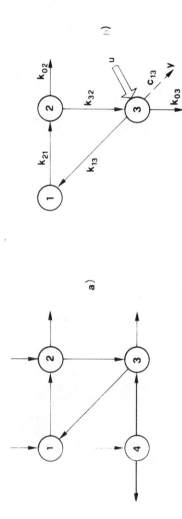

FIG. 4: (a) *A four compartment model of calcium cycle: 1. vegetation; 2. litter; 3. available nutrients; 4. soil and rock minerals.* (b) *The compartmental diagram for the input–output experiment in compartment 3.*

DISPLAY 1: Jacobian matrix $\underset{\sim}{J}_G$ *for the compartment model of Figure 4b.*

$$\begin{bmatrix}
0 & 0 & 0 & 0 & 0 & 1 \\[4pt]
0 & c_{13} & c_{13} & c_{13} & 0 & k_{21}+k_{32}+k_{02} \\[4pt]
0 & k_{32}c_{13}+k_{02}c_{13} & k_{21}c_{13} & k_{21}c_{13} & 0 & k_{21}k_{32}+k_{21}k_{02} \\[4pt]
1 & 1 & 1 & 1 & 1 & 0 \\[4pt]
k_{21}+k_{32}+k_{02} & k_{13}+k_{32}+k_{02}+k_{03} & k_{13}+k_{21}+k_{03} & k_{13}+k_{21}+k_{03} & k_{21}+k_{32}+k_{02} & 0 \\[4pt]
k_{21}k_{02} & k_{13}k_{02}+k_{32}k_{03}+k_{02}k_{03} & k_{13}k_{21}+k_{21}k_{03} & k_{13}k_{21}+k_{21}k_{03} & k_{21}k_{32}+k_{21}k_{02} & 0
\end{bmatrix}$$

considerable amount of formal manipulation as the rank of the
Jacobian matrix related to Markov parameters or to transfer matrix
coefficients is to be evaluated. To avoid the above computation
for surely not identifiable structures, it is convenient to refer
to two necessary conditions, which can be easily tested directly
on the compartmental diagram by evaluating simple quantities as
number of compartments, lengths of paths, etc.

Examples showing the application of the considered conditions
and criteria are provided.

REFERENCES

Bellman, R. and Åström, K. J.   (1970).   On structural identi-
     fiability. *Mathematical Biosciences*, 7, 329-339.

Cobelli, C. and Romanin Jacur, G.   (1975).   Structural identi-
     fiability of strongly connected biological compartmental
     systems. *Medical & Biological Engineering*, 13, 831-838.

Cobelli, C. and Romanin Jacur, G.   (1976a).   Controllability,
     observability and structural identifiability of multi-input
     and multi-output biological compartmental systems. *IEEE
     Transactions on Biomedical Engineering*, 23, 93-100.

Cobelli, C. and Romanin Jacur, G.   (1976b).   On the structural
     identifiability of biological compartmental systems in a
     general input-output configuration. *Mathematical Bio-
     sciences*, 30, 139-151.

Cobelli, C., Lepschy, A., and Romanin Jacur, G.   (1978a).   On
     identifiability problems in biological systems. In *Pro-
     ceedings IV IFAC Symposium on Identification and System
     Parameter Estimation, Part I*, N. S. Rajbman, ed.   North
     Holland, Amsterdam.   525-535.

Cobelli, C., Lepschy, A., and Romanin Jacur, G.   (1978b).
     Structural identifiability of linear compartmental models.
     In *Theoretical System Ecology*, E. Halfon, ed.   Academic
     Press, New York.

Cobelli, C., Lepschy, A., and Romanin Jacur, G.   (1978c).
     Identifiability of compartmental systems and related
     structural properties. *Mathematical Biosciences* (in press).

Cobelli, C., Lepschy, A., and Romanin Jacur, G.   (1978d).   Identi-
     fiability in tracer experiments. *Federation Proceedings* (in
     press).

DiStefano, J. J.   (1976).   Tracer experiment design for unique
    identification of nonlinear physiological systems. *American
    Journal of Physiology*, 230, 476-485.

DiStefano, J. J. and Mori, F.   (1977).   Parameter identifiability
    and experiment design: thyroid hormone metabolism parameters.
    American Journal of Physiology, 233, R134-R144.

Grewal, M. S. and Glover, K.   (1976).   Identifiability of linear
    and nonlinear dynamical systems. *IEEE Transactions on
    Automatic Control*, 21, 833-837.

Hearon, J. Z.   (1963).   Theorems on linear systems. *Annals of
    New York Academy of Sciences*, 108, 36-68.

Jacquez, J. A.   (1972).   *Compartmental Analysis in Biology and
    Medicine*.   Elsevier, Amsterdam.

[*Received August* 1978. *Revised November* 1978]

J. H. Matis, B. C. Patten, and G. C. White, (eds.),
Compartmental Analysis of Ecosystem Models, pp. 125-130. All rights reserved.
International Co-operative Publishing House, Fairland, Maryland.

SIMULATION, DATA ANALYSIS, AND MODELING WITH THE SAAM COMPUTER PROGRAM

MONES BERMAN

Laboratory of Theoretical Biology
National Institutes of Health
Bethesda, Maryland   20014   USA

SUMMARY. SAAM is a general purpose computer program for the analysis of data in terms of dynamic compartmental models. In addition to the numerical solutions, SAAM is designed to deal with the various modeling steps involved with the development of a model and with various experimental procedures involved in the collection of data. Thus, in addition to data fitting, SAAM is oriented towards the modeling methodology as a whole.

KEY WORDS. compartmental models, data fitting, computer programs, modeling methodology.

   SAAM is a general purpose computer program developed for the analysis of data in terms of dynamic models (Berman, Shahn, and Weiss, 1962; Berman, 1965; Berman and Weiss, 1978). It permits simulation and data fitting, and contains various techniques encountered in model building (Berman, Weiss, and Shahn, 1962; Berman, 1978). Although developed primarily for kinetic models in biology the program is of more general utility. It deals best, however, with models that can be described by ordinary (linear and non-linear) differential-difference equations. Such models are frequently referred to as compartmental models. SAAM, however, also solves linear and non-linear algebraic equations and can deal with linear and non-linear regression equations. It is open-ended and can be further expanded to deal with models for which suitable solution algorithms can be provided.

   To appreciate better the power and limitations of SAAM, I'd

like to consider it from a general modeling framework, where various tasks need to be performed. The SAAM program evolved in an effort to carry out some of these tasks routinely.

There is a variety of goals for which modeling may be carried out: description of data, integration of data, testing of hypotheses, formalization of concepts, identification of mechanisms, prediction, diagnosis, etc. The course that one follows in modeling can greatly depend on the goal, and what may be poor modeling with respect to one goal may be perfectly acceptable with respect to another. One therefore needs to be cautious about generalizations.

The most challenging modeling task is: given data on the behavior of a system and other information about a system, identify the mechanisms that underlie these. In this case modeling is usually a cyclic process of stepwise integration of information (data, known physiology, physicochemical laws, hypotheses, etc.) into a consistent framework. In line with this, we can consider two hierarchical levels of modeling. First, given a model structure, simulate and/or adjust its parameters to fit available data, and second, develop strategies of experimentation and data analysis to lead to an appropriate structure of a model.

With respect to the first, the estimation of parameter values and their uncertainties and the testing for uniqueness and consistency of a model need be examined. The data base that one deals with is usually diverse, spotty, and noisy. With respect to the second, the derivations of minimal models subject to known constraints, the validation of models (redundancy and extrapolation), the design of perturbation experiments, to gain new insights into the system and the study of diverse populations for naturally occuring perturbations and the establishment of parameter covariances - need be considered.

SAAM was developed to deal with the various modeling tasks outlined above (Berman, Shahn, and Weiss, 1962; Berman and Weiss, 1978). The major computational tasks include: non-linear differential-difference equation solution algorithms; least squares data fitting; solution interrupts to introduce changes in parameter values; dual solutions for transient and steady states; imposition of constraints between parameters; transformation of models; validation of simpler models; user oriented, simple language with extensive diagnostic capability; and modeling in a BATCH and INTERACTIVE environment.

Given parameter values, p_i, for a compartmental model

(rate constants, initial conditions, etc.), SAAM solves for
f(t) in the differential-difference vector equations

$$f(t) = F(f,t,p) + U(f,t) \tag{1}$$

and computes any user specified functions, G(t) , based on
the parameters and solution values:

$$G(I,t) = g(f,t,p,q) \ . \tag{2}$$

Here, F spells out the form for the differential-difference
equations, U(f,t) is a vector of input functions, g are
user specified relations for the G(I,t) , and q are
additional parameters introduced in G(I,t) . When data are
given, SAAM can calculate and adjust the parameter values,
to obtain a least squares fit of the data, and provides
estimates of statistical measures of confidence for the
parameters p and q and the functions G(I,t) .

SAAM is user oriented and has a simple language for data
input and instructions. For example, the entry of a parameter,
say L(2,1) , implies a first order rate constant for a
transition from state 1 to state 2 and the necessary internal
machinery for this is automatically set up. This parameter
can become more complex by entering with it an operation
@G(I,T) , where @ represents any of the arithmetic operators
(+,-,*,/,**) as well as some special 'ON-OFF' operators, and
G(I,t) may be any specified function. Thus, the rate
constants can be controlled by time dependent or highly nonlinear functions. Instructions entered with data may be applied
to a single datum or to data blocks. The final solutions
are presented both in tabular form and as printer plots.

When executed in CONVERSATIONAL mode there is simple
interactive language for the interrogation of parameter and
solution values, for changing parameter values, for requesting
new solutions, and for the plotting of solutions in graphic
or character mode.

From the Simulation point of view, there are several useful
features:

(1) Parameters values can be specified explicitly or implicitly
     through the simultaneous solution of linear (and some
     non-linear) equations.

(2) Functions to be used during the solution of the differential-
     difference equations or to be calculated subsequently can
     be entered in FORTRAN-like language.

(3) The solution 'interrupt' capability permits the inter-
     ruption of a solution to provide a means for modifying
     parameter values or solution (initial-condition) values.
     This feature also permits multiple solutions of the same
     system under different input conditions or parameter
     perturbations.

(4) A solution dichotomy exists that generates both transient
     and steady state values simultaneously.

     In converging to a least squares fit, various modes of
weighting of data are possible. *A priori* constraints may be
imposed on parameters either by dependence relations, or by
the imposition of variances and/or covariances. Linear
parameters can be identified and linear regressions applied
to them directly, thus speeding convergence. Known solution
functions can be introduced analytically to serve as *forcing
functions* for compartments which they 'feed.' This tends to
decouple subsystems statistically with reduced covariances
between them which greatly improves the rate of convergence
to a least squares fit.

     The solution machinery in SAAM also permits one to carry
out convolution and deconvolutions techniques directly
(Berman, Weiss, and Shahn, 1962; Berman, 1978).

     SAAM is written in FORTRAN and contains about 30,000
statements. With an overlay system it requires about 210K
byte core memory. Versions of the program are available
for the major computers: UNIVAC, IBM, CDC and DEC PDP10. It
is also available for the DEC VAX 11/780 computer which has
a virtual memory operating system and therefore requires
no overlay. Because of core limitations SAAM is restricted
to 25 compartments and 70 parameters.

REFERENCES

Berman, M. (1965). Compartmental analysis in kinetics. In
    Computers in Biomedical Research, Vol. 2, R. Stacy and
    B. Waxman, eds. Academic Press, New York. Chapter VII.

Berman, M. (1978). A deconvolution scheme. *Mathematical
    Biosciences*, 40, 319-323.

Berman, M., Shahn, E., and Weiss, M. F. (1962). The routine fitting of kinetic data to models: a mathematical formalism for digital computers. *Biophysics Journal*, 2, 275-287.

Berman, M. and Weiss, M. F. (1978). *SAAM Manual*. U. S. Department of Health Education and Welfare Publication No. (NIH) 78-180.

Berman, M., Weiss, M. F., and Shahn, E. (1962). Some formal approaches to the analysis of kinetic data in terms of linear compartmental systems. *Biophysics Journal*, 2, 289-316.

[*Received August* 1978. *Revised November* 1978]

J. H. Matis, B. C. Patten, and G. C. White, (eds.),
Compartmental Analysis of Ecosystem Models, pp. 131-144. All rights reserved.
Copyright ©1979 by International Co-operative Publishing House, Fairland, Maryland.

ESTIMATION OF PARAMETERS FOR STOCHASTIC COMPARTMENT MODELS

G. C. WHITE

Group H-12
Los Alamos Scientific Laboratory
Los Alamos, NM 87545 USA

G. M. CLARK

Industrial and Systems Engineering
The Ohio State University
Columbus, OH 43210 USA

SUMMARY. Procedures are described for estimating the rate constants in compartment models when the stochastic variation of the
process (process variation) is assumed large relative to the
sampling variation. The methods are shown to be maximum likelihood if the perturbations to the system are normally distributed.
Estimation for a single compartment is demonstrated with simulated
results to study small sample properties, and with an example.

KEY WORDS. sampling errors, process errors, stochastic compartment, maximum likelihood, random coefficients.

1. INTRODUCTION

The differences between observed and predicted values are
usually assumed due to sampling errors in fitting compartment
models to observed data. That is, the true value of the process
is not observable, so that the observed value has an attached
sampling error. Hence, parameter estimators are chosen such that
the sum of squares of the assumed sampling errors are minimized.

If the observed process also contains some random variation,
denoted here as process variation, a process perturbation at time
t also affects all future observations of the process and hence
generates an autocorrelated sequence of residuals. The model
cannot predict the observed values exactly, not only because there
is some attached sampling error, but also randomness in the real

process. Hence when the sum of squared residuals is minimized, this autocorrelated random error is treated as sampling error, and is not recognized as process variation. The validity of point and interval estimates will depend on the relative magnitude of the two types of variation, and the magnitude of auto-correlation in the residuals. The process variations are not observable, so the investigator must assume that the sampling variation is much larger than the process variation to conclude that the parameter estimates are correct. In the following sections, methods have been developed to give point estimates for the parameters of compartment models when process variation is thought to be the dominant source of variation.

The process error models presented in this paper assume a random rate of movement of material between compartments. Such a random rate would be hypothesized, for example, if one of the forcing functions of the system was a random variable. An example of a random forcing function might be temperature in the environment. The Q10 law states that the rate of oxygen consumption doubles with each 10°C increase in temperature. Hence, if a simple compartmental model of oxygen consumption is hypothesized, with temperature fluctuating randomly, the process models hypothesized in this paper may fit the data.

Kodell and Matis (1976) and Matis and Hartley (1971) have also suggested a method of fitting a stochastic compartment model to data. However, the randomness in their system is assumed due to the discrete nature of the particle movement from one compartment to another. The models hypothesized here allow the rate of transfer to be a random variable.

The methods described will be particularly useful to biologists concerned with estimating the parameters of compartment models of contaminant transport in the environment. Tiwari and Hobbie (1976) have suggested that the random differential equation approach, as opposed to deterministic differential equations, was a better approximation of biological and ecological reality. In this paper maximum likelihood methods of parameter estimation are developed for stochastic compartment models.

2. SINGLE COMPARTMENT RANDOM COEFFICIENT MODEL

Assume a single compartment with concentration $W(0)$ at time $t = 0$. The equation describing the loss of material from the compartment when a random loss coefficient $k(t)$ is assumed is

$$\frac{dW(t)}{dt} = k(t)\, W(t), \tag{1}$$

where $W(t)$ is the concentration at time t. Let $k(t) = k_i$ for $t_{i-1} < t \leq t_i$ where observations of $W(t)$ are taken at times t_i ($i=0,1,\cdots,n$). Note that t_0 need not be equal to zero. Then the solution to equation (1) for any t_i is

$$W(t_i) = W(t_0) \exp\left[\sum_{j=1}^{i} (t_j - t_{j-1}) k_j \right], \tag{2}$$

with recursive solution

$$W(t_i) = W(t_{i-1}) \exp[(t_i - t_{i-1})k_i]. \tag{3}$$

Rearranging equation (3), we solve for k_i:

$$k_i = \log[W(t_i)/W(t_{i-1})]/(t_i - t_{i-1}). \tag{4}$$

Now assume that the k_i are independent and normally distributed with mean κ and variance σ^2. The likelihood of the sample is

$$L(k_1, \cdots, k_n) = \prod_{i=1}^{n} (2\pi\sigma^2)^{-1/2} \exp[-(k_i - \kappa)^2/(2\sigma^2)]. \tag{5}$$

Maximum likelihood estimators of κ and σ^2 are

$$\hat{\kappa} = \sum_{i=1}^{n} k_i/n \quad \text{and} \tag{6}$$

$$\hat{\sigma}^2 = \sum_{i=1}^{n} (k_i - \hat{\kappa})^2/n. \tag{7}$$

The random coefficient model presented suffers from the assumption that $k(t)$ is constant over the interval $t_{i-1} < t < t_i$ when realistically the function $k(t)$ varies randomly about k in the interval.

3.   SINGLE COMPARTMENT ADDITIVE PROCESS ERROR

Assume a single compartment with concentration $W(0)$ at time $t = 0$. Then the equation describing the loss of material from the compartment when a constant loss coefficient k and an additive process error is assumed is

$$\frac{dW(t)}{dt} = kW(t) + e(t), \qquad (8)$$

where $W(t)$ is the concentration at time t, and the $e(t)$ is the process variation or the perturbation in the rate of change of $W(t)$. Assume the compartment is observed at times t_i ($i=0,1,\cdots,n$) and that $t_0 = 0$. For the time period t_{i-1} to t_i, assume $e(t) = a_i$, or that the rate error can be approximated by a constant over the period between observations. Also let a_i ($i=1,\cdots,n$) be random variable normally distributed with mean zero and variance σ_a^2, with all values mutually independent. Then

$$e(t) = a_i \quad \text{for} \quad t_{i-1} \le t < t_i \quad (i=1,2,\cdots,n). \qquad (9)$$

The solution to equations (8) and (9) is known to be of the form

$$W(t) = \exp(kt)f(t) \quad \text{and for} \quad t_0 = 0 \quad \text{is given by}$$

$$W(t) = \exp(kt)[W(t_0) + \int_0^t \exp(-ks)e(s)ds]. \qquad (10)$$

Therefore the compartment concentration at time t is given by

$$W(t) = \exp(kt) \quad \{W(0) - \sum_{i=1}^{p} a_i[\exp(-kt_i) - \exp(-kt_{i-1})]/k$$

$$- a_{p+1}[\exp(-kt) - \exp(-kt_p)]/k\}, \qquad (11)$$

where $p = \max\{i: t_i < t\}$. The last term of (11) is not necessary when $t = t_i$ ($i=1,2,\cdots,m$). Rearranging equation (11), we see that $W(t_i)$ can be expressed in terms of $W(t_{i-1})$:

$$W(t_i) = \exp[k(t_i - t_{i-1})]W(t_{i-1})$$

$$- a_i\{1 - \exp[k(t_i - t_{i-1})]\}/k. \qquad (12)$$

From this equation it is easily seen that an explicit expression for a_i can be obtained:

$$a_i = -k\{W(t_i) - \exp[k(t_i - t_{i-1})]W(t_{i-1})\}$$

$$\div \{1 - \exp[k(t_i - t_{i-1})]\}. \tag{13}$$

A more detailed derivation of equation (13) is given by White (1976). Now assume that the $\{a_i\}$ are independent and normally distributed, $N(0, \sigma_a^2)$. Then the likelihood of the sample is $L(k, a_1, \cdots, a_n)$. Taking the logarithm of L and differentiating with respect to σ_a^2, we see that a maximum likelihood estimate of σ_a^2 is

$$\hat{\sigma}_a^2 = \sum_{i=1}^{n} a_i^2/n. \tag{14}$$

Replacing σ_a^2 by its estimate, we see that the likelihood function is maximized when $\sum_{i=1}^{n} a_i^2$ is minimized. Thus the maximum likelihood estimator of k is $\hat{k}$ such that $\sum_{i=1}^{n} a_i^2$ is minimized. Note that the estimator of initial concentration is just the value observed at $t = 0$. Approximate interval estimates for k can be obtained through the usual nonlinear least square methods. Let

$$\text{Var}(\hat{k}) = \sigma_a^2 \left[\sum_{i=1}^{n} \left(\frac{\partial a_i}{\partial k}\right)^2 \right]^{-1}. \tag{15}$$

An approximate estimate of the $\text{Var}(k)$ is obtained by substituting the estimates for σ_a^2 and k into equation (15). Based on the asymptotic normality of a maximum likelihood estimator, a 95% confidence interval for k is

$$\hat{k} - 1.96\,[\text{Var}(\hat{k})]^{1/2} < k < \hat{k} + 1.96\,[\text{Var}(\hat{k})]^{1/2}. \tag{16}$$

We may extend the additive process error to model to the case where the a_i are assumed independent and normally distributed with mean zero and variance matrix

$$V = \begin{bmatrix} \sigma_1^2 & & & 0 \\ & \sigma_2^2 & & \\ & & \ddots & \\ 0 & & & \sigma_n^2 \end{bmatrix} . \tag{17}$$

To derive an estimator of k, we assume that V can be expressed as

$$V = \Theta \begin{bmatrix} W(t_1)^\phi & & & 0 \\ & W(t_2)^\phi & & \\ & & \ddots & \\ 0 & & & W(t_n)^\phi \end{bmatrix} , \tag{18}$$

where two additional parameters, Θ and ϕ, are introduced. The variance of a_i is assumed to be a power function of $W(t_i)$, with proportionality constant Θ and exponent ϕ.

An explicit expression for the maximum likelihood estimator of Θ is

$$\hat{\Theta} = \frac{1}{n} \sum_{i=1}^{n} \frac{a_i^2}{W(t_i)^\phi} . \tag{19}$$

However we have been unable to obtain explicit expressions for ϕ and k. Thus we increase the parameter space requiring numerical solution from one to two. Given the value of ϕ, we use equation (19) to estimate Θ.

The use of the power function $\Theta W(t_i)^\phi$ to model the variance of a_i provides a very general model.

The two additive process error models are nested, and thus we may use a likelihood ratio test to decide which model is

appropriate. This approach will be illustrated in the example
section.

4. DEVELOPMENT FOR MULTIPLE COMPARTMENTS

The methods described for the single compartment, additive
process error case where the $\{a_i\}$ are iid can be extended to
an m-compartment case. Consider the m differential equations
in matrix notation

$$\underset{\sim}{X}'(t) = \underset{\sim\sim}{KX}(t) + \underset{\sim}{E}(t) \qquad (20)$$

where $\underset{\sim}{X}'(t)$ is the rate of change of the vector $\underset{\sim}{X}(t)$ at time
t, $\underset{\sim}{K}$ is the coefficient matrix, and $\underset{\sim}{E}(t)$ is the vector of
process errors at time t. The error vector is of the same form
as for the one compartment case:

$$\underset{\sim}{E}(t) = \underset{\sim i}{A} \quad \text{for} \quad t_{i-1} < t \leq t_i \quad (i=1,2,\cdots,n), \qquad (21)$$

where $\underset{\sim i}{A}$ is a vector of normally distributed independent process
errors with mean zero and variance matrix $\underset{\sim a}{\Sigma}$. Here $\underset{\sim i}{A}$ and $\underset{\sim j}{A}$
are assumed independent for $i \neq j$. The solution to equation (10)
is known to be (Hirsch and Smale, 1974, p. 99) of the form
$\underset{\sim}{X}(t) = \exp(t\underset{\sim}{K})\underset{\sim}{F}(t)$. Solving for the vector $\underset{\sim}{F}(t)$ with $t_0 = 0$
leads to the solution

$$\underset{\sim}{X}(t) = \exp(t\underset{\sim}{K}) \{\underset{\sim}{X}(0) - \sum_{i=1}^{p} [\exp(-t_i\underset{\sim}{K}) - \exp(-t_{i-1}\underset{\sim}{K})]\underset{\sim}{K}^{-1}\underset{\sim i}{A}$$

$$- [\exp(-t\underset{\sim}{K}) - \exp(-t_p\underset{\sim}{K})]\underset{\sim}{K}^{-1}\underset{\sim p+1}{A} \}. \qquad (22)$$

The previous value of $\underset{\sim}{X}$, $\underset{\sim}{X}(t_{i-1})$, can be substituted into
equation (12) to produce a simpler form for $\underset{\sim}{X}(t_i)$

$$\underset{\sim}{X}(t_i) = \exp[(t_i - t_{i-1})\underset{\sim}{K}]\underset{\sim}{X}(t_{i-1})$$

$$- \{\underset{\sim}{I} - \exp[(t_i - t_{i-1})\underset{\sim}{K}]\}\underset{\sim}{K}^{-1}\underset{\sim i}{A}, \qquad (23)$$

where $\underset{\sim}{I}$ is the identity matrix of dimension m. Note that

$I - \exp[(t_i - t_{i-1})K]$ commutes with K^{-1} because $\exp[(t_i - t_i)K]$ and K^{-1} commute. This is known because

$$\exp[(t_i - t_{i-1})K] = I + \sum_{j=1}^{\infty} (t_i - t_{i-1})^j K^j / j! \qquad (24)$$

with which it is easily seen that K^{-1} commutes. Equation (23) can be rearranged to explicitly solve for A_i

$$A_i = -K\{I - \exp[(t_i - t_{i-1})K]\}^{-1}$$

$$\times \{X(t_i) - \exp[(t_i - t_{i-1})K]X(t_{i-1})\}. \qquad (25)$$

Now assume the A_i are independent and distributed as $N(0, \Sigma_a)$, so that the likelihood of the sample is given by

$$L(K; A_1, \cdots, A_n) = \prod_{i=1}^{n} \frac{1}{(2\pi)^{m/2}(\det\Sigma_a)^{1/2}} \exp(-A_i{}^T \Sigma_a{}^{-1} A_i / 2).$$

$$(26)$$

If Σ_a is known, then we see that the maximum likelihood estimate of K is $\hat{K}$ such that

$$\sum_{i=1}^{n} A_i{}^T \Sigma_a{}^{-1} A_i = \text{Trace}(\Sigma_a{}^{-1} \sum_{i=1}^{n} A_i A_i{}^T)$$

is minimized. If Σ_a is not known, Bard (1974, p. 66) shows that a likelihood function such as equation (26) is maximized when $\log \det \sum_{i=1}^{n} A_i A_i{}^T$ is minimized, and that the maximum likelihood estimator of Σ_a is

$$\hat{\Sigma}_a = (1/n) \sum_{i=1}^{n} A_i A_i{}^T . \qquad (27)$$

The estimator of $X(0)$ is just the compartment values observed
at $t = 0$, because the process is assumed to be observed without
error.

The estimation procedure does not account for the possibility
of multiple eigenvalues of the same value or of complex-valued
systems. Application of the procedures may encounter these problems, and further theoretical development will be required.

5. SMALL SAMPLE PROPERTIES

In Sections 2, 3, and 4 we have developed estimators whose
asymptotic properties are known. The small sample properties of
the random coefficient model presented in Section 2 are also well
understood because the estimator is the simple arithmetic mean of
a sample from a normal population. Now we will consider the
small sample properties of the estimator for the single compartment
additive process error model where the $\{a_i\}$ are assumed iid,
and also look at the robustness of the estimator when the assumption of all process error is not met. Further, we will compare
the results of all process error estimators with the estimator
that assumes all sampling error.

The equation $x_i = 10000 \exp(-0.01t_i) + b_i$ was used to
generate data with sampling errors, with $t_i = 0, 4, 8, \cdots, 200$.
The b_i's were normally distributed, mean zero, $\sigma_b = 400$, and
were added to the function to simulate sampling error. Process
variation data were simulated with equation (12) with $k = -0.01$
and initial value of $x(0) = 10000$. The process was sampled at
$t_i = 0, 4, 8, \cdots, 200$. The a_i's were normally distributed,
mean zero, $\sigma_a = 40$, and were constant between sampling periods.
Data with both process and sampling variation were simulated by
first generating a set of all process variation data, and then
adding a sampling error. Different a_i and b_i values were
used to simulate all process variation data, all sampling variation data, and both process and sampling variation data. Random
normal deviates were generated using the algorithm of Bell (1968).
Three thousand sets of data, each with 50 observations, were
simulated: (1) 1000 with only sampling variation, (2) 1000 with
only process variation, and (3) 1000 with both sampling and
process variation. The data were then fitted by two different
methods: (A) all process variation assumed and (B) all sampling
variation assumed. The estimate of σ_b^2 used was $\sum_{i=1}^{n} [x_i - f(t)]^2/n$,

consistent with the maximum likelihood estimator of σ_a^2. Results are given in Table 1.

Both estimators were found to be unbiased for their respective data types. The all sampling error estimator also appears unbiased for cases which include process error. However, the 95% confidence intervals on $\text{Ave}(\hat{k})$ do not include 0.01 for the all process error estimator for cases including sampling error. Hence the all process error estimator is slightly biased for data that do not fit the model.

Both estimators give biased estimates of σ_a^2 and σ_b^2 respectively. The bias occurs because the maximum likelihood estimators of σ_a^2 and σ_b^2 were used, with n in the denominator. In both cases we see that the bias is removed if the denominators of the all process error estimator of σ_a^2 (equation 14) and the all sampling error estimator of σ_b^2 are replaced with $n - 1$ and $n - 2$ respectively.

Of more concern is the confidence interval coverage for $\hat{k}$ for each estimator. The all process error estimator has very good achieved coverage for the all process error data. Also we would expect the achieved coverage to be even closer to 95% if an unbiased estimate of σ_a^2 was used in calculating the interval. The achieved coverage holds up well when the all process error estimator is applied to data containing sampling error. In Table 1(A) we see that $\text{Ave}(\hat{\sigma}(k))$ is much larger than $\text{SD}(\hat{k})$, indicating that equation (15) overestimates the variance of k when sampling error is present.

The all sampling error estimator has good coverage for the data set containing only sampling error, and would presumably have better coverage if an unbiased estimator of σ_b^2 had been used. The coverage for data containing process error is very poor. Note in Table 1(B) that $\text{Ave}(\hat{\sigma}(k))$ is much less than $\text{SD}(\hat{k})$ for data containing process error, indicating the reason for the poor achieved coverage. When process error is present in the data and the all sampling error estimator is used, the confidence interval is too small.

In summary we see that both estimators provide basically unbiased estimates of k. Only the additive process error estimator provides a confidence interval with the specified or greater coverage. The sampling error estimator results in a precise but wrong interval estimate.

TABLE 1: One thousand data sets, each with 51 observations, were simulated for the three combinations of theoretical variances shown. The true value of k was -0.01. Values in the table represent the mean, standard deviation, or coverage of the parameters estimated by assuming (A) all process variation, or (B) all sampling error, and using the methods derived in the text. The quantity $\hat{\sigma}(k)$ denotes the standard deviation of $\hat{k}$ as calculated from equation (15).

| σ_a | σ_b | $\hat{k}$ | | $\hat{W}(0)$ | | $\hat{\sigma}^2_{(a \text{ or } b)}$ | | $\hat{\sigma}(k)$ | | Achieved |
		AVE	SD	AVE	SD	AVE	SD	AVE	SD	Coverage
				(A)	All	process	variation			
40	0	-.009995	.001104	10000.0	0.0	1558.4	326.6	.001121	.000132	0.944
0	400	-.009932	.000832	9988.5	408.1	19971.6	4780.2	.004002	.000484	1.000
40	400	-.009898	.001389	9999.8	394.7	21498.5	5303.4	.004138	.000546	1.000
				(B)	All	sampling	error			
40	0	-.010074	.001344	10010.5	385.0	86234.2	53658.9	.000190	.000062	0.209
0	400	-.010000	.000275	9997.8	165.8	153911.8	30237.3	.000260	.000026	0.936
40	400	-.010068	.001426	10023.2	423.7	243548.6	73420.9	.000328	.000066	0.368

6.   EXAMPLE

Adams (1976) collected data on the loss of tritium (^{3}H) from a 2-ha Lake Erie marsh. Approximately 11 curies (Ci) of tritium were applied to the marsh unit as tritiated water, HTO. Water samples were taken periodically for the next year and counted for tritium activity. A very simple empirical model which may be fitted to the data is the single compartment model discussed in this paper. Because of incomplete mixing and initially rapid uptake of HTO by sediments, the first five days are not used in the example as the system does not follow constant coefficient kinetics during this period. Thus the example is not perfect, but does serve to illustrate the methods described in the paper.

The results of the different estimation methods applied to these data are presented in Table 2. Two tests on the residuals were performed for each model. Up to $20th$ order autocorrelations were checked with the adequacy of fit test described by Pierce (1971). Also the variance of the first half of the residuals was compared to the variance of the second half of the residuals with an F-ratio. For the heterogeneous variance additive process errors model, the $\{a_i\}$ were standardized by dividing the individual a_i by their associated variances before either of the residual tests were performed.

In Table 2, we see that the only model which is not rejected by either of the tests on the residuals is the heterogeneous variance additive process error model. The likelihood ratio test between the two additive process error models also indicates the inadequacy of the constant variance model.

7.   DISCUSSION

The methods presented provide useful techniques for estimating the parameters of compartment models when sampling error can be assumed trivial relative to the process error. The method is also useful when both sampling and process error are present because an interval estimate with good coverage is obtained.

The additive process error structure we have used to model a stochastic term in the linear differential equations can also be extended to nonlinear differential equations. The solutions are usually difficult to obtain however. A practical application is the Richards' growth curve (Richards, 1959). As yet we have been unable to solve the differential equation. However, allowing a random variation in the rate of growth seems realistic, and would eliminate much of the autocorrelation of residuals which results when growth data are fitted assuming all sampling error.

TABLE 2: Results for four models to estimate the rate coefficient for a single compartment model of HTO loss from a Lake Erie marsh.

Sampling error model

Parameter estimates	$\hat{k} = -0.01015$, $\hat{W}(0) = 2409.3$
Test for autocorrelation of $\{b_i\}$	$P < 0.001$
Test for homogeneity of variance of $\{b_i\}$	$P < 0.001$

Random rate coefficient model

Parameter estimate	$\hat{k} = -0.01001$
Test for autocorrelation of $\{k_i\}$	$P = 0.69$
Test for homogeneity of variance of $\{k_i\}$	$P = 0.001$

Constant variance additive process errors model

Parameter estimate	$\hat{k} = -0.01064$
Test for autocorrelation of $\{a_i\}$	$P = 0.515$
Test for homogeneity of variance of $\{a_i\}$	$P < 0.001$
Max {log likelihood}	$\log L = -219.58$

Heterogeneous variance additive process errors model

Parameter estimates	$\hat{k} = -0.01099$, $\hat{\phi} = 2.64667$
	$\hat{\theta} = 0.38924E{-}5$
Test for autocorrelation of standardized $\{a_i\}$	$P = 0.508$
Test for homogeneity of variance of standardized $\{a_i\}$	$P = 0.670$
Max {log likelihood}	$\log L = -156.47$

Test of constant vs heterogeneous variance additive process errors models	$\chi^2(2) = 126.22$
	$P < 0.001$

ACKNOWLEDGMENTS

This study was partially funded by the Office of Water Resources Research, U. S. Department of the Interior, Project A-038-Ohio, and under contract No. W-7405-ENG-36 between the U. S. Department of Energy and Los Alamos Scientific Laboratory. A portion of this work was done as partial fulfillment of the requirements for the degree of doctor of philosophy for Gary C. White at The Ohio State University.

REFERENCES

Adams, L. W. (1976). *Tritium kinetics in a freshwater marsh.* Ph.D. dissertation, The Ohio State University.

Bard, Y. (1974). *Nonlinear Parameter Estimation.* Academic Press, New York.

Bell, J. R. (1968). Algorithm 334. Normal random deviates (GS). *Communications Association of Computing Machinery,* 11, 498.

Hirsch, M. W. and Smale, S. (1974). *Differential Equations, Dynamical Systems, and Linear Algebra.* Academic Press, New York.

Kodell, R. L. and Matis, J. H. (1976). Estimating the rate constants in a two-compartment stochastic model. *Biometrics,* 32, 377-400.

Matis, J. H. and Hartley, H. O. (1970). Stochastic compartmental analysis: model and least squares estimation with time series data. *Biometrics,* 27, 77-102.

Pierce, D. A. (1971). Distribution of residual autocorrelations in the regression model with autoregressive-moving average errors. *Journal of the Royal Statistical Society, Series B,* 33, 140-146.

Richards, F. J. (1959). A flexible growth function for empirical use. *Journal of Experimental Botany,* 10, 290-300.

Tiwari, J. and Hobbie, J. E. (1976). Random differential equations as models of ecosystems: Monte Carlo simulation approach. *Mathematical Biosciences,* 28, 25-44.

White, G. C. (1976). *A simulation model of tritium kinetics in a freshwater marsh.* Ph.D. dissertation, The Ohio State University.

[*Received April* 1978. *Revised December* 1978]

J. H. Matis, B. C. Patten, and G. C. White, (eds.),
Compartmental Analysis of Ecosystem Models, pp. 145-165. All rights reserved.
Copyright ©1979 by International Co-operative Publishing House, Fairland, Maryland.

STATISTICAL ESTIMATION AND COMPUTATIONAL ALGORITHMS IN
COMPARTMENTAL ANALYSIS FOR INCOMPLETE SETS OF OBSERVATIONS

ROLF E. BARGMANN

Department of Statistics and Computer Science
University of Georgia
Athens, Georgia 30602 USA

SUMMARY. This paper deals with situations where a system can be
observed for a very short time only, so that meaningful extrapolation requires the estimation of elements in the coefficient
matrix of a differential equation system. With a Newton-Raphson
approach to this problem one obtains, in addition to estimates,
asymptotic variance-covariance matrices of these estimates; this
may serve as an indication whether extrapolation on the basis of
available data is valid. Results of sampling studies, based on
successively reduced sets of observations, are presented, and
recommendations are made to help a user to decide when valid conclusions may be drawn for some future point in time as, for
instance, estimating the maximum amount for an intermediate
compartment.

KEY WORDS. compartmental analysis, statistical estimation, confidence intervals, statistical differential equations, Newton-
Raphson on eigenvectors, simulation.

1. INTRODUCTION

1.1 Statement of the Problem. When observations of a system are
available, on one or several variables, at several points in time,
interpolation is usually possible, even extrapolation to a period
immediately following the time of observation is valid, as a rule.
However, where extrapolation far beyond the observed epoch is
required, a mathematical model must be formulated, and the
parameters of the model need to be estimated from the observed
data. The present paper describes a study (Mulherin, 1978) which
attempts to assess the quality of extrapolation, from relatively
short runs under a compartmental model.

In connection with the estimation of specific transition rates of the flow of tracers in an aquarium (Halfon, 1975), a computer program was developed by Bargmann and Halfon (1975) which estimates the coefficients in a system of differential equations:

$$d\hat{y}_{it}/dt = \sum_{j=1}^{n} a_{ij}\hat{y}_{jt} + f_i \qquad (1)$$

$$(i=1,2,\cdots,n), \quad y_{i0} \text{ given},$$

where $\hat{y}_{it}$ is the theoretical amount (or concentration) of a substance in compartment i $(i=1,2,\cdots,n)$ at time t; f_i is a force function (constant rate of inflow into compartment i); and a_{ij} are specific transition rates, i.e., $a_{ij}\hat{y}_{jt}$ represents the rate at which material flows from compartment j into compartment i. The a_{ii}-elements are negative, since $a_{ii}\hat{y}_{it}$ represents the rate of outflow from compartment i. In matrix notation:

$$d\hat{\underline{y}}/dt = A\hat{\underline{y}} + \underline{f}, \quad \hat{\underline{y}}_0 \text{ fixed at time } t=0.$$

Observation vectors (whose elements are the amounts of material in each compartment) are available at m points in time $(0 < t_1 < \cdots < t_m = T)$, and the observed values are assumed to satisfy the model

$$y_{it} = \hat{y}_{it} + e_{it}, \qquad (2)$$

where e_{it} are supposed to be additive errors with zero expectation and variance $\gamma^2 \cdot v_{it}$ [v_{it} known, usually of the form v_i (the same for all time points, but different from compartment to compartment) or v_i/t^2; differences in estimates were slight when choosing one or the other]. The a_{ij} are estimated in such a way that a criterion function

$$\phi = \sum_{i=1}^{n} \sum_{t} w_{it} e_{it}^2 \qquad (3)$$

(usually $\sum_i w_i \sum_t e_{it}^2$) is minimized, where w_{it} or w_i is proportional to the reciprocal of the assumed variance of the error e_{it}. Several computer programs are available for this kind of estimation (e.g., Knott and Reece, 1973) and employ numerical iterative techniques rather than analytical solutions of the differential equation system. For cross-validation, all estimates obtained by our analytical program were calculated once again with the use of the above-mentioned MLAB procedure; results agreed to three or four significant digits.

For our purpose we need not only point estimates of the coefficients, but also estimates of their variances. Hence we wrote a program in which solutions are obtained by the Newton-Raphson technique (the earlier program (Bargmann and Halfon, 1977) used in the tracer-flow system employed the Fletcher-Powell method, and thus had no matrices of second derivatives). As a by-product, at the final stage of iteration, we thus had an estimate of the asymptotic variance-covariance matrix of the estimates of the elements of A. These were compared with sample variance-covariance matrices obtained by sets of ten simulation runs. Agreement was quite satisfactory, and thus the crude 'asymptotic' estimate, the only one available if only a single set of observations is at hand, can serve well for the decision whether or not to extrapolate. The program is listed and documented in Shah (1978); its use in simulation studies has been described in Mulherin (1978). Source programs and computer printouts of several studies were available at the Satellite B (S-10) NATO Advanced Study Institute and ISEP Research Workshop, and were discussed in a work group.

It will be noted that the technique of analysis used in this estimation program is restricted to a model in which the major error component is that due to errors of observation. This seems to be a rather frequent occurrence; in fact, in the tracer flow studies in Halfon (1975) and Bargmann and Halfon (1977), the observed concentrations in each compartment varied so erratically that a smoothing technique had to be applied before even a first guess could be obtained to start the iterations. The computer program handles models of considerable complexity, permitting double and triple roots and complex roots.

In the sense of maximum-likelihood estimation employed in this program, the observed data must be pre-processed in such a way that the resulting errors are additive and can be assumed to be normally distributed. To account for the obvious proportionality of noise to the signal, multiplicative models (log-transformations) were attempted but usually provided a poor fit, because of the relative constancy of the logarithm of concentration in intermediate compartments that have low concentrations for the entire time period. We preferred assigning weights to the data in each compartment. The user may choose a default value in the computer program, which makes the weight for each compartment inversely proportional to the average amount in the compartment, a reasonable approach if the time of observation is long enough to justify the fitting of a gamma distribution to the values. The user has the option to specify weights. Data pre-processing to handle the case of auto-correlated errors has been discussed in Shah (1978). To illustrate the kind of problem which the simulation attempts to approximate, one of several studies is shown which was analyzed by the author in 1966 (see Olson, 1966) by simpler numerical analysis techniques, and which was re-analyzed by the present program.

1.2 Illustration. To study the effect of amount of sensitizer (YSSDS) on the first mitotic division, Olson (1966) observed the proportion of cells remaining in the uninucleate phase (compartment 1), those in the prophase-metaphase (compartment 2), in anaphase (compartment 3), and, finally, in the binucleic state (compartment 4) as observed at time 35(2.5)65 minutes (i.e., at 13 time points) after introducing 60 ml of YSSDS at 34°C. Experimental evidence and estimation from several data sets showed that the first transition begins about 33.5 minutes after sensitizing. The first column for each compartment in Table 1 shows the observed proportions. Plots of these and other data, as shown in Olson (1966) give evidence of inflection points in the first and last compartment, so that the following time-dependent ('aging') model seems indicated:

$$dy_1/dt = -\lambda_1(t-t_0) \cdot y_1$$

$$dy_2/dt = \lambda_1(t-t_0) \cdot y_1 - \lambda_2(t-t_0) \cdot y_2$$

$$dy_3/dt = \lambda_2(t-t_0) \cdot y_2 - \lambda_3(t-t_0) \cdot y_3$$

$$dy_4/dt = \lambda_3(t-t_0) \cdot y_3$$

with $t_0 = 33.5$.

Using the whole set of 13 data points we obtained, by the maximum-likelihood technique, the following estimates of the solution of the system of differential equations:

$$y_1 = \exp\{-.0021875(t-t_0)^2\}$$

$$y_2 = .15464[\exp\{-.0021875(t-t_0)^2\} - \exp\{-.016333(t-t_0)^2\}]$$

TABLE 1:   Observed and predicted proportions of cells.

	Uninucleate			Metaphase		
t	obs	pred(1)	pred(2)	obs	pred(1)	pred(2)
35	.98	.995	.995	.005	.0048	.0047
37.5	.96	.966	.968	.02	.0302	.0278
40	.91	.912	.918	.04	.0634	.0578
42.5	.84	.838	.849	.05	.0883	.0798
45	.77	.749	.765	.07	.0980	.0880
47.5	.67	.651	.673	.089	.0944	.0837
50	.53	.551	.577	.112	.0834	.0753
52.5	.45	.454		.085	.0700	
55	.36	.364		.05	.0562	
57.5	.25	.284		.056	.0419	
60	.19	.215		.05	.0333	
62.5	.13	.159		.03	.0246	
65	.08	.114	(.135)	.021	.0176	(.0179)

	Anaphase			Binucleate		
t	obs	pred(1)	pred(2)	obs	pred(1)	pred(2)
35	.003	.0001	.0001	.012	.0001	.0001
37.5	.01	.0031	.0029	.01	.0011	.0011
40	.01	.0122	.0110	.04	.0127	.0130
42.5	.02	.0220	.0191	.09	.052	.052
45	.02	.0270	.0230	.14	.126	.124
47.5	.021	.0271	.0229	.22	.227	.220
50	.028	.0243	.0206	.33	.341	.327
52.5	.025	.0204		.44	.456	
55	.02	.0165		.57	.564	
57.5	.024	.0129		.67	.660	
60	.01	.0098		.75	.742	
62.5	.01	.0072		.83	.809	
65	.009	.0052	(.0049)	.89	.863	(.842)

$$y_3 = .0438441\exp\{-.0021875(t-t_0)^2\} - .060851\exp\{-.016333(t-t_0)^2\}$$

$$+ .015467\exp\{-.057842(t-t_0)^2\}$$

$$y_4 = 1 - y_1 - y_2 - y_3.$$

The second column in Table 1 for each compartment (headed pred(1)) shows the values as predicted from these equations. The third column in Table 1 (headed pred(2)) displays the predicted values if only the first seven observations are considered. Since the purpose of the present study is to determine to what extent valid inferences can be made from such incomplete sets of data, this reduced set has also been analyzed. The equations for the estimation based on the incomplete set are:

$$y_1 = \exp\{-.0020209(t-t_0)^2\}$$

$$y_2 = .13258[\exp\{-.0020209(t-t_0)^2\} - \exp\{-.0172641(t-t_0)^2\}]$$

$$y_3 = .036409\exp\{-.0020209(t-t_0)^2\} - .048063\exp\{-.0172641(t-t_0)^2\}$$
$$+ .0116538\exp\{-.0648874(t-t_0)^2\}$$

$$y_4 = 1 - y_1 - y_2 - y_3.$$

The estimates of the specific transition rates (twice the exponents) from just the first 7 observations were

$$.004042 \qquad .03453 \qquad .12977.$$

The corresponding standard deviation estimates (asymptotic estimates from just this one set of 7 data) were

$$.000408 \qquad .006562 \qquad .048643.$$

The corresponding estimates of specific transition rates based on the full set of data were

$$.004375 \qquad .03266 \qquad .11568,$$

all less than one standard deviation away from the estimates based on the incomplete sets. This satisfying situation is rather the rule in 'cascade' processes, where the correlations between estimates are low (.324, .134, .084 in the present data set); in more complex systems, where correlations tend to be high, the agreement is not as satisfactory. As another index of the quality of estimation, we may wish to consider the initial half-life ($\sqrt{2\ln 2 / \lambda}$ in this time-dependent process). From the half set (always beginning at $t_0 = 33.5$ minutes) we obtained:

18.5 minutes for uninucleate-metaphase transition,

6.34 minutes for metaphase-anaphase transition, and

3.27 minutes for anaphase-binucleate transition.

These compare in the complete set with

17.8 minutes for uninucleate-metaphase,

6.52 minutes for metaphase-anaphase, and

3.46 minutes for anaphase-binucleate,

a rather small difference. The predicted value for the state of the system at 65 minutes (the last numbers in parentheses in the third columns of Table 1) is equally satisfactory. This example shows how well compartment models fit for certain biological processes, and thus give us some hope of being able to describe complex ecological systems even if we can observe only the very beginning of a process.

2. DESCRIPTION OF THE ALGORITHM

Since a force function can be easily accommodated in a homogeneous system, by addition of an artificial compartment, the following description will be made for the system: $d\hat{\underset{\sim}{y}}_t/dt = \underset{\sim\sim}{A}\hat{\underset{\sim}{y}}_t$; $\hat{\underset{\sim}{y}}_0$ given. Details and derivations are stated in Halfon (1975).

2.1 Initialization. It is well known that the Newton-Raphson technique for several variables is quite efficient, provided that a good initial guess is available. For this purpose we have used, with good success for a long time, a method, essentially due to Prony (1799), of replacing differential equations by difference equations. For the *ith* compartment, let

$$z_{ij} = (y_{it_{j+1}} - y_{it_j})/(t_{j+1} - t_j) \qquad (j=1,2,\cdots,m-1)$$

(recall that m = number of observation vectors). Also let

$$x_{kj} = (y_{kt_j} + y_{kt_{j+1}})/2 \text{ and } z_{ij} = \sum_k a_{ij}x_{kj} + e_{ij};$$

then, in the Prony approximation, we can obtain a first guess of those a_{jk} not assumed to be fixed (e.g., at 0) by multiple regression. Since we deal with an initial guess only, we decided that, where the same coefficient appeared in two different compartments (i.e., where a donor has only one recipient), the simple average of the two regression coefficients was to be used. In systems with unusually large noise (e.g., in Bargmann and Halfon, 1977), previous smoothing of the observed y_{it} by moving averages may be necessary.

2.2 *First Derivatives*. Expressions for first and second derivatives of characteristic roots and eigenvectors of matrices can be
found in some standard textbooks (see, e.g., Porter and Crossley,
1972). Some details for the specific application under consideration are presented in Mulherin (1978) and Shah (1978). Considerable care was taken to insure that intermediate and final results
were correctly programmed.

When the characteristic roots of A are distinct (the sub-
routine that *solves* the differential equation system has provisions
for double and triple roots, but derivatives do not exist when
roots are equal), the solution is $\hat{y}_{it} = \sum_j u_{ij} \exp\{\lambda_j t\}$,
$\sum_j u_{ij} = y_{i0}$ (given), where the λ_j are the characteristic roots
of A (real or complex conjugate pairs) and the u_{ij} are right
eigenvectors, i.e., they satisfy the relation

$$AU = UD_\lambda,$$

where D_λ is the diagonal matrix of characteristic roots. Let
$e_{it} = y_{it} - \hat{y}_{it}$; then

$$\phi = \sum_i w_i \sum_t e_{it}^2 . \tag{4}$$

Hereafter, the symbol b_{ij} will be used to denote a *distinct*
element of A to be estimated. If compartment j is non-
conservative (i.e., if $\sum_{i=1}^n a_{ij} \neq 0$), then $b_{ij} = a_{ij}$. However, if
compartment j empties only to compartments within the system
($\sum_i a_{ij} = 0$), then $\partial\phi/\partial b_{ij} = \partial\phi/\partial a_{ij} - \partial\phi/\partial a_{jj}$, since then a_{jj} is
determined as $-\sum_{i\neq j} a_{ij}$. Thus

$$\partial\phi/\partial b_{k\ell} = -2 \sum_i w_i \sum_t e_{it}(\partial\hat{y}_{it}/\partial b_{k\ell}), \tag{5}$$

$$\partial\hat{y}_{it}/\partial b_{k\ell} = \sum_j (\partial u_{ij}/\partial b_{k\ell})\cdot\exp\{\lambda_j t\}$$
$$+ t \sum_j (\partial\lambda_j/\partial b_{k\ell})\cdot u_{ij}\cdot\exp\{\lambda_j t\}, \tag{6}$$

$$\partial\lambda_j/\partial b_{k\ell} = \begin{cases} u_{\ell j}v_{jk} & (\text{if } \sum_k a_{k\ell} \neq 0) \\ \\ u_{\ell j}v_{jk} - u_{\ell j}v_{j\ell} & (\text{if } \sum_k a_{k\ell} = 0), \end{cases} \qquad (7)$$

where the v_{jk} are the elements of the matrix of left eigenvectors of $\underset{\sim}{A}$, i.e., $\underset{\sim}{V} = \underset{\sim}{U}^{-1}$. (Of course, $\underset{\sim}{V}$ is *not* obtained by a standard inversion program, especially cumbersome if $\underset{\sim}{U}$ is complex-valued, but by the Krylov-sequence technique of solving differential equations, involving only Vandermonde inversion, i.e., synthetic division; Bargmann and Halfon, 1977). Also

$$\partial u_{ij}/\partial a_{k\ell} = u_{ij}\sum_{\substack{\alpha=1\\\alpha\neq j}}^{n} u_{\ell\alpha}v_{jk}/(\lambda_j-\lambda_\alpha) + u_{\ell j}\sum_{\substack{\alpha=1\\\alpha\neq j}}^{n} u_{i\alpha}v_{\alpha k}/(\lambda_j-\lambda_\alpha),$$

$$(8)$$

$$\partial u_{ij}/\partial b_{k\ell} = \begin{cases} \partial u_{ij}/\partial a_{k\ell} & \text{if Comp. } \ell \text{ is non-conservative} \\ \\ \partial u_{ij}/\partial a_{k\ell} - \partial u_{ij}/\partial a_{\ell\ell} & \text{if Comp. } \ell \text{ is conservative.} \end{cases}$$

For second derivatives, $\partial v_{ij}/\partial b_{k\ell}$ is also required; it is

$$\partial v_{ij}/\partial a_{k\ell} = v_{ik}\sum_{\substack{\alpha=1\\\alpha\neq i}}^{n} u_{\ell\alpha}(v_{\alpha j}-v_{ij})/(\lambda_i-\lambda_\alpha). \qquad (9)$$

Again,

$$\partial v_{ij}/\partial b_{k\ell} = \begin{cases} \partial v_{ij}/\partial a_{k\ell} & \text{if Comp. } \ell \text{ is non-conservative} \\ \\ \partial v_{ij}/\partial a_{k\ell} - \partial v_{ij}/\partial a_{\ell\ell} & \text{if Comp. } \ell \text{ is conservative.} \end{cases}$$

2.3 *Second Derivatives.*

$$\partial^2 \phi / \partial b_{k\ell} \partial b_{rs} = 2 \sum_{i=1}^{n} w_i \sum_t (\partial \hat{y}_{it} / \partial b_{rs}) \cdot (\partial \hat{y}_{it} / \partial b_{k\ell})$$

$$- 2 \sum_{i=1}^{n} w_i \sum_t e_{it} (\partial^2 \hat{y}_{it} / \partial b_{k\ell} \partial b_{rs}), \tag{10}$$

where

$$\partial^2 \hat{y}_{it} / \partial b_{k\ell} \partial b_{rs} = \sum_{j=1}^{n} (\partial^2 u_{ij} / \partial b_{k\ell} \partial b_{rs}) \exp\{\lambda_j t\}$$

$$+ t \sum_{j=1}^{n} (\partial u_{ij} / \partial b_{k\ell}) \cdot (\partial \lambda_j / \partial b_{rs}) \cdot \exp\{\lambda_j t\}$$

$$+ t \sum_{j=1}^{n} (\partial^2 \lambda_j / \partial b_{k\ell} \partial b_{rs}) \cdot u_{ij} \cdot \exp\{\lambda_j t\}$$

$$+ t \sum_{j=1}^{n} (\partial \lambda_j / \partial b_{k\ell}) \cdot (\partial u_{ij} / \partial b_{rs}) \cdot \exp\{\lambda_j t\}$$

$$+ t^2 \sum_{j=1}^{n} (\partial \lambda_j / \partial b_{k\ell}) \cdot (\partial \lambda_j / \partial b_{rs}) \cdot u_{ij} \cdot \exp\{\lambda_j t\}. \tag{11}$$

$$\partial^2 \lambda_j / \partial b_{k\ell} \partial b_{rs} = (\partial u_{\ell j} / \partial b_{rs}) \cdot v'_{jk} + u_{\ell j} (\partial v'_{jk} / \partial b_{rs}), \tag{12}$$

where

$$v'_{jk} = \begin{cases} v_{jk} & \text{if Comp. } \ell \text{ is non-conservative} \\ v_{jk} - v_{j\ell} & \text{if Comp. } \ell \text{ is conservative.} \end{cases}$$

Finally,

$$\partial^2 u_{ij}/\partial b_{k\ell}\partial b_{rs} = (\partial u_{\ell j}/\partial b_{rs})\cdot \sum_{\substack{\alpha=1 \\ \alpha\neq j}}^{n} u_{i\alpha}v'_{\alpha k}/(\lambda_j - \lambda_\alpha)$$

$$+ u_{\ell j}\sum_{\substack{\alpha=1 \\ \alpha\neq j}}^{n} \{(\partial u_{i\alpha}/\partial b_{rs})\cdot v'_{\alpha k} + u_{i\alpha}(\partial v'_{\alpha k}/\partial b_{rs})\}/(\lambda_j - \lambda_\alpha)$$

$$- u_{\ell j}\sum_{\alpha\neq j} u_{i\alpha}v'_{\alpha k}\cdot\{(\partial\lambda_j/\partial b_{rs}) - (\partial\lambda_\alpha/\partial b_{rs})\}/(\lambda_j - \lambda_\alpha)^2$$

$$+ \{(\partial u_{ij}/\partial b_{rs})\cdot v'_{jk} + (\partial v'_{jk}/\partial b_{rs})\cdot u_{ij}\}\cdot \sum_{\alpha\neq j} u_{\ell\alpha}/(\lambda_j - \lambda_\alpha)$$

$$+ u_{ij}v'_{jk}\sum_{\alpha\neq j} (\partial u_{\ell\alpha}/\partial b_{rs})/(\lambda_j - \lambda_\alpha)$$

$$- u_{ij}v'_{jk}\sum_{\alpha\neq j} u_{\ell\alpha}\cdot\{(\partial\lambda_j/\partial b_{rs}) - (\partial\lambda_\alpha/\partial b_{rs})\}/(\lambda_j - \lambda_\alpha)^2 \qquad (13)$$

where

$$v'_{\alpha k} = \begin{cases} v_{\alpha k} & \text{if Comp. } \ell \text{ is non-conservative} \\ \\ v_{\alpha k} - v_{\alpha\ell} & \text{if Comp. } \ell \text{ is conservative.} \end{cases}$$

2.4 *Iteration and Approximate Variance Estimates.* Let there be
p distinct elements in the matrix to be estimated by this procedure (i.e., there are p values of the $b_{k\ell}$-type defined after

equation (4)). Beginning with the Prony approximation of these
values, the computer program solves the system of differential
equations, and obtains the value ϕ from equation (4). All values
required in equations (7) and (8) are stored in arrays, then substituted into equation (6) and thus into (5), producing the
gradient vector g with p elements. All required values for

equations (9), (12), and (13) are then obtained, those from (9)
and (12) are stored into internal arrays, whereas those obtained
from (13) are stored in (random access) mass storage, because of
the large space requirement; these are then substituted into (11),
thus into (10), and the (Hessian) matrix of second derivatives,

$\underset{\sim}{H}$, is constructed. At the earlier iterations, the equation system $\underset{\sim}{H}\underset{\sim}{e} = \underset{\sim}{g}$ is solved for $\underset{\sim}{e}$, and the original values $\underset{\sim}{b}$ minus $\underset{\sim}{e}$ represent the new trial for the $b_{k\ell}$ values.

When convergence is reached (in the studies described in this paper, usually three but a maximum of five such iterations were required), i.e., when the elements of $\underset{\sim}{g}$ are very small ($<10^{-6}$) the *inverse* of $\underset{\sim}{H}$ is obtained. For, if the e_{it} are assumed to be normally distributed, with variances inversely proportional to the w_i of equation (4), an estimate of the asymptotic variance-covariance-matrix of the $b_{k\ell}$ is

$$\text{est.var}(\underset{\sim}{b}) = (2\hat{\phi}/df) \cdot \underset{\sim}{H}^{-1} \tag{14}$$

where df = degrees of freedom of error is m (number of observation vectors) times n (number of compartments) minus p (number of elements in $\underset{\sim}{b}$, i.e., number of distinct elements of the model matrix estimated). $\hat{\phi}$ is the value from equation (4) at the final stage.

3. SIMULATION

3.1 Data Generation. Several simulation studies were made in order to assess the quality of the estimates of asymptotic variances, and to obtain some indication whether extrapolation can be recommended. Some of these simulations were performed by discretized transitions (Halfon, 1975), i.e., the deterministic system of differential equations was replaced by a system involving discrete (typically 10,000) elements. To decide whether transition of one element takes place out of compartment i, a uniform random number was generated; if it was below $-a_{ii}\Delta t$ (Δt usually .01 time units), one unit was taken out of the compartment and assigned to one of the recipients in a proportion $a_{ji} / \sum_{j \neq i} a_{ji}$, again by random number lookup. This is by far the simplest approach (the 'process simulation' method) which amounts to a randomized version of the Euler method for the solution of differential equations. It will be recognized, however, that this does not represent the situation assumed in the model (2). However, analysis of such sets by the method described in Section 2 produced no larger or smaller variation of estimates than that based on studies where the simulation had been done by the addition of normally distributed errors.

Other sets were obtained by the addition of random normal numbers to the values obtained from the analytical solution (Mulherin, 1978). Only one of the several sets will be discussed here. For brevity, we choose the four-compartment model:

$$
dy/dt = \begin{bmatrix} -.14 & 0 & 0 & 0 \\ .14 & -.11 & 0 & 0 \\ 0 & .08 & -.07 & 0 \\ 0 & .03 & .07 & 0 \end{bmatrix} \underset{\sim}{y},
$$

where $\underset{\sim 0}{y}^T = [1, 0, 0, 0]$. Fifty observation vectors, for time units .2(.2)10 were obtained from the closed solution. Thus, these 'whole sets' would cover approximately two half-lives of compartment 1 (since compartment 1 has a half-life of $\ln 2/.14 \doteq 5$ time units), a little less than two half-lives of compartment 2, which empties into 3 and 4, and one half-life of compartment 3. Compartment 4 is the final recipient, so the whole system is conservative; values in compartment 4 were subjected to errors of measurement, though, so that the sum of quantities in the four compartments is not 1. Random normal numbers, with standard deviation of .02 in compartments 1 and 2, and .006 in compartments 3 and 4, were added to each deterministic value, with some provision to avoid negative values at earlier time periods in later compartments. For the 'whole set' (t up to 10.0) the coefficient of variation of the noise component amounts to 4% of the average concentration in compartment 1, 7% in 2, 6% in 3, and 20% in 4. Of course, for the half-sets and quarter-sets discussed below, the coefficients of variation are much larger. One typical set (out of ten in this particular run) is reported in Table 2.

To reflect the differences in noise levels, weights of 1, 1, 10, 10 were applied (equation (4)) to the compartments. In one of the analyses, weights of t^2 were used, thus $w_{it} = w_i \cdot t^2$. This yielded, in the restricted sets, a very slight improvement over the time-independent weighting method, and will thus be reported here.

Table 3 shows the estimates of the four parameters $(b_1 = a_{21}, b_2 = a_{32}, b_3 = a_{42}, b_4 = a_{43}$ in the model matrix) for each of the ten 'whole sets' (time .2(.2)10). The estimates of asymptotic standard deviations of the four parameter estimates (equation (14)) are reported in Table 4.

TABLE 2: Sample set (No. 5) generated from compartmental model (15).

t	y_{1t}	y_{2t}	y_{3t}	y_{4t}	t	y_{1t}	y_{2t}	y_{3t}	y_{4t}
.2	.9677	.0109	.00066	.0025	5.2	.5015	.3865	.0876	.0461
.4	.9138	.0783	.0063	.0001	5.4	.4508	.3543	.0914	.0537
.6	.8861	.0642	.0034	.0024	5.6	.4470	.3808	.0985	.0534
.8	.8466	.1010	.0038	.0022	5.8	.4316	.3736	.1038	.0582
1.0	.9014	.1435	.0042	.0016	6.0	.4285	.4225	.1046	.0656
1.2	.8744	.1542	.0076	.0045	6.2	.4516	.4052	.1138	.0710
1.4	.8165	.1774	.0093	.0039	6.4	.4065	.4444	.1186	.0710
1.6	.8102	.1625	.0111	.0066	6.6	.3884	.4345	.1239	.0736
1.8	.8056	.2081	.0169	.0044	6.8	.3931	.3977	.1281	.0797
2.0	.7387	.1682	.0201	.0071	7.0	.3744	.4065	.1302	.0884
2.2	.7365	.2466	.0208	.0107	7.2	.3559	.4257	.1344	.0883
2.4	.7209	.2743	.0252	.0113	7.4	.3773	.4396	.1436	.0923
2.6	.7039	.2549	.0285	.0106	7.6	.3327	.4150	.1486	.0999
2.8	.6804	.2721	.0344	.0144	7.8	.3369	.4063	.1496	.1007
3.0	.6531	.2988	.0367	.0211	8.0	.3129	.4295	.1572	.1027
3.2	.6107	.3036	.0406	.0172	8.2	.3130	.3735	.1580	.1095
3.4	.6132	.2741	.0458	.0220	8.4	.3305	.3750	.1633	.1171
3.6	.6129	.2735	.0477	.0241	8.6	.3205	.4165	.1676	.1217
3.8	.5692	.3330	.0551	.0280	8.8	.3068	.4636	.1689	.1267
4.0	.5749	.3047	.0606	.0290	9.0	.2611	.3867	.1710	.1301
4.2	.5458	.3294	.0650	.0311	9.2	.2417	.3766	.1830	.1383
4.4	.5375	.3889	.0685	.0352	9.4	.2603	.4064	.1831	.1444
4.6	.5146	.3882	.0723	.0388	9.6	.2584	.4121	.1868	.1447
4.8	.4907	.3689	.0770	.0428	9.8	.2492	.3872	.1897	.1471
5.0	.4920	.4103	.0836	.0464	10.0	.2546	.3887	.1941	.1557

TABLE 3: Estimates of transition rates using whole sets.

true	1	2	3	4	5	6	7	8	9	10	mean
.14	.1387	.1397	.1394	.1398	.1401	.1405	.1406	.1413	.1375	.1390	.1397
.08	.0811	.0791	.0809	.0805	.0818	.0793	.0807	.0794	.0812	.0803	.0804
.03	.0300	.0307	.0296	.0300	.0284	.0299	.0288	.0300	.0308	.0297	.0298
.07	.0710	.0667	.0714	.0710	.0760	.0687	.0754	.0690	.0684	.0700	.0708

TABLE 4: Estimates of asymptotic standard deviations using whole sets (each entry should be multiplied by .01).

	1	2	3	4	5	6	7	8	9	Sample S.D.
.060	.066	.061	.064	.062	.050	.061	.058	.055	.053	.108
.162	.171	.162	.169	.169	.131	.162	.149	.150	.142	.091
.160	.169	.160	.167	.165	.129	.160	.147	.148	.140	.074
.628	.675	.628	.657	.645	.515	.638	.585	.575	.552	.297

TABLE 5: Estimates of transition rates using half-sets.

true	1	2	3	4	5	6	7	8	9	10	mean
.14	.1387	.1389	.1397	.1408	.1419	.1383	.1410	.1440	.1410	.1373	.1402
.08	.0768	.0768	.0760	.0812	.0817	.0815	.0814	.0801	.0767	.0847	.0797
.03	.0334	.0339	.0330	.0289	.0268	.0294	.0288	.0284	.0323	.0239	.0299
.07	.0437	.0480	.0377	.0790	.0876	.0864	.0749	.0832	.0454	.1192	.0705

The last column in Table 4 is the sample standard deviation of each estimate for the ten sets. A fact which will become more pronounced in the half-sets and quarter-sets should be noted here: The asymptotic standard deviation estimate is, in every run, somewhat lower than the sample standard deviation based on the sample of ten runs, for the transition rate from compartment 1 to compartment 2, and considerably larger for all the other transitions. Since exact standard deviations of correlated estimates are very difficult to obtain, it is impossible to say, at this time, which one is 'better'. It should be noted that, in live data, only the asymptotic estimate is available, since there will be only one set. The estimates are so good that the coverage by the whole set must be considered quite adequate, even though the maximum of the third (intermediate) compartment was not yet reached. It occurs, for the true system, at 19.49 time units (see Section 4), i.e., extrapolation to twice the observed time period is necessary for the assessment of this important parameter.

3.2 Restriction to Half-Sets and Quarter-Sets. Portions of the same ten sets were re-analyzed, once using the time periods 0.2 to 5.0 ('half-sets') and then using time periods 0.2 to 2.4 ('quarter-sets'). A look at Table 2 shows that, by time 2.4, the last compartment is just beginning to show a monotonic increase; the preponderance of noise up to time 1.8 or 2.0 is quite obvious. In fact, for the quarter sets the coefficient of variation (standard deviation/average response) was 120% for compartment 4. Table 5 is the half-set analog of Table 3, and Table 6 shows the asymptotic standard deviations estimates, followed by the sample standard deviation. Table 7 shows the estimates for the quarter-sets, and is the analog of Table 3 in the whole sets. A solution for the first set could not be found. Again, the estimated asymptotic standard deviations, followed by the sample standard deviation, are shown in Table 8. The quarter sets show the pronouncedly larger values of the estimated asymptotic standard deviations (three or four times the sample estimate) for compartments 2 to 4, and the smaller asymptotic value (1:2) for the estimate of the standard deviation of b_1 ($=a_{21}$).

4. EVALUATION OF SAMPLING EXPERIMENTS

Tables 5 and 6 indicate that the half-sets produce consistently good estimates, even though they represent but one half-life of the first compartment and a small fraction of the process for the following compartments. For the intermediate compartment 2, the maximum value, under the true model, is reached at time 8.0387, thus the half set requires extrapolation even for this value.

TABLE 6: Estimates of asymptotic standard deviations using half-sets (each entry should be divided by 10).

1	2	3	4	5	6	7	8	9	10	Sample S.D.
.0086	.0104	.0087	.0074	.0100	.0085	.0098	.0104	.0090	.0091	.0199
.0534	.0646	.0534	.0481	.0644	.0570	.0626	.0651	.0547	.0646	.0295
.0534	.0646	.0534	.0481	.0643	.0569	.0627	.0650	.0547	.0645	.0323
.4382	.5319	.4411	.3844	.5143	.4567	.4968	.5287	.4499	.5120	.2604

TABLE 7: Estimates of transition rates using quarter sets.

true	2	3	4	5	6	7	8	9	10	mean
.14	.1419	.1430	.1395	.1390	.1395	.1413	.1419	.1330	.1519	.1412
.08	.0807	.0831	.0657	.0863	.0897	.0766	.0719	.0787	.0907	.0804
.03	.0322	.0205	.0438	.0222	.0224	.0385	.0325	-.0454	.0102	.0297
.07	.0978	.2088	-.1962	.1704	.2165	-.0436	-.0397	-.0786	.3836	.0799

TABLE 8: Estimates of asymptotic standard deviations using quarter sets.

2	3	4	5	6	7	8	9	10	Sample S.D.
.00221	.00217	.00149	.00231	.00207	.00252	.00256	.00258	.00234	.00496
.03350	.03662	.01806	.03720	.03412	.03323	.03480	.03632	.04005	.00821
.03351	.03665	.01808	.03722	.03413	.03324	.03480	.03633	.04008	.01178
.56131	.62512	.33025	.60042	.53947	.55305	.61993	.58030	.67022	.18267

In the discussion of quarter-sets it should be noted that the physical model underlying this system would not permit negative values in off-diagonal positions of the matrix, although the Newton-Raphson algorithm does not thus restrict the values. Thus, if the procedure results in negative estimates, such values would be incompatible with the model. As shown in Tables 7 and 8, the quarter sets produce many values of this kind, for the last parameter.

Thus, one index which might give some indication whether extrapolation from incomplete sets is permissible would be the size of the weighted least-squares estimate of the elements in the matrix of coefficients. If they are negative, the model does not fit the assumptions. If they are positive, but so small that approximate 95% confidence intervals (minus 2 Standard Deviations, say) would cover zero, the mathematical solution is inconsistent with the assumptions. The sampling data show, however (compare also the mitosis data in the example in Section 1), that 2 standard deviations is much too rigorous a criterion. As an index for the validity of extrapolation we propose the lower value of an approximate 50% confidence interval (estimate minus estimate of probable error). If this value is inadmissible (negative) for any parameter, extrapolation would not be recommended. Using the asymptotic estimate of the standard deviation (the only one available in real-life data) we find that none of the quarter-sets is reliable for purposes of extrapolation, at least not beyond the second compartment. Set 10 would be the only one coming close to satisfying the criterion of admissibility for extrapolation. If one were to extrapolate from set 10, the following errors would occur in the most important quantity, i.e., the maximum concentration in intermediate compartments:

	Compartment 2		Compartment 3	
	max	at time	max	at time
'True'	.41301	8.0387	.27672	19.49
Whole Set 10	.41160	8.0660	.27730	19.53
Error	−0.3%		+0.2%	
Half Set 10	.41162	8.1707	.21357	16.509
Error	−0.3%		−22.8%	
Quarter Set 10	.44516	8.02	.09881	11.20
Error	+7.8%		−64%	

It is seen that, even for the reasonably well-behaved set 10, prediction of the maximum concentration of compartment 3 from the quarter set is practically useless.

As seen from the results of the half-sets and, for the prediction of compartment 2 even from the quarter sets, observation of a process for a very short time, a virtual snapshot, permits valid extrapolation. Llaurado and Smith (1971) performed a similar study using a deterministic equation which would constitute a compartmental model with one compartment emptying directly into three others, and with no intermediate connections. The noise applied (coefficient of variation approximately 20%) is comparable to ours in the fourth compartment. However, when we look at the half-lives implied in their model (26, 260, and 1527 seconds for the parallel transitions) we note that even their *reduced* set covers practically the whole process (135 half-lives of the first, 14 half-lives of the second, and 2.4 half-lives of the third transition in the one-hour set, non consecutive). By contrast, even our whole set covers only two half-lives of the first, 1.5 half-lives of the second, and one half-life of the third compartment, all consecutive. The stability of estimates in Llaurado and Smith (1971) was thus not surprising. From our results, which were corroborated in four additional simulation runs involving up to 7 compartments in Mulherin (1978) and Shah (1978), we draw the reassuring conclusion that, when the evaluation of estimates of transition rates is done by the Newton-Raphson method, and when approximate values of asymptotic standard deviations of the estimates are thus also available, a recommendation when to extrapolate can be made from even a very short period of observation of a compartmental process. When, for a column of the model matrix designating an intermediate compartment, all estimates minus 0.67 asymptotic standard deviation lead to admissible values, prediction of the maximum amount is very reliable, even if it involves extrapolation to four times the period of observation (as in the prediction of the maximum of compartment 3 from the half-sets).

REFERENCES

Bargmann, R. E. and Halfon, E. (1977). Efficient algorithms for statistical estimation in compartmental analysis; modelling 60-Co kinetics in an aquatic microcosm. *Ecological Modelling*, 3, 211-226.

Knott, G. and Reece, D. (1973). *MLAB - An On-Line Modeling Laboratory*. National Institutes of Health, Bethesda, Maryland.

Llaurado, J. C. and Smith, G. A. (1971). Some observations on the effects of time curtailing on the parameters of decaying exponentials fitted to biomedical data. *International Journal of Bio-Medical Computing*, 2, 265-275.

Mulherin, A. M. (1978). *Extrapolation from noisy data in compartmental analysis*. Ph.D. dissertation, University of Georgia.

Halfon, E. (1975). *A system identification procedure for large-sample ecosystem models*. Ph.D. dissertation, University of Georgia.

Olson, L. W. (1966). *Microtubule morphogenesis in allomyces*. Ph.D. dissertation, University of California, Berkeley.

Porter, B. and Crossley, R. (1972). *Modal Control (Theory and Applications)*. Taylor and Francis, London.

Prony, A. (1799). In *Journal de l'ecole polytechnique*. cahia 2 (an IV). p. 29.

Shah, A. C. (1978). *Computational algorithms for statistical estimation in compartmental analysis*. Ph.D. dissertation, University of Georgia.

[*Received June* 1978. *Revised February* 1979]

STOCHASTIC APPROACHES TO THE COMPARTMENTAL MODELING OF ECOSYSTEMS

INTRODUCTION TO SECTION III

This section presents an overview of the state-of-the-art of stochastic modeling of compartmental systems. The five papers all have one feature in common, namely, they all contain general theoretical arguments justifying some particular stochastic approach to compartment modeling. They also contain, implicitly, general arguments that stochastic models may be justified by data analysis and validation requirements as it becomes apparent that (deterministic) models with sampling or measurement error alone often are not consistent with observed data. Several different approaches to stochastic modeling are outlined, each illustrated by cited or worked-out examples.

The Tiwari paper presents a modeling approach based on stochastic differential equations which are solved by computer simulation. The paper discusses the principle of maximum entropy and of Bayesian inference in the model analysis context. The Matis and Wehrly paper approaches the modeling problem analytically and identifies four common sources of stochasticity which are incorporated with the classical deterministic formulation into a more general model. Purdue presents a survey of stochastic compartment models, reviewing their mathematical theory in some detail, and developing, in particular, the semi-Markov models. The Marcus paper complements the previous semi-Markov development in presenting applications to forest succession and to lead metabolism modeling. Wise, who pioneered the concept of alternative, nonexponential residency time distributions for nonhomogeneous compartments with 'incomplete mixing,' also outlines an example of a semi-Markov model where the causal mechanism is elucidated. The reader should be able to extrapolate from these papers and their examples the present applicability of stochastic compartment models to other ecological modeling problems of interest.

J. H. Matis, B. C. Patten, and G. C. White, (eds.),
Compartmental Analysis of Ecosystem Models, pp. 167-194. All rights reserved.
Copyright ©1979 by International Co-operative Publishing House, Fairland, Maryland.

A MODELING APPROACH BASED ON STOCHASTIC DIFFERENTIAL EQUATIONS, THE PRINCIPLE OF MAXIMUM-ENTROPY, AND BAYESIAN INFERENCE FOR PARAMETERS

J. L. TIWARI

Department of Surgery
School of Medicine
University of California
Los Angeles, California   90024   USA

SUMMARY. This paper presents an outline of a modeling approach based on stochastic differential equations, the principle of maximum entropy, and Bayesian inference for the parameters. Using the principle of maximum entropy, all the existing information about a parameter can be encoded in the form of a probability distribution. These distributions, while using all the available information, are maximally noncommittal with regard to missing information. Any new data that become available in the future can then be incorporated via Bayes theorem. This is particularly useful for the models of ecological systems based on long term investigation.

In the analysis of a relatively simple aquatic ecosystem large numbers of numerical solutions generated by the Monte Carlo simulation procedure gave a range of values of the variables very similar to the data obtained from the field measurements. The stochastic differential equation models enable us to make predictions in terms of means and the associated standard deviations and thus are more useful than the deterministic models for applied problems in ecology, resource management, and environmental science.

KEY WORDS. models, ecological systems, probability distributions, stochastic differential equations, maximum entropy,. Bayesian inference.

1. INTRODUCTION

The increased perception of the serious consequences of the deteriorating environment and ecological systems has generated considerable interest in the experimental as well as mathematical analyses of a wide range of problems in ecology and environmental sciences. The obvious need to predict the effects of man-made changes has also necessitated the search for suitable mathematical models (equations) and methods for estimating and specifying the parameters of these models that would adequately describe the system under consideration.

The basis of selecting a model is, of course, our present knowledge about the time dependent behavior and the interrelation-ship of the relevant biological and ecological processes. The experimental data from these systems provide the basis for specify-ing the parameters of these models. Thus in the development of the mathematical models for prediction one has to deal with the two basic problems: finding appropriate models and specifying the parameters of these models.

The objective of this report is to present a mathematical framework based on stochastic differential equations, the principle of maximum-entropy, and Bayesian inference for the modeling and analysis of applied problems in ecology and resource management. The applications and details of this approach are available in a series of papers by Tiwari and Hobbie (1976a,b) and Tiwari *et al.* (1978a,b). The approach here is essentially numerical and we will not be concerned with the theoretical problems, important as they are, associated with the existence and the uniqueness of the analytical solutions.

In the next Section a brief and general discussion of the adaptive strategy for the selection of models for applied problems will be given. Section 3 will include the rationale for adopting the stochastic differential equation models, basic notations and definitions, and the transition from the deterministic to the stochastic framework of modeling. In Section 4 the principle of maximum-entropy and its application to several ecological problems will be discussed. The updating of the model parameters using Bayesian inference will be taken up in Section 5.

2. THE ADAPTIVE STRATEGY IN THE SELECTION OF A MATHEMATICAL MODEL

The choice of a mathematical model depends, of course, on the nature of the problem and the questions that need to be answered. In biological systems, unlike those in engineering and the physical sciences, the basic mechanisms of only a few processes are relatively

well known. In most situations our knowledge about these systems
is incomplete and thus any particular system can be represented
and described by several models.

Box and Jenkins (1976) have outlined two basic ideas in the
development of mathematical models for predictive purposes. The
principle of parsimony suggests that if there are several models
that could be used to adequately represent the system then we
should select the model with the *smallest possible* number of parameters. Models conceptually interesting but with many parameters
that cannot be estimated from the existing data are of little use
for predictive purposes. The theoretical insight provided by
such models is also of questionable value if the basic mechanism
of the process is not known and there are no experimental data to
compare with the results obtained by the model.

Since most of the biological processes are incompletely
understood, the process of selecting an adequate model is necessarily *iterative* (trial and error). It is frequently possible to
select, based on all the available information, a general class
of models (mathematical functions) with right general properties.
From this general class one can identify the model that will be
considered. The parameters of this model can be estimated from
the experimental data and the predictions can be compared with the
observed values. If the model is adequate and its predictions are
consistent with the data then it can be used for forecasting purposes. If, on the other hand, the model is inadequate then we
can go back to the general class of models and identify another
model. This iterative process is repeated until a suitable model
is obtained (for details see Box and Jenkins, 1976).

3. DETERMINISTIC AND STOCHASTIC DIFFERENTIAL EQUATION MODELS OF ECOSYSTEMS

The basic assumption implicit in the approach outlined in
the following sections is that the differential equations (linear
and nonlinear) provide right general properties so that they can
be used as general class of models for a large number of problems
in ecology and resource management. Linear differential equations
used in modeling the dynamics of the compartmental system (see
Section 2 by Purdue, 1979) can be considered as a special class
selected from this general class of models.

The *system-analysis* type of modeling approach to describe and
study ecological systems in detail for the purpose of resource
management has been used and is being used in a number of situations. The early works of Watt (1963, 1966, 1968), Holling (1966),
and Patten (1971, 1972) provided much of the impetus for the

present widespread application of this technique in ecology. More
recently, it has been used, with various degrees of success, to
construct large scale multi-parameter models of complex ecological
systems studied under the auspices of the International Biological
Program.

In this work, the most general model is the so called *state-
space* model in which the dynamics of the system are described by
a set of differential equations representing the behavior of the
state variables. The right-hand sides of these equations each
contain several terms representing biological and ecological
processes which are influencing the dynamics of the state variables. The functional forms of these processes contain state variables, some parameters, and environmental inputs such as temperature
and light. The estimated values of the parameters and input terms
used in such analyses are, in general, average values, and the
variances around these mean values are not taken into consideration. For relatively small variances this may be the appropriate
procedure and will simplify the modelling work. However, for
large fluctuations and for parameters 'estimated' from few or no
data, a single value may not be a good approximation. Thus the
results of these simulations may not give the average behavior in
which an investigator is interested.

Since all these models are data-based and since the functional
forms of various rate processes are based on experimental observations, a more realistic and appropriate approach to the analysis
of these specific ecological systems would be the one in which
naturally occurring and experimentally observed fluctuations are
taken into consideration. This will enable us to utilize the
maximum amount of information from the experimental data and to
assign a range of values to the parameters in terms of probability
distributions. Using the Monte Carlo simulation technique, a probabilistic description of these quantities can easily be incorporated into the model, and the results of the simulation will give
the mean and variance of each of the variables of interest. This
would be of practical value in any natural resource management
scheme because allowance can thus be made for the expected fluctuations from the average value. A probabilistic description of
the system is also more realisitic and desirable because of our
ignorance and because of uncertainties associated with these
complex natural systems. The advantages of the stochastic models
have been discussed by Matis and Wehrly (1979) and the reader is
referred to Section 2 of this excellent paper.

A probabilistic description of the initial conditions, parameters, and other input variables (e.g., light, temperature, rainfall, etc) is consistent with the observed variability of the data
from the field and laboratory experiments. Furthermore, once we

incorporate all the available information about these quantities, there is no logical reason to arbitrarily adjust and readjust one or more parameters simply to reproduce the observed behavior of the variables, as has been done by many 'modellers.' If the output of the model fails to reproduce the behavior of the varia- bles, the problem may lie in the inadequate conceptualization of the biological processes and their mutual interactions in the form of a mathematical model. The values of the parameters can only be changed if new data suggest such modifications. The results of the computer simulation models based on all the avail- able information, yet inconsistent with the observed data and biological facts, may still provide some insight into the nature of the underlying biological processes and their interactions. These may then help us to modify the experimental procedure or plan entirely new experiments. This valuable role of mathematical modeling in ecology cannot be realized if arbitrary tuning and adjusting of parameters are practiced by pretending to know more than we actually do.

In most situations sufficient data are not available to specify the shape of the probability distributions of initial conditions and parameters. This is either because the experimental and field investigations of an ecosystem are in the early stages or because the lack of funds and other resources limits the scope of the sampling scheme and limits the replication of the exper- iments. The essential consequence of these constraints, from a mathematical modeling point of view, is that we have a limited amount of information at our disposal; yet all the estimates of the parameters must be obtained from this available information.

Furthermore, a systematic study of a large ecological system generally spans several years and seasons. As the investigation progresses new experiments are made, new measurements are taken, and plans for further investigation are formalized in the light of all the information available at that time. If mathematical modeling and analyses are a part of the ecosystem study, and if the results of such models and simulation are to be used as an input in any decision-making process, then parameters and other input quantities of such models should be based on all the exist- ing information, both old and new. In the future, with our grow- ing concern for maintaining the quality of air, water, and other natural environments, it may be possible to organize and pool the information from small scattered studies into some kind of regional data bank. These sets of data may become useful in characterizing parameters and other input variables of a mathema- tical model of a system encompassing a broad geographic region.

These constraints necessitate the adoption of a theoretical framework which allows us to incorporate into the model all the

available information (data) about the system. Furthermore, this
theoretical framework should be open-ended, i.e., it should provide the logical capacity to incorporate the additional information if and when it becomes available. Thus we should have a
mechanism for continuously updating the level of knowledge without
dismantling the existing mathematical structure or disregarding
the previous data.

3.1 Definition and Problem Specification. Let the state of the
system under consideration be defined at any time by a vector X:

$$X = \begin{bmatrix} x_1 \\ x_2 \\ \vdots \\ x_n \end{bmatrix}. \tag{1}$$

The rate of change of the state of the system with respect to time
can be expressed by a vector, $\dot{X}$, of the derivative relations,

$$\dot{X} = dX/dt = \begin{bmatrix} dx_1/dt \\ dx_2/dt \\ \vdots \\ dx_n/dt \end{bmatrix}. \tag{2}$$

Let us define an input vector $U = (u_1, u_2, \cdots, u_n)$ characterizing the environmental variables affecting the behavior of the
system, and let $f = (f_1, f_2, \cdots, f_n)$ be of the vector function defined
on the state space and specifying the set of 'forces' responsible
for the dynamics of the system. We will assume that the derivative
vector $\dot{X}$ is a function of the instantaneous values of the states,
the time t, and the input vector U(t) such that we have a
system of differential equations

$$\dot{X} = f[X(t), U(t), t], \quad X(t_0) = X_0. \tag{3}$$

It should be mentioned here that the differential equation
models of the deterministic compartmental system (see, for example
Section 2 of Purdue, 1979) are special cases of this general formulation.

3.2 Stochastic Differential Equations. The introduction of the
stochastic elements in equation (3) makes $\dot{X}$ a vector of stochastic processes, X(t) $(t \in T)$, whose components are

$X_1(t)$, $X_2(t)$, $\cdots$, $X_m(t)$ [i.e., $X_1 = X(t_1)$, $X_2 = X(t_2)$, $\cdots$, $X_n = X(t_n)$].

Let $x(t)$ ($t \in T$) be a stochastic process defined on a probability space (Ω, B, P), where Ω is an abstract space of points ω, B is a σ-field (Borel field) of observable events in Ω, and P is a probability measure defined for all sets in B. Thus P is a nonnegative, countably additive set function $P(A)$, $A \in B$, with $P(\Omega) = 1$. The nth distribution function of $x(t)$ ($t \in T$) is given by

$$F_n = (x_1,t_1;\ x_2,t_2;\ \cdots;x_n,t_n) = F_{x_1 x_2 \cdots x_n}(x_1,t_1;\ x_2,t_2;\ \cdots;x_n,t_n)$$

$$(n = 1, 2, \cdots). \tag{4}$$

The joint density function associated with (4) is given by

$$f_n(x_1,t_1;\ \cdots;x_n,t_n) = \frac{\partial^n F_n(x_1,t_1;\ \cdots;x_n,t_n)}{\partial x_1 \cdots \partial x_n}. \tag{5}$$

The first moment (mean) μ of $X(t)$ at a given time t is defined by

$$\mu(t) = E[X(t)] = \int_{-\infty}^{\infty} x\, f_1(x,t)\,dx. \tag{6}$$

The second moment (variance) of $X(t)$ is given by

$$\sigma^2(t) = E\{[X(t) - \mu(t)]^2\} = \int_{-\infty}^{\infty} (x - \mu)^2 f_1(x,t)\,dx. \tag{7}$$

There are three levels at which probabilistic arguments can be incorporated in (3). The first and analytically simplest case would be to define the initial condition vector, X_0, as a set of random variables having some well defined probability distributions. These types of equations have been studied extensively in statistical-mechanics, statistical-thermodynamics, problems involving drug administration, and other areas of biology (see Rosen, 1970; Soong, 1973). In this situation the transition from deterministic to stochastic is apparent, and the properties of the solution vector can easily be related to the properties of the initial-condition vector.

In the second case the environmental vector, $U(t)$, can be specified as having some defined probability distribution. Thus $U(t)$ is a vector stochastic process and can be incorporated into (3) either through coefficients of the differential equations or through nonhomogenous terms. In this case our interest is in

predicting the properties and behavior of the solution process
X(t) by knowing the properties of the input vector U(t).

In the third situation the coefficients of the differential
equations may have probabilistic descriptions. Thus, in this
case parameters of the system can be regarded as random variables,
and the behavior of the solution process is determined by and is
a consequence of the randomness in the structure of the system.
(For details see Soong, 1973; Srinivasan and Vasudevan, 1971;
Syski, 1968.)

For most practical applications a knowledge of the mean and
variance is sufficient. Therefore, in computer simulation analyses we will primarily be interested in obtaining the estimates
of the mean and variance, at some prespecified time points, of
each of the state variables of the model.

A stochastic process X(t) is a function on a probability
space (Ω, B, P), and hence it is a function of both t and
ω $(\omega \in \Omega)$. For fixed ω the value of X(t) can be obtained by
varying t. This sequence of random variables $X(t_1), X(t_2), \cdots$,
forms a 'realization' or 'sample function,' $X(\cdot, \omega)$, of the
stochastic process. Each sample of a random system leads to a
deterministic sample differential equation with a unique solution
trajectory, and this collection of trajectories can be investigated
using a measure-theoretic approach. It might be mentioned that
these results are consistent with the mean-square treatment of
the stochastic differential equations (second order stochastic
process). In the computer simulation analysis it is assumed
that the conditions associated with the continuity, differentia-
bility, and existence and uniqueness of the stochastic integral
of (3) are satisfied. (For a detailed description of these
concepts see Astrom, 1970; Soong, 1973.)

4. PARAMETER SPECIFICATION IN TERMS OF MAXIMUM-ENTROPY DISTRIBUTIONS

All the available information and knowledge about an ecosystem can be encoded in the form of probability distributions.
These distributions can then be incorporated into the computer
simulation models to simulate the behavior of ecosystems. The
results would enable us to make inferences about the system based
on maximum degree of information that could be extracted from the
existing data.

The basic problem is to find the probability distributions
which agree with the available data and at the same time are
maximally noncommittal with respect to missing information. In

1957 Jaynes showed that by using Shannon's information theory (Shannon and Weaver, 1949), probability distributions can be constructed on the basis of partial knowledge. He showed that these maximum-entropy probability distributions are the least biased estimates possible from the given information. This concept was further developed and applied successfully by Tribus (1961, 1969) to engineering problems involving decision, design, and reliability.

4.1 Information, Uncertainty, and the Principle of Maximum-Entropy. Let X be a random variable defined on a probability space (Ω,B,P) and taking values $x_1,x_2,\cdots,x_n$ (events) with probabilities $p_1,p_2,\cdots,p_n$, respectively. Let us consider an event x_i with a corresponding probability p_i. Suppose that x_i has actually occurred and that we have knowledge about this event either by experimental measurements or observations or by some other means. This knowledge about x_i is the message, and a numerical measure of the information content, I, is given by

$$I = -K \ln p_i \quad (K > 0),\tag{8}$$

where the constant K specifies the unit of information. This I gives a measure of information after it is received (i.e., after the occurrence of the event). The measure of the expected information in a message before it is received is given by (with K = 1)

$$H = -\sum_i p_i \ln p_i.\tag{9}$$

It is clear from (9) that the greater H, the greater the information needed to specify the state of random variable X. Thus H is a measure of uncertainty concerning the state of random variable X. Since H is a measure of uncertainty, it is called the entropy of the probability distribution or of the random variable having p_i as the probability of its being equal to x_i.

It can be shown that H is positive, increases with increasing uncertainty, and is additive for independent sources of uncertainty (for proofs see the appendix in Jaynes, 1957; also see Tribus, 1961, 1969). In his formulation of statistical mechanics, Jaynes (1957) called H the entropy of the probability distribution p_i.

Thus the terms entropy and uncertainty were considered synonymous, for the quantity H is also the formula for entropy in statistical mechanics.

Although a random-variable description of a parameter is more realistic, the available information is not sufficient to obtain the frequencies. In a number of situations, however, this information is in the form of 'averages,' i.e., some measured functions (means, variances, etc.) of the variable are available. The basic problem, then, is to obtain probability distributions of the random variables (parameters) which agree with the existing information but at the same time are maximally noncommittal with regard to the missing information. This can be achieved by using Jaynes' principle: *The least biased or prejudiced probability distribution is that which maximizes the entropy subject to given information.*

To put this concept in the form of simple mathematical equations, let us consider again random variable X which takes discrete values x_i (i = 1, ···, n) with probabilities p_i.
Furthermore, let us assume that the information about the variable is available in the form of averages:

$$\sum_i p_i = 1, \tag{10}$$

$$\sum_i p_i f_1(x_i) = \bar{f}_1, \tag{11}$$

$$\sum_i p_i f_2(x_i) = \bar{f}_2, \tag{12}$$
$$\vdots$$
$$\sum_i p_i f_k(x_i) = \bar{f}_k, \tag{13}$$

where $p_i = Pr(X=x|\bar{f}_1, ···, \bar{f}_k, D)$, i.e., the probability that the value of the random variable X is equal to x_i given $\bar{f}_1, \bar{f}_2, ···, \bar{f}_k$ and D, where D = all the other known facts about X. Thus Jaynes' principle states that the least biased probability distribution of X is the one which maximizes the entropy H subject to the constraints (10) through (13).

Using the *method of Lagrange multipliers* (Hillier and Lieberman, 1974; Tribus, 1961) we can obtain the least biased probability distribution

$$p_i = \exp(-\lambda_0 - \lambda_1 f_1(x_1) - \lambda_2 f_2(x_2) - ···), \tag{14}$$

where $\lambda_0, \lambda_1, \lambda_2, ···$ are Lagrangian multipliers selected to fit the given data.

It is appropriate to mention here some of the most important and interesting statistical properties of this distribution. It can be shown that this maximum-entropy distribution is a global maximum, not a local saddle point. There are several useful relationships between λ_0 and $\bar{f}_k$. For example,

$$-\frac{\partial \lambda_0}{\partial \lambda_k} = \bar{f}_k, \tag{15}$$

$$\frac{\partial^2 \lambda_0}{\partial \lambda_k^2} = \tag{16}$$

$$\frac{\partial^2 \lambda_0}{\partial \lambda_k \partial \lambda_j} = \text{Cov}(f_k f_j) \tag{17}$$

Since equations (15), (16), and (17) are obtained by differentiating λ_0, it is called the *potential function*. (For derivations and proofs of these equations see Tribus, 1969.)

4.2 Application to Ecosystem Problems. To apply the principle of maximum-entropy let us consider the problem of developing a mathematical model to study and predict the growth and seasonal dynamics of a phytoplankton population in an aquatic system sub= ject to the input of nutrients from an external source. Let x_1 and x_2 be two variables denoting the concentrations of phyto- plankton (cells/liter) and the nutrient (µg/liter). The simplest and most general form of the equation relating the rate of growth of algal cells as a function of nutrient concentration in water can be written as

$$dx_1/dt = w - k_3 \, x_1, \tag{18}$$

$$dx_2/dt = k_4 - k_5 \, w, \tag{19}$$

where $w = k_1 \, x_1 \, [x_2/(x_1 + k_2)]$; $k_1 =$ the maximum rate of growth (division/day); $k_2 =$ the half saturation constant, i.e., the nutrient concentration at which the growth rate is one half the maximum growth rate (µg/liter); $k_3 =$ the death rate; $k_4 =$ rate of nutrient input (µg/liter day); and $k_5 =$ the depletion factor or percent cell composition of the nutrient (µg/cell). This type of equation has been used by several investigators to simulate the growth of phytoplankton; experimental evidence indicate that the

growth rate of some of these species follow Michaelis-Menton
type of curve (Lehman *et al.*, 1975; O'Brien, 1974.)

A study of the seasonal growth of the population requires the
solution of equations (18) and (19). Since we are interested in
the modeling and prediction of a particular system, we must obtain
both the numerical values of the parameters $(k_1$ through $k_5)$

and the initial conditions of x_1 and x_2 from the experimental
data for that particular system. However, in the early stages of
studies we rarely have all the necessary data. Thus at least in
the beginning practically all of the parameter values are derived
from the existing data from similar systems. Such preliminary
modeling is very useful for planning experiments and for examining
various consequences of altering values of the functional relationships and parameters.

Case 1: A literature search usually produces a range of values
from which an inference about possible minimum and maximum values
of a parameter can be made. We can then assume that the actual
value of the parameter can be a number (or numbers) between these
limits. Frequently, an experienced investigator familiar with the
system can also give these interval estimates of the parameters.

Since we do not have any other information (i.e., other functions of the parameter), the only constraint which must be satisfied is

$$\sum_i p_i = 1. \tag{20}$$

Thus the maximum entropy distribution is

$$p_i = e^{-\lambda_0}. \tag{21}$$

From the given constraint we have

$$\lambda_0 = \ln k \tag{22}$$

and hence

$$p_i = 1/k. \tag{23}$$

The entropy is given by

$$H = \ln k. \tag{24}$$

This suggests that if only the upper and lower limits of a parameter are known all the k-values between these limits are equally

probable, and hence the least biased specification of the parameter will be the *uniform distribution*. It is also interesting to note that Laplace's principle of insufficient reason [i.e., *if available evidence does not permit us to favor one of the A's (that is, one of the $A_1, A_2, \cdots, A_n$) over another, then the principle of insufficient reason tells us to assign the same probability to each*] is a special case of Jaynes' formalism (see Tribus, 1961.)

Example. Equations (18) and (19) were numerically analysed by Tiwari and Hobbie (1976b) using the data from O'Brien (1974). The 5 parameters and 2 initial conditions were encoded in the form of uniform distributions (equation (23)) and 10 numerical solutions, each with a different set of values sampled from these distributions, were obtained using Runge-Kutta method with variable t (IBM manual , 1972). The results of the individual solutions as well as the mean and standard deviation of all the 10 solutions were consistent with the observed time-dependent behavior of the phytoplankton population in the lakes.

In the deterministic case in which each parameter and initial condition is specified by a fixed value the results also show the spring population bloom condition. However, one can easily change the shape of the curve by manipulating k_3. Curves obtained for four different values of k_3 while keeping other parameters fixed are shown in Figure 1. With this intentional manipulation of the parameter values one can easily show a slow and continuous growth or slow and continuous decrease of the population. It is quite obvious that even in a simple model with 5 parameters and 2 initial conditions it is possible, by parameter manipulation (the 'tuning and calibration of the model'), to produce any desired result. In doing so, however, we are pretending to know more than we really do and this leads to biased and incorrect results.

Case 2: In the previous case the available information was very general, that is, all we knew about the parameter was its possible upper and lower limits. As the experimental investigation of the ecological system progresses, we accumulate more information and collect data from field and laboratory experiments. From such data the most frequently obtained estimates are averages and standard deviations, generally based on four to five replicates. Although the data may suggest that the parameter is statistically variable, we do not have enough sample values to determine the frequency distribution which can be incorporated into the computer simulation model. Nevertheless, at this stage we have gained more information, and not only do we know the upper and lower limits, but we also have estimates of mean and variance.

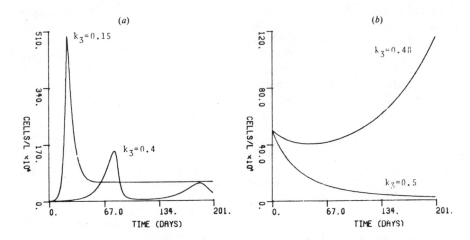

FIG. 1: *Temporal behavior of algal population with various fixed values of k_3. All other parameters are constant.* (k_1 = 0.5, k_2 = 10.0, k_4 = 2.0, k_5 = 2.36 × 10⁻⁷, (x_1 *at* t_0) = 500,000.0, (x_2 *at* t_0) = 150.0) *(Tiwari and Hobbie, 1976b).*

Sometimes, such estimates of mean and variance can be obtained from the published literature on similar and related systems. Based on this information the principle of maximum entropy can be used to obtain the least biased probability distribution of a parameter (Tribus, 1969.)

Let us assume that we have estimates of mean, μ, and variance, σ^2, for a parameter P, whose possible upper and lower limits are P_{max} and P_{min}, respectively. Then the equations of the constraints are

$$\sum_i P_i = 1, \tag{25}$$

$$\sum_i P_i \, P_i = \mu, \tag{26}$$

$$\sum_i P_i \, (P_i - \mu)^2 = \sigma^2. \tag{27}$$

The maximum-entropy distribution is

$$P_i = \exp(-\lambda_0 - \lambda_1 P_i - \lambda_2 P_i^2). \tag{28}$$

The Lagrangian multipliers in equation (28) satisfy the following equations:

$$\lambda_0 = \ln \sum_i \exp(-\lambda_1 P_i - \lambda_2 P_i^2), \tag{29}$$

$$\frac{\partial \lambda_0}{\partial \lambda_1} = -\mu, \tag{30}$$

$$\frac{\partial \lambda_0}{\partial \lambda_2} = -(\mu^2 + \sigma^2). \tag{31}$$

Almost all of the parameters in the models of biological systems have some finite range, and for mathematical convenience we can use a linear transformation to make each range fall between zero and one. With simplification the equation of the constraints can be rewritten as

$$\int_0^1 \exp(-\lambda_0 - \lambda_1 P - \lambda_2 P^2) \, dP = 1, \tag{32}$$

$$\int_0^1 P \exp(-\lambda_0 - \lambda_1 P - \lambda_2 P^2) \, dP = \mu \tag{33}$$

$$\int_0^1 P^2 \exp(-\lambda_0 - \lambda_1 P - \lambda_2 P^2) \, dP = \mu^2 + \sigma^2, \tag{34}$$

for $0 < P < 1$. The problem now is to find the values of λ_0, λ_1, and λ_2 which will satisfy the equality in equations (32), (33), and (34). Tribus (1969) has given an algorithm which can be used to find approximate roots using the Newton-Raphson method.

Some results of the numerical analysis of equation (18) and (19) in which the probability distributions of k_1 and k_2 were generated by equation (22) are available in Tiwari and Hobbie (1976b). All other parameters were sampled from the uniform distribution as outlined in Case 1. The striking difference was a drastic reduction in the variance and almost complete disappearance of the general 'noise' and fluctuations. The confidence interval around the mean was also much narrower, indicating increased precision in the prediction of the behavior of the phytoplankton population. This is one of the necessary consequences of the additional information about k_1 and k_2. As we accumulate more data and get more precise information about other parameters the magnitude of the variance will be further reduced. This points to a more or less obvious conclusion: the accuracy of prediction of the behavior of the variables can be increased by more information about its parameters. This

increased certainty about the parameters will lead to certainty
about the temporal behavior of the system - the deterministic
framework.

5.   BAYESIAN INFERENCE FOR UPDATING THE PARAMETERS

5.1 Notation and Terminology. Let θ be a parameter of a model
of an ecological system under consideration. Let us assume that
our initial information (i.e., that available before any new
experimental measurements are made) is available in the form of a
probability distribution, $f(\theta)$, representing the *prior distribution*
of θ. To get more information about θ, a new set of n obser-
vations $Y = (y_1, y_2, \cdots, y_n)$ is made. Then $f(Y|\theta)$, the like-
lihood function, is determined from the conditional distribution
of Y, given different values of θ. Thus

$$f(\theta)f(Y|\theta) = f(Y,\theta) = f(Y)f(\theta|Y). \tag{35}$$

Given the data Y, the conditional distribution of θ can be
written as

$$f(\theta|Y) = \frac{f(\theta)f(Y|\theta)}{f(Y)}. \tag{36}$$

We can write

$$f(Y) = k^{-1} \begin{cases} \int f(Y|\theta) \ f(\theta) & \text{for } \theta \text{ continuous} \\ \Sigma \ f(Y|\theta) \ f(\theta) & \text{for } \theta \text{ discrete} . \end{cases} \tag{37}$$

(The integral or sum is taken over the admissible range of θ.)
Equation (36) can be rewritten as

$$f(\theta|Y) = k \ f(Y|\theta) \ f(\theta). \tag{38}$$

The equation (38) is known as *Bayes's theorem*. The $f(\theta|Y)$,
called the *posterior distribution* of θ, gives us the information about θ after the addition of the knowledge from the new
data. (For proofs and further details see Box and Tiao, 1973;
De Groot, 1970; Lindley, 1965; Schmitt, 1969; Winkler, 1972).
Thus using Bayes theorem the prior information (represented by the
prior distribution) and the new data (represented by the likelihood
function) can be combined to form the posterior distribution of θ.
The characterization of θ, based on posterior distribution, will
include all the existing information.

5.2 Prior Distribution. The incorporation of the available prior information into prior probability distribution is controversial, for there are no general principles available to translate prior information into a definite prior probability distribution. In the absence of such a principle (or principles) it involves some degree of subjective judgement on the part of the investigator, and thus given information may be translated into different probabilities by different people. If the new information (the likelihood) is based on small sample (relative to prior), then the posterior distribution may be significantly influenced by the prior distribution. This in turn, will affect the behavior of the variables in the model under consideration. On the other hand, if the likelihood dominates the prior, then the posterior distribution will not be significantly changed by the prior distribution. In the intermediate situations, where both prior distribution and the likelihoods have significant influence on the posterior distribution, the choice of a particular prior probability may have important influence on the predictions obtained from the model.

In the mathematical modeling of ecological systems one frequently encounters the situation where for some parameters prior information is the only information available, and parameter specification must be based on this information alone. Furthermore, this prior information is not in the form of frequency data. In some cases, from scattered bits and pieces of information available in the published literature one can 'estimate' an average value and coefficient of variation. In other cases one may be able to establish possible upper and lower limits on a parameter. In the absence of definite rules for translating this kind of information into probabilities, the actual distribution of a parameter used in the model is a matter of subjective judgement. With arbitrary selection of values from this information one can easily produce, from a model, many types of prediction (see Figure 1).

Jaynes (1968) has shown that in many cases it is possible to formulate mathematically the following desideratum: *in two problems where we have the same prior information we should assign the same prior probabilities*. However, some kinds of prior information are too vague and cannot be translated into mathematical terms, and thus if prior information is to be used in a mathematical theory of inference, it must satisfy, as a minimum requirement, the criteria of testable information. (A piece of information I concerning a parameter θ will be called *testable* if, given any proposed prior probability assignment $f(\theta)d\theta$, there is a procedure which will determine unambiguously where $f(\theta)$ does or does not agree with the information I.) Then using the principle of maximum entropy testable information can be converted into prior probability. As discussed in the previous sections,

this principle suggest that the least prejudiced prior probability
distribution is that which maximizes entropy subject to given
prior information. Jaynes' analysis suggests that the principle
may prove to be the final solution to the problem of assigning
discrete prior distributions. For problems involving continuous
distributions, this method can also be applied if the statement
of the problem suggests a definite transformation group which
establishes the invariant measure.

5.3 Likelihood Function. Given the new sample data, Y, the
$f(Y|\theta)$ in equation (38) can be regarded as a function of θ.
Thus, following Fisher (1922), it is called the likelihood function of θ for a given Y (the observed values $y_1, y_2, \cdots, y_n$).

The likelihood function is the second input to the Bayes equation.
It represents the information about the parameter supplied by the
new data. It is this function through which the new data change
the prior information about θ.

The likelihood is equivalently the probability density of
the random variable making up the sample, and thus, to determine
the likelihood we must determine the probability density of the
sample. Based on the characteristics of the sample data, past
experience, the structure of the problem, certain simplifying
assumptions about the data-generating process (the process that
generates the sample information) can be made. With this it may
be possible to represent the data-generating process in terms of
some well-known probability distributions such as Gaussian, exponential, or Poisson (for details see Lindley, 1965; Schmitt, 1969;
Winkler, 1972). Although real-world situations are not exactly
like these theoretical distributions many problems of interest can
be closely approximated by these distributions. It should be mentioned, however, that in the choice of a particular function some
subjective judgement is still involved.

5.4 Applications to Ecological Systems. To examine the usefulness
of Bayesian inference for applied problems in ecology and resource
management, let us consider our simple model (equations (18) and
(19)) of the growth of phytoplankton population in lakes as
influenced by the changes in the nutrient concentrations (e.g.,
phosphorus, nitrogen, etc.). This problem is encountered in water
quality monitoring and management studies and also in impact
statement studies of sewage disposal, reservoirs, etc.

Case 1: Vague Prior Information. In the beginning, we frequently
have no data and can only specify the possible upper and lower
limit of θ. In this case the principle of maximum entropy suggests

that the least biased probability distribution consistent with this information is the uniform distribution where all the values of θ between θ_{max} and θ_{min} are equally probable.

New Data. Suppose n sample values become available after one season of sampling and this new information is to be combined with the prior information (uniform distribution). If we are willing to accept the assumption that the data are independent and come from a Gaussian distribution with unknown mean (μ) and standard deviation (σ) then the likelihood function can be represented by a normal function. Of course, at this stage of analysis this assumption is subjective; but it will simplify the analysis considerably. However, if we look at the pooled data in Figure 2, it is not unreasonable (see Case 2 below). In this case the posterior distribution of θ will be normal, but μ and σ^2 are unknown, and thus now we have to make inferences about these two parameters of the distribution. It has been shown that if $Y = (y_1, y_2, \cdots, y_n)$ is a random sample of size n from $N(\mu, \sigma^2)$ and the prior distribution of μ and $\log \sigma$ are independent and uniform, then given $\overline{Y}$ (the sample mean) and s^2 (the sample variance), the posterior distributions of μ and σ^2 are $t(\overline{Y}, s^2/n, v)$ and $vs\chi^{-2}$, respectively. (For proofs of these theorems, see Box and Tiao, 1973; and Lindley, 1965.) In the above expression $v = n-1 = $ degrees of freedom and χ^{-2} is the inverted χ^2 distribution with v degrees of freedom.

Any new data that become available in the future can be incorporated in this scheme. The present distribution will then be used as the prior distribution. For a numerical example see Tiwari *et al.* (1978a).

Case 2: Prior Information About Mean and Variance. If we have information about possible values of the mean and variance of a parameter the least biased characterization of the parameter (based on the principle of maximum entropy) is in the form of a Gaussian distribution (see Case 2 in Section 4.2). Thus in the absence of any new data this distribution should be used to specify the parameter of the model. On the other hand, if we have some new data, then this distribution should be used as the prior distribution.

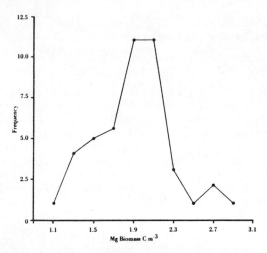

FIG. 2: Frequency distribution of the algal biomass samples at t_0 ($\mu = 1.92$, $\sigma^2 = 0.21$, *and* $n = 42$). *Data from Cayuga Lake, New York (Oglesby and Allee, 1969).*

New Data. Suppose that after several months of experimental investigation we have acquired n sample values of the parameter and these new data are to be combined with the existing probability distribution to update the parameter. If we assume that these sample values come from a Gaussian distribution, then the likelihood can be represented by a Gaussian curve. This likelihood and the prior (also Gaussian) can easily be combined to obtain the posterior distribution.

If a parameter θ has a prior distribution $N(\mu_0, \sigma_0^2)$ and the likelihood function is represented by a normal curve with mean μ and variance σ^2, then the posterior distribution of θ is the Gaussian distribution $N(\mu_n, \sigma_n^2)$, where

$$\mu_n = 1/(w_0 + w_n) \ (w_0\mu_0 + w_n\overline{Y}), \tag{31}$$

$$1/\sigma_n^2 = w_0 + w_n, \tag{40}$$

$$w_0 = 1/\sigma_0^2, \tag{40}$$

and

$$w_n = n/\sigma^2. \tag{42}$$

The mean of the posterior distribution (μ_n) is a weighted average of the prior mean (μ_0) and the sample mean $(\overline{Y})$, w_0 is the reciprocal of the variance of the prior distribution, and w_n is obtained by dividing the sample size by the variance of the data generating process. In most ecological problems this variance is unknown to us. However, for a large sample size it can be approximated by the sample variance (Winkler, 1972.) Thus for a large n, s^2 can be substituted for σ^2 in equation (42).

A numerical example in which the initial condition of x_1 in equations (18) and (19) is updated using this method is given by Tiwari *et al.* (1978a).

Case 3: Sequential Updating. Under the conditions discussed in Case 2, the existing prior distribution can be sequentially updated as new single sample values become available. The present posterior distribution will then become the prior distribution. Any sample value, y_i, can be incorporated by replacing $\overline{Y}$ with y_i and n with 1 in equations (39) and (42), respectively. This type of updating may be very useful in the experimental study of large systems. If the new data do not significantly alter the distribution of a parameter or its effects on the future behavior of the variables, then future collection of data on that parameter may be terminated and the financial and manpower resources can be diverted to measure other quantities or study other processes which are incompletely understood.

Besides these three simple examples there are a number of cases in which these methods can be applied. Some of them are discussed by Tiwari and Hobbie (1978a) and Tiwari, *et al.* (1978b).

6. DISCUSSION

Given a mathemtical model of an ecological system, the problem of specifying the numerical values of the parameters used in the model becomes very important. As shown in Figure 1, the intentional or arbitrary selection of a particular value from the available information can easily produce almost any desired result, especially in a complex model. A general approach using stochastic differential equations, augmented by maximum-entropy and Bayesian inferences for the parameters, can provide a useful tool for modeling and computer simulation analyses of a large class of applied problems in ecology, resource management, and environmental sciences. A decision based on old as well as new

data is particularly necessary and useful in the analysis of
such environmental problems, as the effects of constructing additional power plants on the ecology of the lake, or the effects of
introducing additional quantities of certain chemical or toxic
substances in the air and water. In these cases, new data can be
combined with the old data to obtain the current distributions
of the parameters. The results can thus be examined for the
effects over and above the previous conditions.

Given a prior distribution, a set of sample data can be
analyzed by making different assumptions and using different likelihoods. The effects of different posterior distributions so
obtained can be compared to see if they produce significantly
different predictions. For example, will a parameter produce significant differences in the behavior of the system if it is
specified by a Gaussian distribution rather than a χ^2 or t distribution? This may help us in making decisions about more experiments and data collection.

For applied problems knowledge of the mean and standard
deviation is very useful. It allows us to make predictions in
terms of statistical averages and some appropriate confidence
intervals associated with these estimates. This will also enable
us to test the validity of the model by comparing predicted values
with the data already in hand. The range of the values predicted by
the simulation method can easily be compared with the data
collected in the past.

In a comparative study of deterministic and the corresponding
stochastic differential equation models of a relatively simple
ecological system, Tiwari and Hobbie (1976a) have shown that the
introduction of the random elements into the system produces different behavior. This was observed under a variety of conditions.
This difference in behavior is of course not surprising and can
be expected in a model based on a set of nonlinear differential
equations. Some studies from cellular and macromolecular systems
also indicate this type of difference between stochastic and
deterministic formulations (Smeach and Smith, 1973; Stuart and
Branscomb, 1971). Thus one cannot automatically assume that the
deterministic solution and the mean of the corresponding stochastic
model would give the same result.

The computer simulation results of Tiwari and Hobbie (1976a)
have also shown that the randomness in the system increases its
'stability.' The behavior of the variables in the deterministic
version of the model was very sensitive to small changes in parameter values. For example, if the parameter specifying the respiration rate of algal species was changed from 0.031 to 0.029,
a very sharp increase in the algal biomass was followed by an
equally sharp decrease, leading to final values close to zero.

This in turn influenced the other variables of the system. However, when all the parameters were treated as random variables with Gaussian distributions the values of the respiration parameter ranged between 0.0295 and 0.0334 without any abnormal increase or decrease in the algal biomass. It seems that the random fluctuations in the parameters produce some degree of compensatory effects in which the too large effect of some parameters is balanced by the too small effect of others. It may be that the continuous fluctuations of the parameters generates a type of buffering mechanism in the biological systems.

This observed relationship between variability in the parameter and system stability was based on computer simulation results, and due to the complexity of the system of differential equations, an analytical solution was not possible. However, similar results in nonbiological systems have been obtained by several authors. Analytical as well as experimental studies on the behavior of electrical and mechanical systems suggest that an unstable system can be stabilized through the introduction of random parametric excitation. The resulting random system possesses an equilibrium solution that is almost surely stable (Bogdanoff, 1962). Samuels (1961) using the concept of mean-square stability, has given conditions for the stability of a second-order system with nonwhite parametric excitation. Kozin (1963) has also derived sufficient conditions for the almost sure stability of linear dynamic systems with nonwhite stochastic parameters. Caughey and Gray (1965), employing a Lyapunov type of approach, have obtained sufficient conditions guaranteeing the almost sure stability of linear dynamic systems with stochastic coefficients. Their results are also generalized to include a certain class of nonlinear systems (also see Katz and Krasovskii, 1960; Kozin, 1969; Kozin and Wu, 1973; and Man, 1970). Samuels (1961) has mentioned the possibility of stabilizing a deterministic, linear, unstable system by introducing Gaussian random noise excitation in the system parameters.

REFERENCES

Aström, K. J. (1970). *Introduction to Stochastic Control Theory.* Academic Press, New York.

Bogdanoff, J. L. (1962). Influence on the behavior of a linear dynamical system of some imposed motion of small amplitude. *Journal of the Acoustical Society of America,* 34, 1055-1062.

Box, G. E. P. and Jenkins, G. M. (1976). *Time Series Analysis: Forcasting and Control.* Holden-Day, San Francisco.

Box, G. E. P. and Tiao, G. C. (1973). *Bayesian Inference in Statistical Analysis*. Addison-Wesley, Reading, Massachusetts.

Caughey, T. K. and Gray, A. H. (1965). On the almost sure stability of linear dynamic systems with stochastic coefficients. *Journal of Applied Mechanics, 32,* 365-372.

DeGroot, N. H. (1970). *Optimal Statistical Decisions*. McGraw-Hill, New York.

Fisher, R. A. (1922). On the mathematical foundations of theoretical statistics. *Philosophical Transactions of the Royal Society of London, Series A, 222,* 309-368.

Hillier, F. S. and Lieberman, G. J. (1974). *Operations Research*. Holden-Day, New York.

Holling, C. S. (1966). The strategy of building models of complex ecological systems. In *Systems Analysis in Ecology,* K. E. F. Watt, ed. Academic Press, New York. 195-214.

IBM manual #SH19-7001-2. (1972). Continuous Systems Modeling Program III (CSMP III) Reference Manual. IBM Canada Ltd., Ontario, Canada.

Jaynes, E. T. (1957). Information theory and statistical mechanics. *Physical Review, Series 2, 106,* 620-630.

Jaynes, E. T. (1968). Prior probabilities. *IEEE Transactions on Symtems Science and Cybernetics, 4,* 227-241.

Katz, I. I. and Krasovskii, N. N. (1960). On the stability of systems with random parameters. *Journal of Mathematics and Mechanics, 25,* 809-823.

Kozin, F. (1963). On almost sure stability of linear systems with random coefficients. *Journal of Mathematics and Physics, 42,* 59-67.

Kozin, F. (1969). A survey of stability of stochastic systems. *Automatica, 5,* 95-112.

Kozin, F. and Wu, C. M. (1973). On the stability of linear stochastic differential equations. *Journal of Applied Mechanics, 40,* 87-92.

Lehman, J. T., Botkin, D. B., and Likens, G. E. (1975). The assumptions and rationales of a computer model of phytoplankton dynamics. *Limnology and Oceanography, 20,* 343-364.

Lindley, D. V. (1965). *Introduction to Probability and Statistics From a Bayesian Viewpoint, Part 2: Inference.* Cambridge University Press, London.

Man, F. T. (1970). On the almost sure stability of linear stochastic systems. *Journal of Applied Mechanics,* 37, 541-543.

Matis, J. H. and Wehrly, T. E. (1979). An approach to a compartmental model with multiple sources of stochasticity for ecosystem modeling. In *Compartmental Analysis of Ecosystem Models,* J. H. Matis, B. C. Patten, and G. C. White, eds. Satellite Program in Statistical Ecology, International Co-operative Publishing House, Fairland, Maryland.

O'Brien, W. J. (1974). The dynamics of nutrient limitation of phytoplankton algae: a model reconsidered. *Ecology,* 55, 136-144.

Oglesby, R. T. and Allee, D. J. (1969). *Ecology of Cayuga Lake and the Proposed Bell Station (Nuclear Powered).* Publication No. 27, Cornell University, Ithaca, New York.

Patten, B. C. (1971). *Systems Analysis and Simulation in Ecology, Vol. I.* Academic Press, New York.

Patten, B. C. (1972). *Systems Analysis and Simulation in Ecology, Vol. II.* Academic Press, New York.

Purdue, P. (1979). Stochastic compartmental models: A review of the mathematical theory with ecological applications. In *Compartmental Analysis of Ecosystem Models,* J. H. Matis, B. C. Patten, and G. C. White, eds. Satellite Program in Statistical Ecology, International Co-operative Publishing House, Fairland, Maryland.

Rosen, R. (1970). *Dynamical Systems Theory in Biology. Vol. I: Statility Theory and Its Applications.* Wiley, New York.

Samuels, J. C. (1961). Theory of stochastic linear systems with Gaussian parameter variations. *Journal of the Acoustical Society of America,* 33, 1782-1786.

Schmitt, S. A. (1969). *Measuring Uncertainty: An Elementary Introduction to Bayesian Statistics.* Addison-Wesley, Reading, Massachusetts.

Shannon, C. E. and Weaver, W. (1949). *The Mathematical Theory of Communications.* University of Illinois Press, Urbana.

Smeach, S. C. and Smith, W. (1973). A comparison of stochastic and deterministic models for cell membrane transport. *Journal of Theoretical Biology*, 42, 157-167.

Soong, T. T. (1973). *Random Differential Equations in Science and Engineering*. Academic Press, New York.

Srinivasan, S. K. and Vasudevan, R. (1971). *Introduction to Random Differential Equations and Their Applications*. Elsevier, New York.

Stuart, R. N. and Branscomb, E. W. (1971). Quantitative theory of *in vitro* lac repressor: Significance of repressor packaging. I. Equilibrium considerations. *Journal of Theoretical Biology*, 31, 313-329.

Syski, K. (1968). Stochastic differential equations. In *Modern Equations*. T. L. Saaty, ed. McGraw-Hill, New York 346-456.

Tiwari, J. L. and Hobbie, J. E. (1976a). Random differential equations as models of ecosystems: Monte Carlo simulation approach. *Mathematical Biosciences*, 28, 25-44.

Tiwari, J. L. and Hobbie, J. E. (1976b). Random differential equations as models of ecosystems. II. Initial conditions and parameter specifications in terms of maximum entropy distributions. *Mathematical Biosciences*, 31, 37-53.

Tiwari, J. L., Hobbie, J. E., and Peterson, B. J. (1978a). Random differential equations as models of ecosystems. III. Bayesian inference for parameters. *Mathematical Biosciences*, 38, 247-258.

Tiwari, J. L., Hobbie, J. E., Reed, J. P., Stanley, D. W., and Miller, M. C. (1978b). Some stochastic differential equation models of an aquatic ecosystem. *Ecological Modelling*, 4, 3-27.

Tribus, M. (1961). *Thermostatistics and Thermodynamics*. Van Nostrand, New York.

Tribus, M. (1969). *Rational Descriptions, Decisions, and Designs*. Pergamon, New York.

Watt, K. E. F. (1963). Dynamic programming, "look ahead programming," and the strategy of insect pest control. *Canadian Entomologist*, 95, 525-536.

Watt, K. E. F. (1966). The nature of systems analysis. In *Systems Analysis in Ecology*, K. E. F. Watt, ed. Academic Press, New York.

Watt, K. E. F. (1968). *Ecology and Resource Management*. McGraw-Hill, New York.

Winkler, R. L. (1972). *An Introduction to Bayesian Inference and Decisions*. Holt, Reinhart, and Winston, New York.

[*Received February* 1979]

J. H. Matis, B. C. Patten, and G. C. White, (eds.),
Compartmental Analysis of Ecosystem Models, pp. 195-222. All rights reserved.

AN APPROACH TO A COMPARTMENTAL MODEL WITH MULTIPLE SOURCES OF STOCHASTICITY FOR MODELING ECOLOGICAL SYSTEMS

J. H. MATIS and T. E. WEHRLY

Institute of Statistics
Texas A & M University
College Station, Texas   77840   USA

SUMMARY. This paper presents a stochastic approach for the compartmental modeling of ecological systems. The utility of compartmental modeling and the need for stochasticity in such models are briefly discussed. After a brief introduction of various sources of stochastic variability, a unified model is presented which combines these individual sources for the one-compartment case. It is shown that, in general, the mean of the stochastic model is not the same as the deterministic model evaluated at the mean rate. Also, although several stochastic formulations have the same mean value, it is shown that the underlying chance mechanism is identifiable from the covariance structure of the model. The theoretical appeal of the stochastic model, combined with the tractability of its first two moments, make it an important technique for modeling ecological systems.

KEY WORDS. compartmental analysis, stochastic models, ecosystem models.

1.   INTRODUCTION

This paper outlines a new approach to the development of a broad class of compartmental models with multiple sources of stochasticity. The paper is divided into three parts. The first part discusses our concept about the role of stochastic compartmental models in ecosystem modeling. The need for such models is pointed out together with a summary of their advantages and disadvantages. The second part briefly reviews the principal

stochastic compartmental models contained in the current
literature and comments on how the assumptions of the existing
stochastic models interface with the practical requirements of
general ecosystem modeling. The final part develops a new
unified stochastic framework and proposes a general compartmental
modeling strategy.

2.   THE ROLE OF STOCHASTIC COMPARTMENTAL MODELS IN ECOLOGY

2.1 The Case for Stochastic Models in Ecosystem Modeling.   The
deterministic versus stochastic question in ecosystem modeling
has been considered in detail by many authors, and several such
reviews are included in other volumes of the Satellite Program
in Statistical Ecology series. Although there has been much
philosophical controversy on whether or not there is an absolute,
inherent determinism in nature, that question seems unimportant
in the world of practical modeling. Most leading observers
agree that, in our present state of knowledge, the 'real world'
is stochastic in practice. Poole (1979) concludes that
"the only determinism in the real world is that nothing is certain"
and hence that "idealistically all ecological prediction problems
are stochastic." Solomon (1979) also infers that stochastic
models are more realistic. Gold (1977) states that "every real
system must be considered to be subject to uncertainties of one
type or another" and Tiwari *et al.* (1978) provide a succinct
summary by noting that "the existence of variability, at every
level of organization, is one of the most fundamental characteristics of all living systems."

In addition to this general consensus that the real world of
ecological prediction is stochastic, there is a general awareness
that the stochastic and the deterministic approaches may differ
not only in their theoretical foundations but also in their
practical predictions. Poole (1979) indicates that the 'average
behavior' of the stochastic model may not be equivalent to its
deterministic analog. Tiwari *et al.* (1978) come to the same
conclusion and then explicitly point out some of the differences
with a simulation study. O'Neill (1979a) observes certain
situations where the application of the deterministic model can
yield "misleading and erroneous results."

If, then, it is generally agreed that 1) the real world
is stochastic and 2) predictions from a deterministic model
may be misleading, one may well ask whether the deterministic
model has any utility in ecosystem modeling or, looking at it
from another perspective, question why most of the current
modeling effort is deterministic. O'Neill (1979a) justifies
the deterministic approach on the ground of it being "simpler

to implement and more efficient to utilize." Poole (1979)
also notes the pragmatic value of the simpler analytical (or
numerical) solution of the deterministic model and also develops
a rationale for its scientific appropriateness in providing
simpler predictions which, even if not fully realistic, may be
readily compared to the available data.

It should be pointed out that these considerations apply to
stochastic ecosystem models in general. The above arguments
are reviewed subsequently as applied specifically to compartmental
models.

2.2 The Case for Compartmental Models in Ecosystem Modeling.
The deterministic compartmental model implicitly assumes that the
dynamic behavior of an ecosystem may be modeled by a system of
differential or difference equations. The development in this
paper, as indeed in most of compartmental theory, is further
restricted to systems of linear first-order differential
equations.

The question of the general utility of this compartmental
approach to ecosystem analysis has been considered by several
authors. Regier and Rapport (1978) survey the principal
paradigms in ecology and conclude that the compartmental-flow
approach is a promising paradigm not only in theory but also
for the practical analysis of ecosystem models. Patten (1975)
considers several arguments to support the specific hypothesis
that "ecosystems are nominally linear in their large-scale
dynamic characteristics." O'Neill (1979b), after a thorough
review in another paper in this volume, makes the following
conclusions: (1) that the chief advantage of compartmental
analysis is the impressive "number and variety of mathematical
techniques available for (the) analysis" which "argue strongly
for its adoption," (2) that the chief disadvantage of classical
compartmental analysis is its restrictions by the assumption of
linearity and (3) that, on the balance, the "compartmental model
should continue to find useful application in ecosystem science"
under certain conditions.

It should be pointed out that the above assessments of the
utility of compartmental modeling are almost completely limited
to a consideration of deterministic compartmental models. We
subsequently consider arguments specifically applicable to
stochastic compartmental models.

*2.3 The Case for Stochastic Compartmental Models in Ecosystem
Modeling.* The basic thesis of this research has two inter-
related parts - firstly that the usefulness of the compartmental

approach is greatly enhanced through a general stochastic
formulation and, secondly, that the stochastic approach is
particularly powerful in compartmental modeling. Let us
consider now some of the general properties of the stochastic
compartmental models.

Mathematical Tractability. As noted previously, a chief
advantage of using deterministic compartmental models for
describing ecosystems is the mathematical tractability of
such models. It should be pointed out that, as a rule, the
stochastic approach preserves this mathematical tractability in
one of several ways. First of all, there are some classes of
stochastic solutions which are immediate generalizations of the
deterministic solutions. Hence, the stochastic solutions can
be obtained and many of their consequences derived with relative
ease. But there is another important sense in which a
stochastic model preserves or even enhances the tractability of
a problem. Most deterministic compartmental models, in an
attempt to be realistic, consist of a large number of directly
interrelated endogenous variables and a host of other exogenous
variables which also influence the system behavior. Thus, one
approximates the detailed causal mechanisms producing the
dynamic behavior with a large, linear deterministic model. The
size of these models usually rules out an analytical solution,
and they are instead typically studied by computer simulation.
Sometimes, an attractive alternative procedure is to replace the
complex deterministic model with a simpler stochastic representation where the detailed causal mechanism of the former is
supplanted by the probabilistic variation of the latter. In
other words, one may deliberately introduce simplifications in
the model conceptualization with the consequential 'errors in
the equations' to yield an analytically tractable stochastic
model from which valid statistical inferences can be made, in
principle, on the operation of the complex deterministic
system. The reduced, analytical stochastic solution may be
as realistic and yet much more amenable to analysis than its
deterministic counterpart.

Turner (1964) provides a simple example of this procedure.
Consider prediction from an n-compartment model with solution

$$\eta = \alpha_0 + \alpha_1 e^{-\lambda_1 t} + \alpha_2 e^{-\lambda_2 t} + \cdots + \alpha_n e^{-\lambda_n t}, \tag{1}$$

where η is the amount in a compartment at time t.

The model has $(2n + 1)$ parameters and is intractable for
parameter estimation when $n \geq 3$. Turner's solution to this

problem is to replace the 'discrete spectrum' of (α_i, λ_i) by a continuous approximation, say, for example, the gamma distribution for λ with the ordinate of the gamma, $f(\lambda)$, approximating the scaled α_i. Taking the expectation of η, one then has the four parameter model

$$E(\eta) = \alpha_0 + \beta(t-\mu)^\delta, \tag{2}$$

where μ and δ are parameters of the gamma distribution and β is a scaling parameter. This new model is tractable for estimation and prediction.

The Restriction of Linear Kinetics. As pointed out previously, the tractability of deterministic compartmental analysis is largely a result of its limitation to linear kinetics. This is, of course, a severe theoretical constraint even though the practical consequences of this constraint are controversial, as mentioned previously. The practical considerations of tractability also limit the stochastic compartmental models to linear rates. However, as reviewed by Purdue (1979), the linear stochastic compartmental model can adequately approximate some complex nonlinear systems. Faddy (1977) shows this in some detail for a nonlinear epidemic process. Faddy's results, indicating the excellent approximation which may be present for any single component, together with Patten's (1975) arguments on the general robustness of the model when the individual components are combined into whole system, tend to suggest that a linear stochastic compartmental model may be a rich framework to approximate nonlinear ecosystems. Further research is now in progress to investigate this robustness by numerical simulation.

Generality. A strong argument for the stochastic compartmental model which is typically not noted explicitly is its generality. The deterministic model can be produced as a limiting case of all the common stochastic formulations. Hence, in principle, one might initially conceptualize a very broad stochastic formulation with multiple sources of stochasticity. It is unlikely in any single ecosystem modeling problem that all sources of stochasticity would have a measurable effect. However, it seems important to start with a framework where all potential sources are represented and to then discard those that are negligible in a particular application. If, indeed, all the sources of stochasticity are minor, one would end up with a deterministic model.

In summary, the stochastic compartmental model may have

three desirable properties: it may be mathematically relatively
tractable, it may approximate nonlinear systems well, and it is
a more general formulation than a deterministic model. Of
course, it may also be more 'realistic.' A final advantage
is in the ultimate empirical consequence of making an 'incorrect'
decision on the choice of a model. Paraphrasing Tsokos and
Tsokos (1976), if one assumes a stochastic model when in fact
the system is deterministic, very little, if anything, has
been lost. However, if one assumes a deterministic model when
in fact the system is stochastic, the results can be very
misleading.

3. AN OVERVIEW OF PRESENT STOCHASTIC COMPARTMENTAL MODELS

If one is inclined for either theoretical or practical
reasons to use a stochastic formulation of a compartmental model
to describe an ecosystem, one must carefully choose among the
stochastic formulations available. Although there may be great
merit in principle in using a stochastic formulation, it, of
course, does not follow that all stochastic models are worthwhile
for every application.

Cobelli and Morato (1978) note that there historically have
been two principal classes of stochastic compartmental models.
One approach incorporates stochasticity into the system via a
probabilistic transfer mechanism through the use of multidimensional migration-death processes with either constant or
time-varying transfer coefficients. These are called the
'particle' models by Eberhardt *et al.* (1976). The other approach
has been to assume a deterministic transfer mechanism with a
random variable rate coefficient. Interestingly, the sets of
researchers in each class have been mutually exclusive and each
class has proudly advertised itself as using 'the' stochastic
compartmental model. We will now review briefly the development
of each of these branches separately.

3.1 Models with a Stochastic Transfer Mechanism. Our primary
objective is to examine various properties of this class of
models and hence, for the sake of simplicity, we will restrict
the present considerations, for the most part, to models with no
more than n=2 compartments. Although the properties of interest
can be more clearly demonstrated with this reduced set, it
should be pointed out that explicit solutions of the problem
exist for much larger n. Purdue (1979) reviews this class of
models in detail and may be consulted for such higher dimensional
solutions.

Let $X_i(t)$ denote the number of particles in compartment

i at time t and let $k_{ji}(t)$ $(i,j=0,1,2;i \neq j)$ be the transition intensity coefficient from compartment i to compartment j at time t where 0 denotes the system exterior. These coefficients thus define the following six elementary transitions in a small time increment Δt:

Pr[particular unit in i goes to j in $(t,t+\Delta t) | X_1(t),X_2(t)]$

$= k_{ji}(t)\Delta t + o(\Delta t)$ $(j=0,1,2;i=1,2;i \neq j)$, and

Pr[compartment i gains a unit in $(t,t+\Delta t) | X_1(t),X_2(t)]$

$= k_{i0}(t)\Delta t + o(\Delta t)$ $(i=1,2)$.

Schematically one has the system below

Consider first the simplest case, say Case 1, where $X_1(0)$ and $X_2(0)$ are fixed, though perhaps unknown, parameters and where the rate coefficients are constant over time, i.e. $k_{ji}(t) = k_{ji}$ for all $t \geq 0$. Then it is well-known (see e.g. Matis, 1970; Purdue, 1979) and can be established from first principles that the vector $[X_1(t),X_2(t)]$ is distributed as the sum of two independent trinomial and a Poisson random vector. For most practical applications, it is sufficient to examine only the first and second moments which, in this case, are easily shown to have the forms

$$E[X_i(t)] = \sum_{\ell=1}^{2} X(0)p_{i\ell}(t) + \delta_i(t) \qquad (i=1,2) \quad ,$$

$$Var[X_i(t)] = \sum_{\ell=1}^{2} X(0)p_{i\ell}(t)[1-p_{i\ell}(t)] + \delta_i(t) \qquad (i=1,2) \quad , \quad (3)$$

$$Cov[X_1(t), X_2(t)] = - \sum_{\ell=1}^{2} X(0)p_{1\ell}(t)p_{2\ell}(t) \quad ,$$

where $p_{i\ell}(t)$ is the probability that a particle starting initially in compartment ℓ is in compartment i at time t and $\delta_i(t)$ is a parameter of the Poisson distribution. These parameters for the present simple assumptions are of the form

$$p_{i\ell}(t) = \sum_{m=1}^{2} b_{i\ell m} e^{-\lambda_m t} \qquad (i=1,2;\ell=1,2) \quad , \tag{4}$$

$$\delta_i(t) = c_{i0} + \sum_{m=1}^{2} c_{im} e^{-\lambda_m t} \qquad (i=1,2) \quad , \tag{5}$$

where the c_{ij}, $b_{i\ell m}$, and λ_m may be involved functions, as given explicitly in previous references, of the k_{rs} coefficients. Note that under these assumptions one has the so-called sums of exponentials model.

A natural extension of the problem, call it Case 2, is to allow time dependent rates with the fixed initial values $X_1(0)$ and $X_2(0)$. Raman and Chiang (1973) and Faddy (1976, 1977) note that the random vector $[X_1(t), X_2(t)]$ has the same distributional form as Case 1. Therefore one has the same moment structure as in (3); however, the explicit expressions for the $p_{i\ell}(t)$ and $\delta_i(t)$ parameters become much more involved in this case and are given explicitly in Cardenas and Matis (1974,1975a,b) and in Epperson and Matis (1979) for certain compartmental configurations.

A somewhat different formulation, say Case 3, is to approach the problem as a semi-Markov process as pioneered by Weiner and Purdue (1977) and Marcus and Becker (1978). This is an elegant approach incorporating arbitrary residency-time distributions within each compartment. The semi-Markov formulation again has the same probability structure as before with the form of the moments as given by (3), although the $p_{i\ell}(t)$ and $\delta_i(t)$ as given in the above references are different generalizations of the sums of exponentials models. As an illustration, Matis (1973) considers a simple model where the residency-time distributions in each compartment are integral gamma distributions, whereupon the $p_{i\ell}(t)$ in (3) have positive powers of time, i.e.

$$p_{i\ell}(t) = \sum_{m=1}^{2} e^{-\lambda_m t} \sum_{h=1}^{g_m} c_{i\ell mh} t^{h-1} .$$

It is instructive to examine for the above three cases the coefficients of variation of the compartments, denoted CV_i (i = 1,2); and defined as

$$CV_i(t) = \{Var[X_i(t)]\}^{\frac{1}{2}} \{E[X_i(t)]\}^{-1} .$$

As noted in Thakur *et al.* (1973), and frequently rediscovered, it follows from (3) that

$$Var[X_i(t)] \leq E[X_i(t)]$$

for i=1,2 and for any t>0 regardless of the form of $p_{ij}(t)$ and $\delta_i(t)$. Therefore one can show that

$$CV_i(t) \leq \{E[X_i(t)]\}^{-\frac{1}{2}} .$$

The practical significance of this is that for any ecosystem application where one is concerned with counts of molecules or particles, $E[X_i(t)]$ is extremely large whereby the coefficient of variation for these models is minute. However, the observed coefficients of variation in most ecosystem applications are not at all negligible, as predicted by these models. Therefore, although the variation due to the stochastic transfer mechanism probably exists to some degree in all applications, the models with *only* this one source of variability have limited utility in general ecosystem modeling. It should be noted, however, that the above models have been very useful in population dynamics and in other problems where the counts are typically small.

One generalization, say Case 4, of the above models is the inclusion of another source of stochastic variability through random initial $X_1(0)$ and $X_2(0)$. Cardenas and Matis (1975a) give the moments for this case as

$$E[X_i(t)] = \sum_{\ell=1}^{2} \mu_i p_{i\ell}(t) + \delta_i(t) \quad , \quad \text{and}$$

$$\text{Var}[X_i(t)] = \sum_{\ell=1}^{2} \mu_i p_{i\ell}(t)[1-p_{i\ell}(t)] + \delta_i(t) + \sigma_{11}p_{i1}^{2}(t)$$

$$+ 2\sigma_{12}p_{i1}(t)p_{i2}(t) + \sigma_{22}p_{i2}^{2}(t) \quad ,$$

where μ_1, μ_2, σ_{11}, σ_{22}, and σ_{12} are the means, variances, and covariance, respectively, of $X_1(0)$ and $X_2(0)$. The last three terms are a quadratic form in $p_{i1}(t)$ and $p_{i2}(t)$ and are non-negative. Note that this model is not necessarily subject for finite t to the previous difficulty of a small coefficient of variation. Since uncertain (random) initial conditions are also a likely occurrence in ecosystem modeling, it seems important to include this source of stochasticity in many ecosystem modeling applications.

3.2 *Models with a Random Rate Coefficient.* The principal alternative stochastic formulation of a compartmental model envisages a deterministic system with a random vector of initial amounts, $[X_1(0), X_2(0)]$, and/or a random vector of rates $\{k_{ji}: i, j = 0, 1, 2; i \neq j\}$. Any particular replicate of the experiment is based on a particular realization of these random vectors but, once the realization is determined, the behavior of the system is deterministic for that replicate. The objective is again to determine the joint distribution and the moments of $[X_1(t), X_2(t)]$.

Soong (1971) is apparently the first to provide a mathematical analysis for this approach to compartmental analysis by using a Liouville-type theorem to obtain a general formulation for the joint density of $[X_1(t), X_2(t)]$. However, as he notes, one could also obtain the same results by applying the well-known algebraic transformation-of-variables procedures to the distributions of $[X_1(0), X_2(0)]$ and $[k_{ji}]$ to derive the distribution of $[X_1(t), X_2(t)]$. This latter procedure is more useful when the main objective is to obtain the moments, and we will illustrate its use in Section 4.

Soong also illustrates an important property of the closed two-compartment system below.

For this particular model, he assumes that the initial values are fixed, but that the rate vector $[k_{21}, k_{12}]$ has a specified truncated bivariate normal distribution. He then shows numerically in this particular case that the "resulting behavior of the stochastic model, even in an average sense, cannot be predicted adequately by deterministic approaches where the rate constants are chosen deterministically."

Soong and Dowdee (1974) consider the problem of parameter estimation for this stochastic formulation. Their approach is based on the deterministic models parameterized in the form

$$X_i(t) = \sum_{\ell=1}^{2} b_{i\ell} e^{-\lambda_\ell t} \quad (i=1,2) \ . \tag{6}$$

Let us assume that the λ_ℓ are real (although the method does allow complex conjugates) and that the random variables $b_{i\ell}$ and $\exp(-\lambda_\ell t)$ are mutually uncorrelated for all t . It is then easy to show that

$$E[X_i(t)] = \sum_{\ell=1}^{2} E[b_{i\ell}] M_{\lambda_\ell}(-t) \quad (i=1,2) \ , \tag{7}$$

where $M_{\lambda_\ell}(\cdot)$ is the moment generating function of λ_ℓ . [Soong and Dowdee assume that $b_{i\ell}$ and λ_{ℓ} are mutually uncorrelated, however (7) does not necessarily follow from their assumption.] The data averaged over two or more replicates may then be fitted to (7). As an illustration, Soong and Dowdee apply this to the previous closed two-compartment model where the non-zero eigenvalue λ of the system is assumed to have a gamma distribution with parameters ω and ν. Letting $\mu_{i\ell}$ denote $E[b_{i\ell}]$ and noting that the second eigenvalue is 0, one has the regression model

$$E[X_i(t)] = \mu_{i1} + \mu_{i2}(1+\omega t)^{-\nu} \qquad (8)$$

whence one can estimate the parameters μ_{i1}, μ_{i2}, ω, and ν.

Since the variables $b_{i\ell}$ and $\exp(-\lambda_{\ell}t)$ are usually rather involved functions of the k_{ji} rates, the assumption that they are mutually uncorrelated is very difficult to satisfy in practice. However, Soong and Dowdee simulate several compartmental models and suggest that estimation with model (7) is somewhat robust against departures from this assumption.

Campello and Cobelli (1978) note that the assumption of mutual uncorrelatedness is a severe constraint if the initial values are known, whereupon $\Sigma_{\ell} b_{i\ell}$ for a given i, must be a constant. Therefore, they relax the assumption of mutual uncorrelatedness somewhat. They also give an expression for the variance of $X_i(t)$ in this model and describe in detail an estimation procedure. Tsokos and Tsokos (1976) derive an expression for the probability distribution of a two-compartment open model, and they show numerically that a deterministic solution differs from the mean of the stochastic solution when the rates in the latter have a trivariate normal distribution. As indicated previously, Tiwari et al. (1978) also obtain a difference in the model predictions numerically for a variety of conditions. Other papers which develop this approach are Soong (1972) on optimal design in this model, Dowdee and Soong (1974) on estimation, and Chuang and Lloyd (1974) which applies these results to a chemotherapy model.

As applied to ecosystem modeling, the models with
uncertainties in the rate coefficients provide alternative
regression functions, $E[X_i(t)]$ for $i=1,\cdots,n$, which have
been shown in some specified cases to differ from the deterministic function, $X_i(t)$. However, they are all based on a model
which has no inherent uncertainty once the initial conditions
(amounts and/or rates) are determined.

4. A UNIFIED STOCHASTIC COMPARTMENTAL MODEL

It is clear from this brief overview that each of these
principal types of stochastic models has extended the basic
deterministic model to include some stochastic variation, and each
type of model has been successfully applied to various modeling
problems. However, the theoretical argument that every realistic
problem would include both types of variability, although in
differing amounts, seems irrefutable. Our approach therefore is
two-fold. The first part is to combine these and other sources
of stochasticity into a single, unified model structure and the
second is to identify the effects of the individual sources by
using the covariance structure of the model. The objective is
to commence any particular modeling application with a very
general formulation and then to eliminate any negligible effect
either by *a priori* reasoning and/or by statistical hypothesis
testing with the data.

4.1 Model Nomenclature. Consider now some nomenclature for the
sources of variability. The previous sources of stochasticity
may be categorized into two classes: one class, denoted by P ,
is the stochasticity associated with individual 'particles' and
the other class, denoted by R , is associated with replicates
of a whole experiment. To establish this partitioning, note
that the formulation in Section 3.1 concerning the stochastic
transfer mechanism is a property associated with the behavior
of individual particles. Let us denote this stochasticity as P1 .
We now introduce P2 as a source of stochasticity where the rate
coefficient of each particle is a random variable. Such
stochasticity in the particle transfer rate would arise naturally
in a population of heterogeneous particles where there is some
variation, for example, in the physical shape or in the chemical
composition of the particles. It should be noted that this effect
has not been incorporated explicitly into either of the previous
formulations of a stochastic compartmental model. Note also that
although P1 and P2 stochasticity both lead to a random
residency time of a particle within a compartment, they are
different chance mechanisms which could (and most likely do)

appear together.

The R stochasticity is clearly divisible into two principal types. Let R1 denote the randomness in the initial amount or count of substance and let R2 denote a random rate coefficient where, for any particular experiment, the rate coefficient which is common to all units is a realization of some random variable. Note that these broad definitions may be applied to ecosystem models whether the variable $X(t)$, which we are ultimately interested in modeling, is continuous, as in the usual deterministic formulation, or discrete, as in the stochastic particle formulation.

In developing an overall model notation, it is useful also to categorize a model according to whether it has constant or time-varying rates. Let C denote a model where, if it is a deterministic model, all the transition rates are constant or where, if it is a stochastic model, all the hazard rates are constant with the resulting exponential residency times. Let T denote a system with one or more time varying rates. The above nomenclature is summarized in Table 1.

TABLE 1: Summary of nomenclature.

C - constant (time invariant) transition or hazard rates

T - time-varying transition or hazard rates

P1 - probabilistic transfer of the individual particles

P2 - random rate coefficient for the individual particles

R1 - random initial amount or count in a replicate

R2 - random rate coefficient for an individual replicate
     which is common to all particles within the replicate

We now propose a notation wherein a compartmental model is designated by the letter C or T followed by the specification of particular P and/or R stochasticity, if such is present, with each source separated by a slash. With this notation, C would denote the classical deterministic model with constant coefficients which leads to the sums of exponentials models. The models of Section 3.1 with the stochastic transfer mechanism would be classified as C/P1 for case 1, T/P1 for cases 2 and 3, and T/P1/R1 for case 4 since this case includes a random initial count. The models of Section 3.2, as they are usually applied, would be denoted as C/R2 or C/R1/R2 .

4.2 *Probability Distributions and Moments for the Unified Models.*

It is relatively easy to indicate the symbolic solution for the unified formulation, but it should be pointed out that the explicit solution, which is needed for fitting to the data, may still be very difficult to obtain in practice. The considerations may be divided into two cases based on whether or not the model includes P1 stochasticity, i.e. a random transfer mechanism.

Let us consider first the case without P1 stochasticity. In this case, once the random vector

$$\underset{\sim}{\theta}^T = [X_1(0),X_2(0),k_{ji} \quad \text{for} \quad i,j=0,1,2,i \neq j]$$

is specified, the conditional vector $[X_1(t),X_2(t)|\underset{\sim}{\theta}]$ is fixed (deterministic) for all t . It is therefore conceptually simple to derive the joint density function of $[X_1(t),X_2(t)]$ from the joint density of $\underset{\sim}{\theta}$ by the classical transformation-of-variables procedures (see e.g. Hogg and Craig, 1970, chap. 4). The result, of course, is identical to that from Soong's approach.

On the other hand, when one has P1 stochasticity, the conditional vector $[X_1(t),X_2(t)|\underset{\sim}{\theta}]$ is random for all t>0 . A conceptually simple approach in this case is to first determine the conditional density of $[X_1(t),X_2(t)|\underset{\sim}{\theta}]$ from the literature previously cited and to then derive the unconditional density of $[X_1(t),X_2(t)]$ as a mixture of density functions, whereby one has

$$f_{X_1(t),X_2(t)}(x_1,x_2) = E_{\underset{\sim}{\theta}}[f_{X_1(t),X_2(t)}(x_1,x_2|\underset{\sim}{\theta})] \tag{9}$$

In most practical applications, the derivation of the first and second cumulants of the variables is a primary objective, and the solution of the joint probability density function is only of secondary importance. Therefore, it is important to note that it is usually much simpler to obtain the desired moments directly, through the use of conditional expectations, rather than to obtain them from the explicit form of the joint density functions. The direct approach uses the following formulas involving conditional expectations (see e.g. Parzen, 1962, chap. 2).

$$E[g(X_1,X_2)] = E_\theta\{E[g(X_1,X_2|\theta)]\} \quad \text{and} \quad (10)$$

$$Var[g(X_1,X_2] = Var_\theta\{E[g(X_1,X_2|\theta)]\} + E_\theta\{Var(X_1,X_2)|\theta)\} .(11)$$

Note that the considerations in this section can be easily generalized to n-dimensional systems.

5. SOME ONE-COMPARTMENT EXAMPLES OF THE UNIFIED MODEL

The scope and richness of the unified models and most of the fundamental operations can be demonstrated within the framework of the simple one-compartment system where the fundamental concepts are relatively uncluttered by the algebraic detail. Let us consider the two basic one-compartment models and derive some moments and properties.

5.1 Some Moments of a One-compartment System with an Initial Load. It is assumed in this model that an initial amount of substance is introduced into a compartment and that the substance leaves the compartment through linear kinetics. This basic model and its multicompartment generalizations have been widely used in ecosystem modeling to describe the dynamics of tracers (see e.g. O'Neill, 1979b) and of contaminants (see e.g. Eberhardt *et al.*, 1976). An extensive development of some stochastic models for this system is contained in Matis and Tolley (1979) which also includes the distribution theory for the models. We focus here only on the means of $X(t)$, $\mu(t)$, and on the covariances of $X(t)$ and $X(u)$ for $t \leq u$, denoted by $\sigma^2(t,u)$. Let μ_0 and σ_0^2 be the mean and variance of the initial amount, X_0 , for R1 stochasticity and let μ_k and σ_k^2 be moments of the rate k for P2 or R2 stochasticity. There are twelve main stochastic models for this system with different combinations of stochasticity as follows:

Model C. This is the deterministic model with no stochasticity. Therefore, it is assumed that the initial amount, X_0 , and the rate, k , are fixed, which in turn leads to the well-known solution

$$X(t) = X_0 e^{-kt} \quad (t \geq 0) . \tag{12}$$

Models C/R1, C/P2, C/R2, C/P2/R1, *and* C/R1/R2. All of these models are mixtures of Model C where X_0 or k or both are random variables. The mean values obtained by substituting into (10) are as follows:

for Model C/R1,

$$E[X(t)] = E_{X_0} [X_0 e^{-kt}] = \mu_0 e^{-kt} ; \tag{13}$$

for Models C/P2 and C/R2,

$$E[X(t)] = E_k [X_0 e^{-kt}] = X_0 M_k (-t) , \tag{14}$$

where $M_k(\theta)$ is the moment generating function, $E_k[e^{k\theta}]$, of the random variable k ;

and for Models C/P2/R1 and C/R1/R2,

$$E[X(t)] = E_{k,X_0} [X_0 e^{-kt}] = \mu_0 M_k (-t) . \tag{15}$$

The covariances for these models, as given by Matis and Tolley, are listed in Table 2.

Model C/P1. This stochastic model assumes a known initial count, say X_0 , of independent units, and a fixed rate k. This implies that $X(t)$ is binomial with mean

$$E[X(t)] = X_0 e^{-kt} \tag{16}$$

and the vector $[X(t), X(u)]$ is chain binomial with covariance

$$\sigma^2(t,\mu) = X_0 e^{-ku} [1 - e^{-kt}] \quad (t \leq u) . \tag{17}$$

TABLE 2: *Covariance*, $\sigma^2(t,u)$, *for various stochastic models.*

Model	Covariance
C	0
C/R1	$\sigma_0^2 \exp\{-k(t+u)\}$
C/P2	$X_0[M(-t-u) - M(-t)M(-u)]$
C/R2	$X_0^2[M(-t-u) - M(-t)M(-u)]$
C/P2/R1	$\mu_0[M(-t-u) - M(-t)M(-u)] + \sigma_0^2 M(-t)M(-u)$
C/R1/R2	$\mu_0^2[M(-t-u) - M(-t)M(-u)] + \sigma_0^2 M(-t-u)$
C/P1	$X_0 \exp(-ku)[1-\exp(-kt)]$
C/P1/R1	$\mu_0 \exp\{-ku\}[1-\exp\{-kt\}] + \sigma_0^2 \exp\{-k(t+u)\}$
C/P1/P2	$X_0[M(-u) - M(-t)M(-u)]$
C/P1/R2	$X_0[M(-u) - M(-t-u)] + X_0^2[M(-t-u) - M(-t)M(-u)]$
C/P1/P2/R1	$\mu_0[M(-u) - M(-t-u)] + \sigma_0^2 M(-t)M(-u)$
C/P1/R1/R2	$\mu_0[M(-u) - M(-t-u)] + \mu_0^2[M(-t-u) - M(-t)M(-u)]$ $+ \sigma_0^2 M(-t-u)$

Models C/P1/P2, C/P1/R1, C/P1/R2, C/P1/P2/R1, *and* C/P1/R1/R2.
All of these models are mixtures of model C/P1 and their moments
can be found by manipulation of (16) and (17). Since the mean
for C/P1 in (16) is the same as the mean for C in (12), it
follows that (13) is also the mean of C/P1/R1, (14) the mean of
both C/P1/P2 and C/P1/R2, and (15) the mean of C/P1/P2/R1 and
C/P1/R1/R2. The covariances, however, are all different as is
apparent in Table 2.

5.2 *Some Properties of a One-compartment System with an Initial
Load.* This simple stochastic system has some profound implications.
Note that although several stochastic formulations share the same
mean value function, the covariance for each formulation is unique.
It is therefore possible, in principle, to identify the model
from the covariance matrix of time series data. This identification should be aided considerably by capitalizing on the
characteristic form of each stochastic effect in the covariances.
The numerical differences of the covariances are illustrated
for certain assumed distributions and parameters in the Matis
and Tolley paper.

Another feature of these models is the relative simplicity
of the regression function with the random rate. In all such
models with P2 or R2 stochasticity, one would fit the
regression model

$$\mu(t) = c M_k(-t)$$

to the (averaged) data. The moment generating function has a
benign form for many rich families of distributions; the gamma
distribution of k , for example, has the form

$$M_k(-t) = (1+\beta t)^{-\alpha} \qquad (18)$$

which is easily fitted to data. The gamma, the uniform, and
mixtures of the gamma and/or the uniform distributions provide
an excellent framework to approximate any continuous distributions
of k.

It has often been observed, as noted previously, that the
stochastic regression function may differ from its deterministic
counterpart. This is typically demonstrated numerically, however
in this simple example one can demonstrate this analytically.
This is immediately apparent in comparing the regression
function in (18), which arises from a gamma distribution of k ,
to the corresponding exponential decay function of the deterministic model. Indeed one can show in general by using
Jensen's Inequality (see e.g. Matis and Tolley, 1979) that, for

the present one-compartment system, one has

$$M_k(-t) = E_k[\exp(-kt)] > \exp[-tE(k)]   (t>0)   \tag{19}$$

for any non-trivial distribution of k. The implication is that, for this system, the expected value of a model with a random rate coefficient will always exceed the expected value of a model with a fixed coefficient evaluated at the mean rate. This brings to question the validity of the exponential decay model whenever the particles (or replicates over which the data is averaged) have any variation which results in a random rate. As mentioned before, many feel that such variation is inherent in all natural systems.

Finally, we note that this model may be generalized to non-constant hazard rates, or equivalently to non-exponential residency times, with the same qualitative results, as shown in Matis and Wehrly (1979). As an illustration, let us assume that the residency times follow a Weibull distribution with hazard rate $h(t) = (\lambda\gamma t^{\gamma-1})k$, where k is now the random scale parameter. If k is assumed to have a gamma distribution among units or replicates, then the regression function is

$$E[X(t)] = X_0[1+\beta\lambda t^{\gamma}]^{-\alpha}   \tag{20}$$

Note that (18) is a special case of the above with $\lambda=1$ and $\gamma=1$.

5.3 Some Moments of a One-compartment System with Continuous Input. Consider now the dynamics of a matter through a compartment which is initially devoid of the substance but which has a continuous input (and output) over time. Let us assume for the deterministic Model C that the input occurs with constant rate λ and the output rate parameter is μ, as illustrated in the schematic below

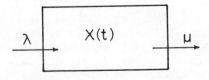

It can be shown that the deterministic solution which solves its

equation $\dot{X}(t) = \lambda - \mu X(t)$ with $X(0) = 0$ is

$$X(t) = \lambda(1-e^{-\mu t})/\mu \ . \tag{21}$$

The stochastic model $C/P1$ assumes the analogous Poisson entries with rate λ and an exiting hazard rate of μ . It is relatively easy to show in this case that the random count, $X(t)$, has a Poisson distribution with parameter $\lambda(1-e^{-\mu t})/\mu$ (see e.g. Purdue, 1979). Hence, the regression function for Model $C/P1$ is also

$$E[X(t)] = \lambda(1-e^{-\mu t})/\mu \ . \tag{22}$$

Let us now incorporate $P2$ and $R2$ stochasticity into the model, whereupon λ and μ are random rates. We first consider the specific case where λ and μ are independent gamma distributed random variables, say $\lambda \sim gamma(\alpha,p)$ with $E(\lambda) = \alpha p$ and $\mu \sim gamma(\beta,q)$ with $E(\mu)=\beta q$. Then the expected value of $X(t)$ for Models $C/P2$, $C/R2$, $C/P1/P2$ and $C/P1/R2$ may be found by substituting (21) or (22) into (10) to yield, for $t>0$,

$$E[X(t)] = E_{\lambda,\mu}[\lambda(1-e^{-\mu t})/\mu]$$

$$= E[\lambda]E[(1-e^{-\mu t})/\mu] \tag{23}$$

$$= \begin{cases} \alpha p[1 - (\beta t+1)^{-(q-1)}]/[\beta(q-1)] & (q>1) \\ \alpha p[\ln(1+\beta t)]/\beta & (q=1) \end{cases} \ . \tag{24}$$

Consider also the variances of several of the stochastic models to illustrate (11). The variance of $X(t)$ for Model $C/R2$ with the independent gamma distributions may be obtained by substituting (21) into (11). For the case $q>2$, this yields

$$Var[X(t)] = Var_{\lambda,\mu}[\lambda(1-e^{\mu t})/\mu]$$

$$= E_\lambda(\lambda^2)E_\mu[(1-e^{-\mu t})^2/\mu^2]$$

$$- \{\alpha p[1 - (\beta t+1)^{-(q-1)}]/\beta(q-1)\}^2$$

$$= \frac{\alpha^2 p}{\beta^2 (q-1)} \left\{ \frac{p+1}{q-2} \left(1 - 2(1+\beta t)^{-(q-2)} \right. \right.$$

$$\left. + (1+2\beta t)^{-(q-2)} \right)$$

$$\left. - \frac{p}{q-1} [1 - (1+\beta t)^{-(q-1)}]^2 \right\} . \qquad (25)$$

The variance of $X(t)$ for Model C/P1/R2 may be obtained by substituting (22) and $Var[X(t)|\lambda,\mu] = \lambda(1-e^{-\mu t})/\mu$ into (11). For the case $q>2$, one has the sum of (23) and (25) as follows:

$$Var[X(t)] = Var_{\lambda,\mu}[\lambda(1-e^{-\mu t})/\mu] + E_{\lambda,\mu}[\lambda(1-e^{-\mu t})/\mu]$$

$$= \frac{\alpha p}{\beta(q-1)} \left\{ \frac{\alpha(p+1)}{\beta(q-2)} \left(1 - 2(1+\beta t)^{-(q-2)} \right. \right.$$

$$\left. + (1 + 2\beta t)^{-(q-2)} \right)$$

$$\left. - \frac{\alpha p}{\beta(q-1)} [1 - (1+\beta t)^{-(q-1)}]^2 \right.$$

$$\left. + [1 + (1+\beta t)^{-(q-1)}] \right\} . \qquad (26)$$

It is immediate in comparing these variances that (26) contains a variance component due to the addition of P1 stochasticity to the model.

5.4 *Some Properties of a One-compartment System with Continuous Input.* The basic properties noted in the compartmental system of Section 5.2 are manifested in this system also. Perhaps the property of greatest interest due to its implications in practice is the comparison between the regression functions of the stochastic models with random rates and the solution of the deterministic model evaluated at the mean rates. Matis and Wehrly (1979) show that since $g(\mu) = (1-\exp(-\mu t))/\mu$ is a convex function, one has from Jensen's Inequality the relationship

$$E[g(\mu)] > g[E(\mu)] . \qquad (27)$$

When (27) is evaluated and multiplied by $E(\lambda)$, where λ is independent of μ , one has

$$E_{\lambda,\mu}[\lambda(1-e^{-\mu t})/\mu] > E(\lambda)[1-e^{-tE(\mu)}]/E(\mu) \qquad (28)$$

which implies that the expected value of this stochastic model will *always exceed* the deterministic model evaluated at the mean rate provided λ and μ are independent.

This relationship is easily shown in the special case of independent gamma distributions in Section 5.3; where, for all $t>0$, (23) and (24) exceed (21) evaluated at the mean rates of αp and βq. Note in particular that the equilibrium value of the deterministic model is λ/μ which upon substituting the mean rates yields $\alpha p/(\beta q)$, whereas the expected value of the stochastic model with P2 or R2 stochasticity has limit $\alpha p/[\beta(q-1)]$ when $q>1$. For the case $q=1$, i.e., an exponential distribution of μ , one has the novel phenomenon of $E[X(t)]$ increasing without bound over time.

Two other properties are obvious in this model. Firstly, the stochastic hypothesis may lead to useful closed form solutions, as in (23) which is the consequence of the seemingly plausible assumption of the gamma distributions. Secondly, although several stochastic models may share the same expected value, the individual sources of stochasticity are identifiable from the variances, as illustrated by (25) and (26), or from the more general covariances.

6.   A SUMMARY OF THE APPROACH

The present research has identified some of the principal sources of stochasticity in a compartmental system and has combined them into a unified model structure. In general, the approach would also call for other potential sources to be included upon their identification. It has been demonstrated that the combined model may have convenient closed form analytical solutions, at least in the one compartment case, which can be fitted to data for prediction, parameter estimation and hypothesis testing. Moreover, as indicated, the approach generalizes readily to certain non-exponential residency time distributions and to certain multiple compartment systems.

It is clear from the above considerations that the regression functions of the stochastic and deterministic models may differ substantially. If the hazard rate is any non-trivial random variable which is conditionally fixed over time, the regression function for the stochastic model cannot yield the classical sums of exponentials function. More generally, if the hazard rate is a function which varies over time and some of its parameters are random variables, then the deterministic solution is not the mean of the stochastic model. In the case of a continuous input, even the mean of the equilibrium distribution may differ substantially from the deterministic equilibrium point, as illustrated in the simple example.

The present approach also points out that the covariance structure of a stochastic model may be used for model identification. Heretofore the covariances have been derived primarily to optimize the estimation of parameters. However, in a model with multiple sources of stochasticity, the covariances may be used, in principle, to identify the individual sources of variability in an analysis-of-variance-like procedure. Current research on such techniques is promising and will be reported separately.

ACKNOWLEDGEMENTS

We are indebted to Dr. William Grant of Texas A&M University for many helpful discussions. This research was supported in part by Development Award 1 K04 BM00033-4 from the National Institute of General Medical Science to the first author.

REFERENCES

Campello, L. and Cobelli, C. (1978). Parameter estimation of biological stochastic compartmental models - An application. *IEEE Transactions on Bio-Medical Engineering*, 25, 139-146.

Cardenas, M. and Matis, J. H. (1974). On the stochastic theory of compartments: Solution for n-compartment systems with irreversible, time-dependent transition probabilities. *Bulletin of Mathematical Biology*, 36, 489-504.

Cardenas, M. and Matis, J. H. (1975a). On the time-dependent reversible stochastic compartment models: I. The general two compartment model. *Bulletin of Mathematical Biology*, 37, 505-519.

Cardenas, M. and Matis, J. H. (1975b). On the time-dependent reversible stochastic compartment model: II. A class of n-compartment systems. *Bulletin of Mathematical Biology*, 37, 555-564.

Chuang, S. N. and Lloyd, H. H. (1974). Analysis and identification of stochastic compartmental models in pharmacokinetics: Implication for cancer chemotherapy. *Mathematical Biosciences*, 22, 57-74.

Cobelli, C. and Morato, L. (1978). On the identification by filtering techniques of a biological n-compartment model in which the rate parameters are assumed to be stochastic processes. *Bulletin of Mathematical Biology*, 40, 651-660.

Dowdee, J. W. and Soong, T. T. (1974). On random least-square analysis. *Journal of Mathematical Analysis and Applications*, 46, 447-462.

Eberhardt, L. L., Gilbert, R. O., Hollister, H. L., and Thomas, J. M. (1976). Sampling for contaminants in ecological systems. *Environmental Science and Technology*, 10, 917-925.

Epperson, J. and Matis, J. H. (1979). On the distribution of the general irreversible n-compartmental model having time-dependent transition probabilities. To appear in *Bulletin of Mathematical Biology*.

Faddy, M. J. (1976). A note on the general time dependent compartmental model. *Biometrics*, 32, 443-448.

Faddy, M. J. (1977). Stochastic compartmental models as approximations to more general stochastic systems with the general stochastic epidemic as an example. *Advances in Applied Probability*, 9, 448-461.

Gold, H. J. (1977). *Mathematical Modeling of Biological Systems*. Wiley, New York.

Hogg, R. V. and Craig, A. T. (1970). *Introduction to Mathematical Statistics*. Macmillan, New York.

Marcus, A. H. and Becker, A. (1977). Power laws in compartmental analysis: II. Numerical evaluation of semi-Markov models. *Mathematical Biosciences*, 35, 27-45.

Matis, J. H. (1970). *Stochastic compartmental analysis: Model and least squares estimation from time series data*. Ph.D. dissertation, Texas A&M University.

Matis, J. H. (1973). Gamma time-dependency in Blaxter's compartmental model. *Biometrics*, 28, 597-602.

Matis, J. H. and Tolley, H. D. (1979). Compartmental models with multiple sources of stochastic variability: The one-compartment, time invariant hazard rate case. *Bulletin of Mathematical Biology*, in press.

Matis, J. H. and Wehrly, T. E. (1979). Compartmental models with multiple sources of stochastic variability: The one-compartment case with continuous infusion and time-varying hazard rates. Manuscript.

O'Neill, R. V. (1979a). A review of stochastic modeling in ecology. In *Systems Analysis of Ecosystems*, G. Innis, ed. Satellite Program in Statistical Ecology, International Co-operative Publishing House, Fairland, Maryland.

O'Neill, R. V. (1979b). A review of linear compartmental analysis in ecosystem science. In *Compartmental Analysis of Ecosystem Models*, J. H. Matis, B. C. Patten, and G. C. White, eds. Satellite Program in Statistical Ecology, International Co-operative Publishing House, Fairland, Maryland

Parzen, E. (1962). *Stochastic Processes*. Holden-Day, San Francisco.

Patten, B. C. (1975). Ecosystem linearization: An evolutionary design problem. *American Naturalist*, 109, 529-539.

Poole, R. W. (1979). Ecological models and the stochastic-deterministic question. In *Scientific Modeling and Quantitative Thinking with Examples in Ecology*, C. S. Holling, G. P. Patil, D. Solomon, and D. Simberloff, eds. Satellite Program in

Statistical Ecology, International Co-operative Publishing House, Fairland, Maryland.

Purdue, P. (1979). Stochastic compartmental models: A review of the mathematical theory with ecological applications, In *Compartmental Analysis of Ecosystem Models*, J. H. Matis, B. C. Patten, and G. C. White, eds. Satellite Program in Statistical Ecology, International Co-operative Publishing House, Fairland, Maryland.

Raman, S. and Chiang, C. L. (1973). On a solution of the migration process and the application to a problem in epidemiology. *Journal of Applied Probability*, 10, 718-727.

Regier, H. A. and Rapport, D. J. (1978). Ecological paradigms, once again. *Bulletin of the Ecological Society of America*, 59, 2-6.

Solomon, D. (1979). On a paradigm for mathematical modeling. In *Scientific Modeling and Quantitative Thinking with Examples in Ecology*, C. S. Holling, G. P. Patil, D. Solomon, and D. Simberloff, eds. Satellite Program in Statistical Ecology, International Co-operative Publishing House, Fairland, Maryland.

Soong, T. T. (1971). Pharmacokinetics with uncertainties in rate constants. *Mathematical Biosciences*, 12, 235-243.

Soong, T. T. (1972). Pharmacokinetics with uncertainties in rate constants: II. Sensitivity analysis and optimal dosage control. *Mathematical Biosciences*, 13, 391-396.

Soong, T. T. and Dowdee, J. W. (1974). Pharmacokinetics with uncertainties in rate constants. III. The inverse problem. *Mathematical Biosciences*, 19, 343-353.

Thakur, A. K., Rescigno, A., and Schafer, D. E. (1974). On the stochastic theory of compartments: II. The multi-compartment systems. *Bulletin of Mathematical Biology*, 35, 263-271.

Tiwari, J. L., Hobbie, J. E., Reed, J. P., Stanley, D. W., and Miller, M. C. (1978). Some stochastic differential equation models of an aquatic ecosystem. *Ecological Modelling*, 4, 3-27.

Tsokos, J. O. and Tsokos, C. P. (1976). Statistical modeling of pharmacokinetics systems. *Journal of Dynamic Systems Measurement and Control (Transactions of the American Society of Mechanical Engineers)*, 98, 37-43.

Turner, M. E. (1964). Mean exponential regression. *Biometrics*, 19, 183–186.

Weiner, D. and Purdue, P. (1977). A semi–Markov approach to stochastic compartmental models. *Communications in Statistics*, A6, 1231–1243.

[Received August 1978. *Revised December* 1978]

J. H. Matis, B. C. Patten, and G. C. White, (eds.),
Compartmental Analysis of Ecosystem Models, pp. 223-260. All rights reserved.
Copyright ©1979 by International Co-operative Publishing House, Fairland, Maryland.

STOCHASTIC COMPARTMENTAL MODELS: A REVIEW OF THE MATHEMATICAL THEORY WITH ECOLOGICAL APPLICATIONS

PETER PURDUE

Department of Statistics
University of Kentucky
Lexington, Kentucky   40506   USA

SUMMARY. In this paper we give a unified review of some of the models and methods of stochastic compartmental analysis. Applications to a number of ecological problems are given with particular emphasis on epidemic models and radiocalcium turnover.

KEY WORDS. compartments, Markov population processes, semi-Markov processes, radiocalcium turnover, epidemics.

1. INTRODUCTION

Compartmental models have been used to model biological systems for many years. Atkins (1969) attributes the first such application to a 1923 paper by Hevesey (1923) in which radioactive lead was used to study the uptake and loss of lead by the roots of various plants. The first application to animal studies is also attributed to Hevesey and co-workers (1924). Much of the early work was in the area of drug and tracer kinetics, starting with Teorell (1937). Excellent accounts of this work may be found in the books by Atkins (1969), Jacquez (1972), Gibaldi and Perrier (1975), and in the many references contained therein.

Almost all of the work referenced above is concerned with deterministic systems and makes heavy use of differential equations. While the deterministic theory describes the average or ideal behavior of a system, models examining the deviations from such ideal behavior are also needed. Bartholomay (1958a, b) was among the earliest authors to incorporate stochastic behavior into his models. Today there is a well developed theory of stochastic

compartmental systems. In ecological systems modelling, there
are certainly strong arguments for the use of stochastic
processes. One such argument is that, even if nature is completely deterministic, the complexity of an ecological system is
outside either our theoretical understanding or descriptive tools.
This lack of complete knowledge can be handled using probabilistic methods.

Matis and Gerald (1977) give a number of examples of the use
of compartmental analysis in the study of some ecological
systems. Two other relevant early papers concerned with the
circulation of carbon in nature and which use a deterministic
compartmental approach are Craig (1957) and Eriksson and Welander
(1956). Ericksson and Welander used a 4-compartment system which
was extended to a 5-compartment system by Craig. Bryant (1969)
used a stochastic model for the movement of aphids among plants.
A field was assumed to contain n+1 similar plants. To study
the migration between any two plants, Bryant used a 3-compartment
system with each plant being a compartment and the ground being
the third one. This model can be represented diagramatically as:

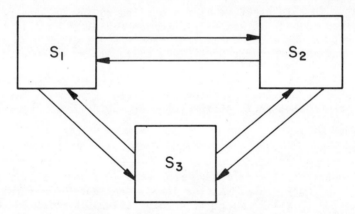

where S_i is plant i (i=1,2) and S_3 is the ground. A brief
outline of the remainder of the paper is this. In Section 2 the
deterministic theory is reviewed. Section 3 contains a discussion
of the homogeneous Markov population process approach to the
stochastic theory of compartments and is followed in Section 4 by
the semi-Markov process approach. Section 5 is a study of systems
with time dependent parameters. In Section 6 some applications
to problems in the theory of epidemics are given.

2.   DETERMINISTIC COMPARTMENTAL SYSTEMS

The fundamental assumption in compartmental analysis is that
a system may be divided into distinct components called compart-

ments which interact by exchanging material. There may be input
from the surrounding environment to the system and the system
may also leak material into the environment (open systems). If
there are no exchanges with the outside environment then the
system is a closed system. We consider now a system comprising
n compartments labelled C_1, C_2,$\cdots$,C_n. In the deterministic
case the flow rate of material from C_i to C_j is denoted by
$F_{ij}(t)$ and the amount of material in C_i at time t is denoted
by $X_i(t)$. The mass-balance equation for each compartment is

$$\frac{dX_i(t)}{dt} = \sum_{j=0}^{n}{}' F_{ji}(t) - \sum_{j=0}^{n}{}' F_{ij}(t) \quad (i=1,2,\cdots,n) \tag{1}$$

where Σ' means summation over all indicated values of j with
$j \neq i$ and where the first term on the right gives the inflow and
the second term the outflow. F_{0i} and F_{i0} represent interactions with the environment.

    The case which has received most attention is the time
homogeneous, linear donar controlled system which requires

$$F_{ij}(t) = a_{ij} X_i(t) \quad (i=1,\cdots,n),$$

where a_{ij} is the transfer rate between C_i and C_j. For this
case equation (1) becomes [· denotes differentiation],

$$\dot{X}_i(t) = F_{0i} + \sum_{j=1}^{n}{}' a_{ji} X_j(t)$$

$$- (a_{i0} + \sum_{j=1}^{n}{}' a_{ij})X_i(t) \quad (i=1,2,\cdots,n). \tag{2}$$

This can be written in matrix form as

$$\dot{X}(t) = A X(t) + B, \tag{3}$$

where $X(t) = [X_1(t),\cdots,X_n(t)]^T$, $B = [F_{01},\cdots,F_{0n}]$, and
A is the n×n matrix with $A_{ij} = a_{ji}$ ($i \neq j$) and with diagonal
entries

$$A_{ii} = 1 - (a_{i0} + \sum_{j=1}^{n}{}' a_{ij}).$$

From here the story is simple. The steady-state solution, X_0, is given by $-A^{-1}B$ assuming the existence of A^{-1}. Every solution to (3) converges to X_0 as $t \to \infty$, provided that the homogeneous form of (3) is asymptotically stable, i.e., that the solution to

$$\dot{Y}(t) = A\, Y(t), \tag{4}$$

which is $e^{At}Y(0)$, converges to 0 as $t \to \infty$. Equation (4) is stable if the eigenvalues of A have negative real parts. Finally, since $a_{ii} \geq \sum_{j \neq i} a_{ij}$, the matrix A is diagonally dominant and this ensures the required stability.

For the open system with no input and assuming that A has distinct eigenvalues $\lambda_1, \cdots, \lambda_n$ with associated eigenvectors $A_1, \cdots, A_n$ the solution at any time t is given by

$$X(t) = \sum_{i=1}^{n} C_i\, A_i\, e^{\lambda_i t}.$$

For more details on this system and its many special cases which are of interest see Jacquez (1972).

Ecosystems are temperature dependent and this in turn means that the flows in the compartmental models of such systems are also temperature and hence time dependent. We can build this into the model by using time dependent transfer rates and it seems appropriate to use periodically varying functions. Using the same notation as in equation (1), we let

$$F_{ij}(t) = \alpha_{ij}(t)\, X_i(t) \qquad (i=1, \cdots, n;\ t \geq 0),$$

where $\alpha_{ij}(t+\tau) = \alpha_{ij}(t)$ for all $t \geq 0$. Here τ is the common period of the rate functions. Again, in matrix form we have

$$\dot{X}(t) = A(t)\, X(t) + B(t). \tag{6}$$

The solution to equation (6) is not nearly as simple as for equation (3). Ideally the solution should display the same periodic behavior as the rate functions but this need not be the case. Mulholland and Keener (1974) and Aronsson and Kellogg

(1978) discuss this type of problem and present some solutions. For example Aronsson and Kellogg show for a 'leaky' system that under mild conditions on $A(t)$, every solution must go to 0. Also, under the same conditions, they show that (6) has a unique periodic solution. The whole question of differential equations with periodic coefficients is difficult to handle, but for a good discussion see Yakubovich and Starzhinskii (1975).

This completes a rather brief review of the deterministic case; in the next section we begin the study of stochastic compartmental systems.

3. HOMOGENEOUS MARKOV COMPARTMENTAL SYSTEMS

A stochastic compartmental system may be visualized in a number of ways. Perhaps the most immediate is to replace the transfer rates in the deterministic case by random variables. This leads to an equation of the type

$$\dot{X}(t) = K\, X(t), \tag{7}$$

where K is a random matrix satisfying almost surely the same conditions as the matrix A in the deterministic case. We will not discuss this approach any further in this paper; the interested reader may refer to Soong (1972) and Chuang and Lloyd (1974).

Another approach, and the one which we follow in this paper, is to consider a set of compartments which are connected by a probabilistic transfer mechanism with sure rate functions. This approach is examined, amongst others, in Thakur, Rescigno, and Schafer (1972, 1973), Matis and Hartley (1971), Cardenas and Matis (1974, 1975), Purdue (1974, 1975), Weiner and Purdue (1977a, b), and Faddy (1976, 1977). We examine in this section such models when the system is Markov. In the next section we will discuss in detail the semi-Markov case.

3.1 Linear Markov Population Processes. Let there be k compartments into which particles are introduced and among which particles move according to a Markov chain. Particles may also be permitted to leave the system. Using the notation of Kingman (1969), we can give a formal mathematical description of the model. Let $\underset{\sim}{n} = (n_1, \cdots, n_k)$ be a k-vector of non-negative integers and let C^k be the set of all such k-vectors. We are interested in a continuous time Markov chain with state space C^k and with transition rate from $\underset{\sim}{n}$ to $\underset{\sim}{m}$ denoted by $q(\underset{\sim}{n},\underset{\sim}{m})$ and with $q(\underset{\sim}{n}) = \sum_{\underset{\sim}{m} \neq \underset{\sim}{n}} q(\underset{\sim}{n},\underset{\sim}{m})$

denoting the overall rate of flow out of state $\underset{\sim}{m}$. This Markov chain will be called a Markov population process if for any $\underset{\sim}{n}$ the only values of $\underset{\sim}{m}$ for which $q(\underset{\sim}{n},\underset{\sim}{m})$ is non-zero are those with

$$m_i = n_i+1, \qquad m_\alpha = n_\alpha \qquad (\alpha \neq i)$$

$$m_i = n_i-1, \qquad m_\alpha = n_\alpha \qquad (\alpha \neq i)$$

$$m_i = n_i-1, \qquad m_j = n_j+1, \qquad m_\alpha = n_\alpha \qquad (\alpha \neq i,j).$$

Writing e_i for the vector with all components zero except for 1 in the ith position we have

$$q(\underset{\sim}{n}, \underset{\sim}{n} + \underset{\sim i}{e}) = \alpha_i(\underset{\sim}{n}), \qquad q(\underset{\sim}{n}, \underset{\sim}{n} - \underset{\sim i}{e}) = \beta_i(\underset{\sim}{n}),$$

$$q(\underset{\sim}{n}, \underset{\sim}{n} - \underset{\sim i}{e} + \underset{\sim j}{e}) = \gamma_{ij}(\underset{\sim}{n}) \qquad (i \neq j).$$

For compartmental systems we impose the further restriction that $\alpha_i(\underset{\sim}{n}) = \alpha_i(n_i)$, $\beta_i(\underset{\sim}{n}) = \beta_i(n_i)$, and $\gamma_{ij}(\underset{\sim}{n}) = \gamma_{ij}(n_i,n_j)$. In fact for the linear, donar controlled compartmental system we have

$$\alpha_i(n_i) = \alpha_i, \qquad \beta_i(n_i) = \beta_i n_i \qquad (i=1,2,\cdots,k),$$

$$\gamma_{ij}(n_i,n_j) = \gamma_{ij} n_i \qquad (i,j=1,2,\cdots,k; \; i \neq j).$$

The characteristic feature of the linear system is that particles enter the system according to a Poisson process at a rate $\alpha = \alpha_1 + \alpha_2 + \cdots + \alpha_k$ and then move among the compartments independently according to a continuous time Markov chain. By taking other choices for $\alpha_i(\underset{\sim}{n})$, $\beta_i(\underset{\sim}{n})$, and $\gamma_{ij}(\underset{\sim}{n},\underset{\sim}{m})$ we can incorporate the effects of crowding and other social interactions into a compartmental system.

We can describe the linear system in a slightly different way. Particles enter the system according to k independent Poisson processes, the ith process feeding into compartment i at rate α_i. On entering a compartment the particle stays a length of time which is exponential with mean $1/\sigma_i$ and then transfers to compartment j with probability Θ_{ij} or leaves the system with probability Θ_{i0} where

$$\sum_{j=0}^{k,} \Theta_{ij} = 1.$$

Let $X_i(t)$ be the number of particles in compartment i at time t with $\underset{\sim}{X}(t) = [X_1(t), \cdots, X_k(t)]$. Then $\{\underset{\sim}{X}(t); t \geq 0\}$ is a (linear) Markov population process. It is easy to see that

$$\alpha_i = \alpha_i, \quad \beta_i = \sigma_i \Theta_{i0} \quad (i=1, \cdots, k),$$

$$\gamma_{ij} = \sigma_i \Theta_{ij} \quad (i, j=1, \cdots, k; \ i \neq j).$$

The interesting problems are to find (a) the joint distribution of $X_1(t), \cdots, X_k(t)$, (b) the distribution of $X(t) = X_1(t) + \cdots + X_k(t)$, and (c) the long-run or steady state behavior of the system. Toward this, let us introduce two new k-dimensional vectors, $Y(t)$ and $Z(t)$, where $Y_i(t)$ is the number of particles which were present in the system at $t=0$ and which are in compartment i at t, and $Z_i(t)$ is the number of particles in compartment i at t which entered the system in $(0,t]$. We can then write

$$X(t) = Y(t) + Z(t). \tag{8}$$

To proceed further we let $P(t,j)$ be the probability that a particle at time t after its entry into the system is in compartment j; clearly $P(t,j)$ is governed by the mechanism by which particles move through the system. If we let

$$M(t,j) = \alpha \int_0^t P(u,j) \, du,$$

we can easily show (Kingman, 1969) that (a) the random variables $Z_1(t), \cdots, Z_k(t)$ are independent and (b) $Z_i(t)$ has a Poisson distribution with mean $M(t,j)$ $(j=1, 2, \cdots, k)$.

The next problem is how to handle the $Y_i(t)$ variables. For this, suppose we let $Y'_{\ell i}(t)$ be the number of particles which were in compartment ℓ at $t=0$ and which are in i at t. Then we have

$$Y_i(t) = \sum_{\ell=1}^{k} Y'_{\ell i}(t), \tag{9}$$

with $Y'_{\ell i}(t)$ $(\ell=1,2,\cdots,k)$ being binomial with parameters $X_1(0)$, $P_{\ell i}(t)$ where $P_{\ell i}(t)$ is the probability that a particle which is in ℓ at time 0 is in i at time t.

The major difficulty with this approach is to find $P_{ij}(t)$ and $P(t,j)$. (c) The steady state distribution of the system is easy to determine. In fact,

$$\lim_{t\to\infty} \Pr[X(t) = n] = \sum_{i=1}^{k} e^{-\lambda_i} \lambda_i^{n_i}/n_i!,$$

where $\lambda_i = \alpha/\sigma_i$ $(i=1,2,\cdots,k)$.

From the above results we can write down the mean and variance of the number in any compartment. We get, letting $X_\ell(0) = N_\ell$ $(\ell=1,\cdots,k)$,

$$E[X_i(t)] = \sum_{\ell=1}^{k} N_\ell P_{\ell i}(t) + M(t,i),$$

$$\mathrm{Var}[X_i(t)] = \sum_{\ell=1}^{k} N_\ell P_{\ell i}(t)[1-P_{\ell i}(t)] + M(t,i). \tag{10}$$

All that now remains is to determine the $P_{ij}(t)$ terms. We note that

$$P(t,i) = \sum_{\ell=1}^{k} (\alpha_\ell/\alpha) P_{\ell i}(t). \tag{11}$$

By the definition of the process, as $h \to 0$,

$$P_{ij}(h) = \sigma_i \Theta_{ij} h + o(h) \qquad (i,j=1,2,\cdots,k;\ i\neq j).$$

If a new compartment, say 0, is added and represents the 'outside' of the system then $P_{0,0}(h) = 1$ and

$$P_{i0}(h) = \sigma_i \Theta_{i0} h + o(h) \qquad (i=1,2,\cdots,k).$$

Hence if we let V denote the intensity matrix then the Kolmogorov equations are

$$P'(t) = VP(t) = P(t)V,$$

with solution,

$$P(t) = e^{Vt}. \tag{13}$$

In the case where V has distinct eigenvalues we can proceed thus: Let $\lambda_0, \cdots, \lambda_k$ be the eigenvalues of V and write V as

$$V = T \, \mathrm{diag}(\lambda_0, \cdots, \lambda_k) T^{-1} \tag{14}$$

and then

$$P(t) = T \, \mathrm{diag}(e^{\lambda_0 t}, \cdots, e^{\lambda_k t}) T^{-1}.$$

A good example of the use of this method is provided by Bryant (1969) in his study of the plant-to-plant migration of aphids. He uses a closed 3-compartment system and solves equation (13) explicitly. The initial condition used is that compartment 3 is empty at $t=0$. Then

$$\Pr[X_i(t) = \nu]$$

$$= \sum_{\gamma=0}^{\nu} \binom{N_1}{\gamma} P_{1i}^{\gamma}(t) [1-P_{1i}(t)]^{(N_1-\gamma)}$$

$$\times \binom{N_2}{\nu-\gamma} P_{2i}^{(\nu-\gamma)}(t) [1-P_{2i}(t)]^{(N_2-\nu+\gamma)} \qquad (i=1,2).$$

3.2 Some Other Markov Compartmental Systems. In the above discussion it was assumed that the individual particles all behaved independently, an assumption which seems to be made in almost all compartmental systems. As one possible way of removing this independence we suppose that there is a maximum rate at which particles may leave a compartment. This may be caused by exits from a compartment being possible at only certain points on the boundary. Such problems have been studied in Queueing Theory under the heading of Jackson networks. We discuss a simple example. As before particles enter the system at compartment i according to a Poisson process with rate α_i $(i=1,\cdots,k)$. However the rate at which particles may leave compartment i $(i=1,2,\cdots,k)$ depends upon the number present in the following way

$$\Pr[\text{particle leaves in } (t,t+h)\,|\,X_i(t) = m]$$

$$= \begin{cases} m\,\sigma_i & (0 \le m \le n_i), \\ n_i\sigma_i & (m \ge n_i). \end{cases}$$

As before, when a particle leaves compartment i it goes to j with probability Θ_{ij} or leaves the system with probability

$$\Theta_{i0} = 1 - \sum_{\ell=1}^{k} \Theta_{i\ell}.$$

This is identical with the network of waiting lines as studied by Jackson (1957). The main concern here will be with the steady state behavior of the system. Let Γ_m be the average arrival rate into compartment m from all sources, then we must have

$$\Gamma_m = \alpha_m + \sum_{\ell} \Theta_{\ell m} \Gamma_k \qquad (m=1,2,\cdots,k).$$

The main result for this model is given by the following. Let

$$P_i^{(m)} = \begin{cases} P_0^{(m)} \left(\dfrac{\Gamma_m}{\sigma_m}\right)^i /i!, & (0 \le i \le n_m) \\[3ex] P_0^{(m)} \left(\dfrac{\Gamma_m}{\sigma_m}\right)^i \dfrac{(n_m)^{n_m-i}}{n_m!} & (n_m \le i). \end{cases} \qquad (15)$$

Then

$$\lim_{t\to\infty} \Pr[\underline{X}(t) = \underline{v}] = \sum_{\ell=1}^{k} P_{v_\ell}^{\ell}.$$

The surprising thing is that the system behaves, in so far as steady state is concerned, as if its compartment were independent. For further results on Jackson networks, see Malamed (1977).

Another interesting compartmental system arises when the possibility of 'clearing' the system is considered. Consider a field into which a pest migrates at a certain rate and suppose that at certain points in time we eliminate all of the pests in the field. What is the size of the pest population at any time t?

In the absence of clearing, each pest has a random lifetime in field which is exponential with mean $1/\mu$. As we saw earlier for a simple one compartment model, if $X(t)$ denotes the number of particles present at t and if $X(0) = 0$, then

$$\Pr[X(t) = n] = \frac{[q(t)]^n e^{-q(t)}}{n!} ,$$

where $q(t) = (\lambda/\mu)1-e^{-\mu t}$. We now suppose that at times $\{S_0, S_1, S_2, \cdots\}$, the compartment is cleared out and we let $\{S_0, S_1, S_2, \cdots\}$ be a renewal process with inter-renewal times governed by the distribution function F. The renewal function associated with F, which gives the expected number of renewals in $[0,t]$, is

$$R(t) = \sum_{n=0}^{\infty} F^{(n)}(t),$$

where $F^{(n)}$ is the n-fold convolution of F with itself. Let $Z(t)$ be the number of pests in the cleared system at t. Formally,

$$Z(t) = \begin{cases} N(t) & (0 \leq t < S_1), \\ N(t-S_n) & (S_n \leq t < S_{n+1}). \end{cases}$$

Let $P_n(t) = \Pr[Z(t) = n/Z(0) = 0]$. Feldman and Curry (1977), working with queueing systems, obtained the following results for this model:

(a) $P_n(t) = \frac{1}{n!} \int_0^t [1-F(t-s)][q(t-s)]^n e^{-q(t-s)} dR(s).$

(b) $P_n = \lim_{t \to \infty} P_n(t) = \frac{1}{Mn!} \int_0^{\infty} [1-F(t)] q^n(t) e^{-q(t)} dt,$

where M is the mean of F.

(c) The steady state mean population size, L, is given

by $L = \frac{1}{M} \int_0^\infty [1-F(t)]q(t)dt.$

These integrals may be found in closed form for certain types of renewal processes. If the clearing times occur at times determined by a Poisson process with rate γ then $L = \lambda/(\gamma+\mu)$; while if the inter-renewal times are from a uniform distribution on (a-b, a+b), then

$$L = \frac{\lambda}{a\mu} \{a - \frac{1}{\mu} + (e^{\mu b} - e^{-\mu b})e^{-\mu a}/2b\mu^2\}.$$

This finishes our survey of the time homogeneous Markovian compartmental system. The approach taken, namely a direct random variable technique avoids the use of generating functions which are used by most authors. Of course one of the major disadvantages of the time homogeneous Markovian model is that the time spent in each compartment is an exponential random variable which depends only upon the compartment occupied by the particle. It seems desirable to have a more general set-up in which in-compartment delays are general random variables and the time spent in a compartment may depend upon the next compartment to be visited. Such a degree of generality is introduced by using semi-Markov processes. We outline this approach in the next section.

4. SEMI-MARKOV COMPARTMENTAL MODELS

Weiner and Purdue (1977) presented a detailed account of the semi-Markov compartmental model. Marcus and Becker (1977) treated certain two and three compartment models. In this section we will give a brief review of this work.

We consider a system which consists of k compartments. Particles may enter the system through any one of the compartments. The stream of particles entering the ith compartment is a Poisson process with rate λ_i; the k Poisson processes are assumed to be independent. On entering the system, a particle's progress through the system is assumed to be a semi-Markov process (see Cinlar, 1969, for basic definitions and properties). A particle which leaves the system will be said to be in compartment k+1. When a particle leaves the system, it may not reenter. Finally we assume that particles behave independently within the system.

For a given particle, let X_n denote the compartment entered at its nth transition within the system and let T_n be the epoch of this transition. We then have, by assumption,

$$Pr[X_{n+1} = j, T_{n+1} - T_n \leq t \,|\, X_0, X_1, T_1, \cdots X_n = i, T_n]$$

$$= Pr[X_{n+1} = j, T_{n+1} - T_n \leq t \,|\, X_n = i]$$

$$= A_{ij}(t) \qquad\qquad (1 \leq i \leq k+1; \; 1 \leq j \leq k+1).$$

The matrix $A(t)$ with $(i,j)th$ element $A_{ij}(t)$ is a semi-Markov matrix and will be called the semi-Markov matrix associated with the system. We will assume that

$$A_{ij}(t) = \alpha_{ij} F_i(t) \qquad (1 \leq (i,j) \leq k+1).$$

Here $F_i(t)$ is the distribution function of the length of time spent in compartment i and α_{ij} is the conditional probability of entering j given that a particle leaves i. Since $(k+1)$ is an absorbing state we have,

$$\alpha_{k+1,j} = \begin{cases} 0 & (j=1,\cdots,k), \\ \\ 1 & (j=k+1). \end{cases}$$

Consider now a particle which enters the system at time 0 and let J_t denote the compartment it is in at time t. We define $\tilde{P}_{ij}(t) = Pr[J_t = j \,|\, J_0 = i]$ and $B(t) = \mathrm{diag}\{F_1(t), \cdots, F_{k+1}(t)\}$. Then we have (Cinlar, 1969)

$$\tilde{P}_{ij}(t) = \int_0^t [1 - F_j(t-x)] \, R_{ij}(dx),$$

where $R_{ij}(x)$ is the $(i,j)th$ element of

$$R(t) = I + \sum_{n=1}^{\infty} A^{*(n)}(t),$$

where I is the $(k+1)$-dim. identity matrix and $A^{*(n)}(t)$ is defined by

$$A^{*(1)}(t) = A(t), \qquad A^{*(n)}(t) = A * A^{*(n-1)}(t),$$

and

$$A*B_{ij}(t) = \int_0^t \sum_{\ell=1}^{k+1} A_{i\ell}(t-x) \, B_{\ell j}(dx).$$

This theorem enables us to find the distribution function of the time spent in the system by a particle. Let $G(t)$ be the distribution function (d.f.) of the time spent in system; then $G(t)$ is given by

$$G(t) = \frac{1}{\lambda} \sum_{i=1}^k \lambda_i \, \tilde{P}_{i,k+1}(t),$$

where $\lambda = \lambda_1 + \lambda_2 + \cdots + \lambda_k$.

We let $X(t)$ denote the number of particles in the system at time t and $N(t)$ denote the number of particles which entered the system in $(0,t]$. We will assume that $X(t) = 0$ a.s. Then, for $t > 0$,

$$\Pr[X(t) = n \,|\, X(0) = 0]$$

$$= [\lambda \int_0^t [1-G(x)]dx]^n \, \exp\{-\lambda \int_0^t [1-G(x)]dx\}/n! \quad (n=0,1,\cdots).$$

We now consider the contents of the ith compartment at time t. Let $X_i(t)$ be the number of particles in compartment i at time t. Finally let $\tilde{P}_j(t) = \Pr(J_t = j)$. Then (a) For each $t > 0$, $X_j(t)$ has a Poisson distribution with parameter

$$\lambda \int_0^t \tilde{P}_j(x)dx \quad (j=1,\cdots,k).$$

(b) For each $t > 0$, $X_1(t),\cdots,X_k(t)$ are independent random variables. In order to find the $\lim P[X(t) = n]$ as $t \to \infty$, we note that we must evaluate $\int_0^\infty [1-G(x)]dx$. This is, of course, the mean first passage time to state $k+1$, which we will denote by μ. Now we let μ_i denote the mean first passage time to $k+1$ starting in state i and let μ_i' denote the mean time spent in compartment i. Letting $\underset{\sim}{\mu} = (\mu_1 \cdots \mu_k)^T$, $\underset{\sim}{\mu}' = (\mu_1' \cdots \mu_k')^T$, and

$$\Phi = \begin{bmatrix} \alpha_{11} & \cdots & \alpha_{1k} \\ \cdot & \cdots & \cdot \\ \cdot & \cdots & \cdot \\ \alpha_{k1} & \cdots & \alpha_{kk} \end{bmatrix},$$

we have (Ross, 1970), $\mu = (I-\Phi)^{-1} \mu'$, where $[I-\Phi]^{-1}$ exists since the maximal eigenvalue of Φ is less than 1. Clearly now $\mu = \frac{1}{\lambda} \sum_{i=1}^{k} \lambda_i \mu_i$ and we see,

$$\lim_{t \to \infty} \Pr[X(t) = n] = e^{-\mu\lambda} \frac{(\lambda\mu)^n}{n!} \quad (n=0,1,2,\cdots).$$

We can use this result to determine the mean occupation time of the system. The system is occupied if there is at least one particle present in the system. Let μ_0 denote the mean occupation time. Then $\mu_0 = [\exp(-\lambda\mu) - 1]/\lambda$.

The main difficulties with implementing the above scheme is the evaluation of R and μ. We illustrate how this can be done in the following example.

In this example we consider the open, two-compartment reversible system with independent Poisson arrivals which was studied by Purdue (1975). Upon entering the system a particle remains in the compartment it entered for a random amount of time before transferring to the other compartment or leaving the system. The semi-Markov matrix associated with this system is

$$A(t) = \begin{bmatrix} 0 & \alpha_{12}F_1(t) & \alpha_{13}F_1(t) \\ \alpha_{21}F_2(t) & 0 & \alpha_{23}F_2(t) \\ 0 & 0 & F_3(t) \end{bmatrix} .$$

Diagrammatically we have:

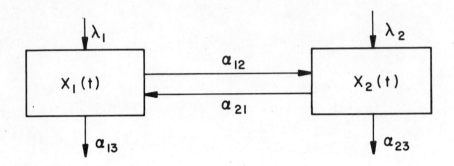

An easy computation shows that

$$R(t) = \begin{bmatrix} H(t) & \alpha_{12}F_1*H(t) & r_{13}(t) \\ \alpha_{21}F_2*H(t) & H(t) & r_{23}(t) \\ 0 & 0 & F_3(t) \end{bmatrix},$$

where $H(t) = \sum_{n=0}^{\infty} (\alpha_{12}\alpha_{21}F_1*F_2)^{(n)}(t)$, and $r_{13}(t)$, $r_{23}(t)$ are not yet determined. We shall see below that it is not necessary to obtain explicit expressions for $r_{13}(t)$ and $r_{23}(t)$. The matrix $B(t)$ is again diagonal with ith diagonal $F_i(t)$ $(i=1,2,3)$. It follows then that $\tilde{P}(t)$ is given by

$$\begin{bmatrix} (1-F_1)*H(t) & (1-F_2)*\alpha_{12}F_1*H(t) & 1-P_{11}(t)-P_{12}(t) \\ (1-F_1)*\alpha_{21}F_2*H(t) & (1-F_2)*H(t) & 1-P_{21}(t)-P_{22}(t) \\ 0 & 0 & 1 \end{bmatrix}.$$

Note that by using the fact that $\sum_{j=1}^{3}\tilde{P}_{ij}(t) \doteq 1$ $(i=1,2)$, we were able to obtain $\tilde{P}_{13}(t)$, $P_{23}(t)$ without obtaining an explicit expression for $r_{13}(t)$, $r_{23}(t)$.

We have then that

$$G_1(t) = 1 - [(1-F_1) + \alpha_{12}(1-F_2)*F_1] * H(t)$$

$$= (\alpha_{13}F_1 + \alpha_{12}\alpha_{23}F_1*F_2) * H(t),$$

$$G_2(t) = 1 - [(1-F_2) + (1-F_1)*\alpha_{21}F_2] * H(t)$$

$$= [\alpha_{23}F_2 + \alpha_{21}\alpha_{13}F_1*F_2] * H(t),$$

$$G(t) = \sum_{i=1}^{2} \lambda_i G_i(t)/\lambda.$$

Also, $\quad \tilde{P}_1(t) = \frac{\lambda_1}{\lambda} (1-F_1)*H(t) + \frac{\lambda_2}{\lambda} (1-F_1)*\alpha_{21}F_2 * H(t),$

$$\tilde{P}_2(t) = \frac{\lambda_1}{\lambda} (1-F_2)*\alpha_{12} F_1*H(t) + \frac{\lambda_2}{\lambda} (1-F_2)*H(t)$$

$$P_j(t) = \int_0^t \tilde{P}_j(t-x)dx/t \quad (j=1,2).$$

To obtain $\underset{\sim}{\mu}$ note that

$$\Phi = \begin{bmatrix} 0 & \alpha_{12} \\ \alpha_{21} & 0 \end{bmatrix}.$$

Thus $\qquad \underset{\sim}{\mu} = (I-\Phi)^{-1} \underset{\sim}{\mu}' = \dfrac{1}{1-\alpha_{12}\alpha_{21}} \begin{bmatrix} 1 & \alpha_{12} \\ \alpha_{21} & 1 \end{bmatrix} \begin{bmatrix} \mu_1' \\ \mu_2' \end{bmatrix}$

$$= \frac{1}{1-\alpha_{12}\alpha_{21}} \begin{bmatrix} \mu_1' + \alpha_{12}\mu_2' \\[2ex] \mu_2' + \alpha_{21}\mu_1' \end{bmatrix} .$$

The semi-Markov approach presented here provides a framework within which one can answer many questions arising in compartmental analysis. One appealing aspect of this method is that it does not require the solution of systems of differential equations. One of the difficulties of this approach is that it may be difficult to write down explicitly the $R(t)$ matrix when dealing with reversible systems with several compartments. However, as shown in the example and in Weiner (1976), $R(t)$ can be determined rather easily for many familiar models.

An interesting system where explicit results may be obtained is the Mammillary system. This has been used in an ecological context by Aronsson (1972) in a mark-recapture model where animals are marked in a central area and recaptured in several adjacent areas. Matis, Cardenas, and Kodell (1974) report the use of a similar model to study the migration of Boll-Weevils from a central zone into adjacent fields. As shown in Weiner (1976), this model is a special case of the semi-Markov compartmental system. Some of the interesting aspects of the set-up will be outlined here.

The mammillary model consists of one central compartment and k peripheral compartments. These $k+1$ compartments will hereafter be called the system. Particles remain in the central compartment for a random amount of time before leaving the system or transferring to one of the k peripheral compartments. A particle in one of the peripheral compartments remains there a random amount of time before leaving the system. A particle which has entered the central compartment and has left the system (either directly from the central compartment or through a peripheral compartment) is said to be in the $k+2$ compartment. Once a particle has left the system it may not reenter it. Further, it is assumed the particles behave independently. We need the following definitions:

$G_i(t)$ = distribution function of the amount of time a given particle remains in the ith peripheral compartment before leaving the system $(i=1,\cdots,k)$;

$G_{k+1}(t)$ = distribution function of the amount of time a given particle remains in the central compartment before leaving the system or entering one of the peripheral compartments;

α_i = probability that a given particle makes a transition to the $i th$ peripheral compartment given that it leaves the main compartment;

α_{k+1} = probability that a given particle leaves the system given that it leaves the main compartment;

$P_i(t)$ = probability that a given particle in the main compartment at time 0 is in the $i th$ compartment at time t $(i=1,\cdots,k+2)$;

$X_i(t)$ = number of particles in the $i th$ compartment at time t $(i=1,\cdots,k+2)$.

This model is analyzed for the case where $X_{k+1}(0) = N$ and $X_i(0) = 0$ $(i=1,\cdots,k,\ k+2)$.

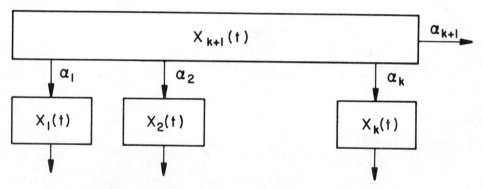

Since the particles behave independently and since at any point t in time each of the N particles must be in one of the $k+2$ compartments it follows that $X(t) = [X_1(t),\cdots,X_{k+2}(t)]$ has a multinomial distribution with parameters $[N, P_1(t),\cdots,P_{k+2}(t)]$. To find $P_i(t)$ $(i=1,\cdots,k)$, we first note that a particle must make a transition at some time $x \leq t$ to the $i th$ compartment then remain there until time t. Thus

$$P_i(t) = \int_0^t [1 - G_i(t-x)]\alpha_i dG_{k+1}(x) \quad (i=1,\cdots,k),$$

$$P_{k+1}(t) = 1 - G_{k+1}(t), \quad \text{and} \quad P_{k+2}(t) = 1 - \sum_{i=1}^{k+1} P_i(t).$$

Further it is well known that

$$E[X_i(t)] = NP_i(t), \quad Var[X_i(t)] = NP_i(t)[1-P_i(t)],$$

$$Cov[X_i(t),X_j(t)] = -NP_i(t)P_j(t) \quad (i,j=1,\cdots,k+2; \ i \neq j).$$

Matis, Cardenas, and Kodell (1974) have examined this model for the case when the waiting time for a given particle in the *ith* compartment is exponential, $(i=1,\cdots,k+1)$. Their model is a special case of the above.

Next, we examine

$$X^*(t) = \sum_{i=1}^{k+1} X_i(t),$$

which is the number of particles in the system at time t. Then

$$Pr[X^*(t)=k] = \binom{N}{k} [\bar{S}(t)]^k [S(t)]^{N-k} \quad (k=0,1,\cdots,N),$$

where

$$\bar{S}(t) = 1-S(t) = \alpha_{k+1}G_{k+1}(t) = \sum_{i=1}^{k} \alpha_i \int_0^t G_i(t-x)dG_{k+1}(x).$$

S(t) is the distribution function of the time spent in the system by any particle. If we let w = time to first emptiness of the system, then $Pr(w \leq t) = Pr[X^*(t)=0] = [S(t)]^N$.

Matis, Kodell, and Cardenas (1975) found an upper bound for the probability that the $\max_{i \leq i \leq k} \sup_{t \leq 0} \{X_i(t)\}$ is larger than some threshold value. We generalize that bound to the non-exponential case. Consider the closed mammillary model where once a particle enters a peripheral compartment it remains there. For this model let $T_i(t)$ be the number of particles in compartment i at time t $(i=1,\cdots,k+2)$. Then $T_i(t)$ is bounded, non-decreasing, and converges almost surely to T_i', where T_i' is a random variable having a binomial distribution with parameters (N, α_i) $(i=1, \cdots, k)$. Further, $T_{k+2}(t)$ converges almost surely to a random variable T'_{k+2} having a binomial distribution with parameters (N, α_{k+1}) and $T_{k+1}(t)$ converges almost surely to a random variable T_{k+1} which takes on the value zero with probability one. It follows then that $T' = (T'_1, \cdots, T'_{k+2})$ is distributed as a multinomial random variable with parameters $(N, \alpha_1, \cdots, \alpha_k, 0, \alpha_{k+1})$.

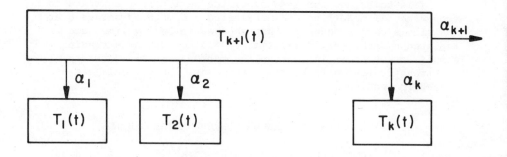

Next, let $M' = \max\{T_i : i=1, \cdots, k\}$. Then

$$\Pr[M' \geq m] \leq \sum_{i=1}^{k} \Pr[T'_i \geq m].$$

Recalling that T'_1 has a binomial(N, α_i) distribution, we have

$$\Pr[M' \geq m] \leq \min\left\{1, \sum_{i=1}^{k} \sum_{j=m}^{N} \binom{N}{j} (\alpha_i)^j (1-\alpha_i)^{N-j}\right\}. \qquad (16)$$

If we let $X'_i = \sup_t\{X'_i(t)\}$ and $Y' = \max\{X'_i\}$, then we note that $\Pr[T'_i > m] \geq \Pr[X'_i > m]$ $(i=1, \cdots, k)$, which implies $\Pr[M' \geq m] \geq \Pr[Y' \geq m]$. Thus (16) is an upper bound for $\Pr[Y' \geq m]$.

Note that if for some $\alpha > 0$, $k_1 > 0$, we let

$$\alpha_i = \frac{\alpha}{k\alpha + k_1\alpha} = \frac{1}{k + k_1} \quad (i=1, \cdots, k),$$

$$\alpha_{k+1} = \frac{k_1\alpha}{k\alpha + k_1\alpha} = \frac{k_1}{k + k_1},$$

we see that the bound found in Matis, Cardenas, and Kodell (1975) is a special case of the above results.

This concludes our account of the semi-Markov approach to compartmental modelling. As indicated the main advantage over the simple Markov model is that exponential delay times are replaced by arbitrary delay times. Another way of extending the simple Markov model is to keep the Markov assumption but to allow the transfer routes to be time dependent. Such models are discussed in Section 5.

5. TIME DEPENDENT MARKOV COMPARTMENTAL SYSTEMS

In this section we return to the Markovian assumption but allow all transition rates to be time dependent. This is motivated for the same reasons as outlined in Section 2 for the deterministic case. The results here are based mainly on Faddy (1977, 1976), Cardenas and Matis (1974, 1975a, b), Epperson and Matis (1978), Brown and Ross (1969), and Collings and Stoneman (1976).

As in the homogeneous case, let there be k compartments into which particles are introduced and among which particles move according to a time dependent continuous time Markov chain. Once in the system, the particles behave independently of one another. We suppose that particles enter compartment i according to an inhomogeneous Poisson process with rate function $\lambda_i(t)$ and that the entry processes are independent. A particle in compartment i at t will move to compartment j in $(t, t+h)$ with probability $\nu_{ij}(t)h + o(h)$; a particle in C_i at t will leave the system entirely in $(t, t+h)$ with probability $\mu_i(t)h + o(h)$. We let $X_i(t)$ denote the number of particles in C_i at t and $\underset{\sim}{X}(t) = [X_1(t), \cdots, X_k(t)]$. The development from here on is remarkably similar to the simpler time homogeneous case. Let $Y_i(t)$ be the number of original (i.e., in system at $t=0$) particles which are in C_i at t and let $Z_i(t)$ be the number of new particles present in C_i at t. Then, using the obvious vector notation,

$$X(t) = Y(t) + Z(t)$$

where, for each t, the vectors $Y(t)$ and $Z(t)$ are independent. To find the distributions of the entries in the Y and Z vectors we introduce the function $P_{ij}(t_0,t)$ which is the probability that a particle which is in C_i at t_0 is in C_j at time t. As pointed out by Faddy, these satisfy the Kolmogorov forward equation,

$$\frac{dP(t_0,t)}{dt} = P(t_0,t)\ V(t), \tag{17}$$

where $P(t_0,t)$ is the k×k matrix with entries $P_{ij}(t_0,t)$ and $V(t)$ has $(i,j)th$ coefficient $v_{ij}(t)$ and

$$v_{ii}(t) = -\left\{ \sum_{j=1}^{k}{}' v_{ij}(t) + \mu_i(t) \right\}.$$

The major problem is solving (17); we return to this point later. We will now show that the solution to (17) gives a simple way of finding the needed distribution functions. To begin, let us consider the $Y(t)$ vector.

Let $Y_{ij}^0(t)$ be the number of particles which were in C_i at t=0 and which are in C_j at t. Then we can write

$$Y_j(t) = \sum_{i=1}^{k} Y_{ij}^0(t), \tag{18}$$

and also, for each i, the random variables $Y_{i1}^0(t), \cdots, Y_{ik}^0(t)$ have multinomial distribution given by

$$Pr[Y_{i1}^0(t) = r_1, \cdots, Y_{i,k}^0(t) = r_k]$$

$$= \frac{N_i!}{N_i-r} \left[1 - \sum_{j=1}^{k} P_{ij}(t) \right]^{N_i-r} \prod_{j=1}^{k} \frac{[P_{ij}(t)]^{r_j}}{r_j!}, \tag{19}$$

where $r = r_1 + \cdots + r_k$ and N_i is the (non-random) number of particles in C_i at $t=0$.

For the remaining terms, let $Z_{ij}^0(t)$ be the number of new particles which entered the system in $(0,t]$ by way of C_i and which are in C_j at t. Again, we can write

$$Z_j(t) = \sum_{i=1}^{k} Z_{ij}^0(t);$$

but now, using the same argument as in Weiner and Purdue (1977) or the argument in Faddy (1977), we see that for each j, $Y_{ij}^0(t)$ $(i,j=1,2,\cdots,k)$ are independent Poisson random variables with means

$$\int_0^t \lambda_i(x) \, P_{ij}(x,t) dt.$$

As an example we can write down an expression for the mean and variance of $X_j(t)$ very easily. These are given by

$$E[X_j(t)] = \sum_{i=1}^{k} \{N_i P_{ij}(t) + \int_0^t \lambda_i(x) \, P_{ij}(x,t) dx\},$$

$$Var[X_j(t)] = \sum_{i=1}^{k} \{N_i P_{ij}(t)[1-P_{ij}(t)]$$

$$+ \int_0^t \lambda_i(x) \, P_{ij}(x,t) dx\}. \tag{20}$$

Matis and his students have explicitly evaluated these expressions for certain special compartmental systems. However the expressions are much too long and complicated to write down here.

We will now consider in a little more detail the single compartment system with time dependent rates. In this case let $\lambda(t)$ be the entry rate function and $\mu(t)$ the death rate function. Actually, instead of only allowing a single particle to enter at any time we can, following Brown and Ross (1969) allow particles to arrive in batches so that the probability that a batch which arrives at t is of size r is $P_t(r)$ and that each particle which enters the system at time t has an in-system time with a distribution function G_t. As before we let $X(t)$ denote the number in system at t and we will now make the assumption that $X(0) = 0$. Under these assumptions,

$$\Pr[X(t)=k] = \sum_n P\{Z_1 + \cdots + Z_n = k\}e^{-m(t)} \frac{[m(t)]^n}{n!} , \qquad (21)$$

where $m(t) = \int_0^t \lambda(x)dx$ and $\{Z_i\}$ are independent identically distributed, each with the distribution $\Pr(Z_1=j)$ given by

$$\frac{1}{m(t)} \int_0^t \sum_{r=j}^{\infty} P_x(r) \binom{r}{j} [1-G_x(t-x)]^j [G_x(t-x)]^{r-j} dm(x). \qquad (22)$$

Or, to put it more simply, $X(t)$ has a compound Poisson distribution with Poisson parameter $m(t)$ and jumps distributed according to (22). A single compartment goes through periods when it is empty and periods when it is not empty. There is a simple expression for the long run emptiness probability in the case $m(t) = \lambda t$, $G_t = G$, and $P_t = P$ which is given by

$$\lim_{t \to \infty} \Pr[X(t)=0] = e^{-\lambda M} \quad \text{where} \quad M = \int_0^{\infty} \sum_{r=0}^{\infty} [1-G^r(x)]P(r)dx.$$

There is a nice interpretation for this. If r is the size of a batch and if $Y_1, \cdots, Y_r$ represent the lengths of time spent by each particle in the system, then $M = E[\max(Y_1, \cdots, Y_r)]$ as would be expected. For further information on systems of this type, the paper by Brown and Ross (1969) should be consulted and, in the case where the batches are of size 1, the paper by Collings and Stoneman (1976) has some interesting material.

We will end this section with a brief discussion of another way of introducing parameter fluctuations. We consider a one compartment system with input and death rate functions $\lambda(t)$ and $\mu(t)$ which are themselves stochastic processes. Specifically we assume that $\lambda(t)$ and $\mu(t)$ are influenced by an extraneous environmental process which is an m-state, irreducible continuous time Markov chain. Whenever this extraneous process is in state i, the input rate is λ_i and the death rate is μ_i. As an illustration of the way to analyze such a process we will look at the amount of time spent in the system by any one particle. Suppose that a particle enters the system at time 0 while the environmental process is in state i. Let J_t denote the environmental state at t and let $P_{ij}(t) = \Pr[T>t, J_t=j \mid J_0=i]$, where T is the time spent in the compartment. As stated earlier, $\{J_t; t \geq 0\}$ is a Markov chain; let $\underset{\sim}{P}$ denote the Markov matrix governing the successive states visited and let $1/\sigma_i$ be the mean time spent in state i (i=1,2,$\cdots$,m). Using the Markovian nature of the process it is easy to see that

$$P'_{ij}(t) = (\sigma_j + \mu_j) P_{ij}(t) + \sum_{k=1}^{m} P_{ik}(t) P_{jk} \sigma_k. \tag{23}$$

Letting $M = \text{diag}(\mu_1, \cdots, \mu_m)$ and $\Delta_0 = \text{diag}(\sigma_1, \cdots, \sigma_m)$, this can be written as

$$P'(t) = - P(t)[M + \Delta_0 - \Delta_0 P]. \tag{24}$$

The solution to this equation is

$$P(t) = \exp[-(M + \Delta_0 - \Delta_0 P)t]. \tag{25}$$

So we see that $\Pr(T>t \mid J_0=i)$ is the *ith* entry in the vector $P(t)\underset{\sim}{e}$ where $\underset{\sim}{e}$ is an $(n \times 1)$-dimensional vector of 1's. The mean length of time spent in the compartment is easily obtainable. Let $\beta_i = E(T \mid J_0=i)$ and let $\underset{\sim}{\beta} = (\beta_1, \cdots, \beta_m)^T$; then

$$\underset{\sim}{\beta} = [M + \Delta_0 - \Delta_0 P]^{-1} \underset{\sim}{e}. \tag{26}$$

As an illustration suppose there are only two environmental states and that the environmental process alternates between these two states. Then we have,

$$\beta_1 = \frac{\sigma_1 + \sigma_2 + \mu_2}{\mu_1\mu_2 + \mu_1\sigma_2 + \mu_2\sigma_1} \; , \qquad \beta_2 = \frac{\sigma_1 + \sigma_2 + \mu_1}{\mu_1\mu_2 + \mu_1\sigma_2 + \mu_2\sigma_1} \; .$$

Other results are also known for this system but will appear in a future paper.

Thus far our goal has been to give a review of some of the methods of attack and results for some compartmental systems. In the next section we will show how to apply some of these results.

6. APPLICATIONS

Compartmental models have been used extensively in many different areas. However, traditionally the term has been used to mean only *deterministic* models, as for example in Wise's (1971) "The evidence against Compartments."

As we have seen earlier in this paper a *deterministic* multi-compartment system is mathematically equivalent to a system of linear differential equations with constant coefficients. The solutions to such systems are then mixtures of exponentials and non-negative integer powers of time. However as Marcus (1975) and Wise (1968, 1971, 1974), among others, point out this is inconsistent with observations on many systems which exhibit clearance curves with either negative powers of time, $at^{-\alpha}$, or gamma curves, $at^{-\alpha}e^{-\beta t}$. We will explain here how such results *can* be accommodated using a stochastic compartmental approach, specifically a semi-Markov compartmental system.

Let us first of all briefly compare deterministic, Markovian, and semi-Markovian formulations of a compartmental system. In the deterministic case each particle spends exactly the same amount of time in every compartment as any other particle and a fixed fraction of the particles leaving compartment i must go to j. In the Markovian case a particle spends an Exponential length of time in a compartment and if the exchange rate from i to j is λ_{ij}, the mean length of stay in compartment i is $1/\lambda_i$ where $\lambda_i = \sum_{j \neq i} \lambda_{ij}$, and the particle goes to j with probability $P_{ij} = \lambda_{ij}/\lambda_i$. But, the important thing here is that the modeller has to be prepared to accept *only* exponential waiting times. This seems to be a major disadvantage and will not get us out of the problems raised by Marcus and Wise. The great versatility of the semi-Markov approach is that *any* waiting time distributions can be used for the delay in a compartment. As we will show this allows us to construct compartmental systems consistent with the observation mentioned by Marcus and Wise.

We will illustrate these remarks by discussing a semi-Markov formulation of Wise's results on clearance curves for bone-seeking isotopes. We will also demonstrate the use of time dependent Markov systems in some epidemic problems.

6.1   A Model for the Turnover of Radiocalcium. After ingestion, by injection, swallowing, or breathing-in, of a tracer, its concentration-time curve can be followed for many hours or days. Wise (1968, 1974) points out that such curves, in the case of bone-seeking isotopes, are known to fit power functions $At^{-\alpha}$, or gamma curves $At^{-\alpha}e^{-\beta t}$, or indeed two different negative power curves with a short transition period in between.

In Wise (1974) there is an example given for the specific-activity curve for ^{45}Ca in blood after intravenous injection.

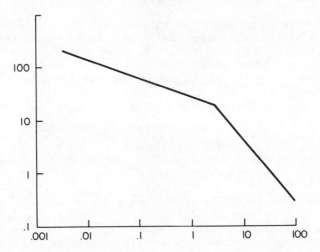

We will describe here how Wise's model for the turnover of radio-calcium fits into a semi-Markov compartmental model.

We take a 4-compartment model where the compartments correspond to:

Compt. 1: Plasma   Compt. 2: Non-Plasma
Compt. 3: Bone   Compt. 4: Excretory Organs.

These are linked according to:

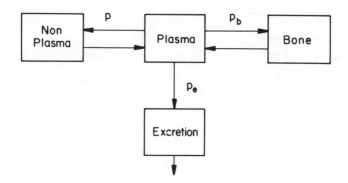

A tracer particle which enters the plasma stays there a random length of time and then with probability p_b enters the bone compartment, with probability $p = 1 - p_e - p_b$ enters the non-plasma compartment. The amount of time spent in bone is much greater than the time spent in either plasma or non-plasma.

Suppose that at $t=0$ a quantity Q_0 of tracer is placed in compartment 1. This then is a semi-Markov compartmental system where there is an impulse at $t=0$ and no further additions to the system. The main problem is to specify the distribution functions for the delays in compartments 1, 2, and 3. Because of the nature of the bone/calcium interaction, any tracer which enters the bone compartment is essentially lost to the system for a long time. Hence for the short term behavior of the system we can look upon any particle which enters compartment 3 as being lost to the plasma-non-plasma system. With this in mind the lifetime of a particle in the plasma-non-plasma part of the system may be written as

$$T = X_0 + \sum_{i=1}^{N} (X_i + Y_i),$$

where: N is the total number of plasma-non-plasma cycles made by a particle before it leaves the system; X_0 is the time spent in compartment 1 just prior to entering compartment 3 or 4; and $X_i + Y_i$ is the total time spent in plasma and non-plasma during the ith plasma-non-plasma cycle $(i=1, \cdots, N)$.

Also, $Pr(N=n) = p^n(1-p)$ and all the above random variables are independent. This corresponds to equation (5) in Wise (1968). By choosing distribution functions for X_i and Y_i corresponding to $1st$ passage probabilities for random walks, Wise shows how the observed activity curves of the form $\beta^{1-\alpha} t^{-\alpha} e^{-\beta t}/(-\alpha)!$ can arise.

By using the results for semi-Markov compartmental systems we can also extend Wise's results to the case where there is a continuous input of radionuclide into the system. In Wise (1974) details are given as to how to handle the bone-non-bone long term behavior of the system. For information on a similar system for lead metabolism see Marcus (1979).

6.2 Application to Epidemic Models. McClean (1976) uses a compartmental approach to study a hierarchial population model having Poisson recruitment. Raman and Chiang (1973) apply a time dependent Markov compartmental system to a problem in the epidemiology of Leprosy. Renshaw (1973) seems to have introduced the idea of using a linear compartmental system to approximate a more complex non-linear system. A similar method was used by Faddy (1977) in a study of the general stochastic epidemic. In this section we will describe briefly the uses of a compartmental approach in epidemic theory.

(i) An approximate solution to the general stochastic epidemic. The results here are due to Faddy (1977). The basic theory of the general stochastic epidemic is as follows. A population is divided into two groups called susceptibles and infectives. Let $X(t)$ denote the number of susceptibles at time t and $Y(t)$ the number of infectives and assume $X(0) = m$ and $Y(0) = n$. The dynamics are such that in $(t, t+h]$ a susceptible may become infected with probability $Y(t)h + o(h)$ while an infective may be removed from the population with probability $\rho Y(t) + o(h)$, both being independent of all other members of the population. This then is a Markov population process with 2 compartments and with

$$\alpha_1(x,y) = 0, \qquad \alpha_2(x,y) = 0,$$

$$\beta_1(x,y) = 0, \qquad \beta_2(x,y) = y,$$

$$\gamma_{12}(x,y) = xy, \qquad \gamma_{21}(x,y) = 0.$$

Diagrammatically,

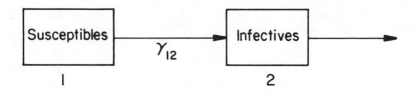

The problem is that this is *not* a linear donar controlled system; however, it is time homogeneous. In order to approximate this system, Faddy replaced the above time homogeneous system by a certain non-homogeneous two compartment model. The idea is to replace Y(t) by an appropriate deterministic function.

There is a deterministic version of the general stochastic epidemic which is governed by the differential equations,

$$\frac{dx}{dt} = -xy, \qquad \frac{dy}{dt} = xy-\rho y, \tag{27}$$

where x(t) and y(t) are the number of susceptibles and infectives respectively. Using the solution to equation (27) we get an approximating, linear, non-homogeneous system with transition rate from susceptible to infective given by y(t) while the other rates remain as above. So, in the notation of Section 5,

$$\Pr[X(t) = r, Y_1(t) = s_1]$$

$$= \frac{m!}{r!s!(m-r-s)!} [P_{11}(t)]^r [P_{12}(t)]^s [1-P_{11}(t)-P_{12}(t)]^{m-r-s}, \tag{28}$$

$$\Pr[Y_2(t)=s_2] = \frac{n!}{s_2!(n-s_2)!} [P_{22}(t)]^{s_2} [1-P_{22}(t)]^{m-s_2}, \tag{29}$$

where $Y_i(t)$ (i=1,2) is the number of particles in compartment 2 at time t which started in compartment i at time 0. Also,

$$P_{11}(t) = \frac{x(t)}{m}, \quad P_{12}(t) = \frac{y(t) - ne^{-\rho t}}{m}, \quad P_{22}(t) = e^{-\rho t}. \tag{30}$$

Faddy examined, using simulation methods, how good this approximate solution was. When $(\rho/m)^n$ was very small the approximation was shown to be excellent. See Faddy for a further discussion of the accuracy of the method.

(ii) The simple stochastic epidemic. The simple stochastic epidemic is similar to the stochastic case but does not allow any individuals to leave the system. An interesting way of modelling this, exactly, using a compartmental system will now be described. Let N be the total number of individuals in the population and let $X(t)$ denote the number of infectives present at t with $X(0) = 1$. The only permissible transition is that the number of infectives increases by 1. And this happens in $(t,t+h]$ with probability $X(t)[N-X(t)]h + o(h)$. Consider now the N compartment, catenary system,

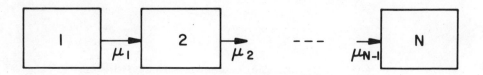

where $\mu_i = i(N-i)$, and suppose there is exactly one particle in compartment 1 at t=0. Let $X_i(t)$ denote the number of particles in compartment i at t; clearly $X_i(t)$ is either 0 or 1. Then the event that $X_i(t)$ is the same as the event that $X(t) = i$, and hence $Pr[X(t)=i] = Pr[X_i(t)=1]$. Let T_i be the amount of time spent in compartment i by the particle. Now we can use the technique developed in Section 3; however, due to the special nature of the process we have a simpler way of doing things here. Let T_i be the amount of time spent in state i by the particle let $S_j = T_1 + \cdots + T_j$. Then S_j is the time at which compartment i is entered and $X_i(t) = 1$ if and only if $S_i \leq t < S_{i+1}$. So,

$$Pr[X_i(t)=1] = Pr(S_{i-1} \leq t) = Pr(S_i \leq t) \quad (2 \leq i \leq N-1), \tag{31}$$

$$Pr[X_1(t)=1] = Pr(S_1 > t), \quad Pr[X_N(t)=1] = Pr(S_{N-1} \leq t).$$

So the whole solution depends upon obtaining an expression for the sum of independent but not identically distributed exponential random variables.

Bartholomew (1973) indicates that, for n independent, exponential random variables with parameters $\mu_1, \cdots, \mu_n$,

$$Pr(T_1 + \cdots + T_n > t) = (-1)^{n-1} \sum_{j=1}^{n} e^{-\mu_j t} \prod_{\substack{i=1 \\ i \neq j}}^{n} [\mu_j/(\mu_j - \mu_i)].$$

Using this we can easily write down an expression for the desired problems. Billard, Lacayo, and Langberg (1978) give explicit expressions for Pr[X(t)=i].

7. SUMMARY

The main goal of this paper was to give a unified review of some of the models and methods of stochastic compartmental analysis. We have deliberately emphasized an approach which does not use generating function arguments in the belief that thereby a clear picture of the overall behavior of the process emerges. As Jacquez (1972) points out, there are three steps in applying compartmental models to specific systems. The first step is to develop a plausible compartment model. The second step is the solution of the resulting mathematical equations. The final step is the acquisition of the data to estimate the model parameters. Somewhere between steps two and three, the problem of the derivation of good estimators must be tackled. All of these problems are extremely important. However the second step is probably the most well developed of the four steps. Clearly, much work needs to be done in the last two steps; for, without this, the mathematical theory is sterile.

REFERENCES

Atkins, G. L. (1969). *Multicompartment Models in Biological Systems*. Methuen, London.

Aronsson, G. and Kellogg, R. B. (1978). On a differential equation arising from compartmental analysis. *Mathematical Biosciences*, 38, 113-123.

Bartholomay, A. F. (1958a). On the linear birth and death processes of biology as Markov chains. *Bulletin of Mathematical Biophysics*, 20, 97-118.

Bartholomay, A. F. (1958b). Stochastic models for chemical reactions: I. Theory of the unimolecular reaction process. *Bulletin of Mathematical Biophysics,* 20, 175-190.

Bartholomew, D. J. (1973). *Stochastic Models for Social Processes.* Wiley, New York.

Billard, L., Lacayo, H., and Langberg, N. (1978). A new look at the simple epidemic process. *Florida State University Statistics Report M454.*

Brown, M. and Ross, S. M. (1969). Some results for infinite server Poisson queues. *Journal of Applied Probability,* 6, 604-611.

Bryant, E. H. (1969). A model for plant-to-plant movement of aphids: A new approach. *Researches on Population Ecology,* 11, 34-39.

Cardenas, M. and Matis, J. H. (1974). On the stochastic theory of compartments. *Bulletin of Mathematical Biology,* 36, 489-504.

Cardenas, M. and Matis, J. H. (1975a). On the time dependent, reversible stochastic compartment model: I. The general two compartment system. *Bulletin of Mathematical Biology,* 37, 505-519.

Cardenas, M. and Matis, J. H. (1975b). On the time dependent, reversible stochastic compartment model: II. A class of n compartment systems. *Bulletin of Mathematical Biology,* 37, 555-564.

Chuang, S. N. and Lloyd, H. H. (1974). Analysis and identification of stochastic compartment models in pharmacokinetics: Implications for cancer chemotherapy. *Mathematical Biosciences,* 22, 57-74.

Cinlar, E. (1969). Markov renewal theory. *Advances in Applied Probability,* 1, 123-187.

Collings, T. and Stoneman, C. (1976). The $M|M|\infty$ queue with varying arrival and departure rates. *Operations Research,* 24, 760-774.

Craig, H. (1957). The natural distribution of radiocarbon and the exchange time of carbon dioxide between atmosphere and sea. *Tellus,* 9, 1-17.

Epperson, J. O. and Matis, J. H. (1978). *On the distribution of the general irreversible n-compartmental model having time-dependent transition probabilities.* Unpublished manuscript.

Eriksson, E. and Welander, P. (1956). On a mathematical model of the carbon cycle in nature. *Tellus,* 8, 165-175.

Faddy, M. J. (1976). A note on the general time dependent compartment model. *Biometrics,* 32, 443-448.

Faddy, M. J. (1977). Stochastic compartmental models as approximations to more general stochastic systems with the general stochastic epidemic as an example. *Advances in Applied Probability,* 9, 448-461.

Gibaldi, M. and Perrier, P. (1975). *Pharmacokinetics.* Marcel Dekker, New York.

Hevesey, G. (1923). The absorption and translocation of lead by plants. *Biochemical Journal,* 17.

Hevesey, G., Christiansen, I. A., and Lomholt, S. (1924). Rescherches, par une methode radiochimique, sur la circulation du dismuth dans l'organisme. *Comptes Rendus Hebdomadaires des Seances, Academie des Sciences, Paris,* 178.

Jackson, J. R. (1957). Networks of waiting times. *Operations Research,* 5, 518-521.

Jacquez, J. A. (1972). *Compartmental Analysis in Biology and Medicine.* Elsevier, Amsterdam.

Kingman, J. F. C. (1969). Markov population processes. *Journal of Applied Probability,* 6, 1-18.

McClean, S. I. (1976). A continuous time population model with Poisson recruitment. *Journal of Applied Probability,* 13, 348-354.

Melamed, B. (1977). *Characterizations of Poisson traffic streams in Jackson networks.* Department of Industrial and Operations Engineering, University of Michigan Technical Report #77-2.

Marcus, A. H. (1975). Power laws in compartmental analysis. Part I: A unified stochastic model. *Mathematical Biosciences,* 23, 337-351.

Marcus, A. H. and Becker, A. (19). Power laws in compartmental analysis. Part II: Numerical evaluation of semi-Markov models. *Mathematical Biosciences* (to appear).

Marcus, A. H. (1979). Semi-Markov compartmental models in ecology and environmental health. In *Compartmental Analysis of Ecosystem Models*, J. H. Matis, B. C. Patten, and G. C. White, eds. Satellite Program in Statistical Ecology, International Co-operative Publishing House, Fairland, Maryland.

Matis, J. H. and Gerald, K. B. (1979). Concepts and methods of compartmental models analysis in ecological work. In *Compartmental Analysis of Ecosystem Models*, J. H. Matis, B. C. Patten, and G. C. White, eds. Satellite Program in Statistical Ecology, International Co-operative Publishing House, Fairland, Maryland.

Matis, J. H. and Hartley, H. O. (1971). Stochastic compartment analysis: model and least squares estimation from time series data. *Biometrics*, 27, 77-102.

Matis, J. H., Cardenas, M., and Kodell, R. L. (1974). On the probability of reaching a threshold in a stochastic mammillary system. *Bulletin of Mathematical Biology*, 36, 445-454.

Matis, J. H., Cardenas, M., and Kodell, R. L. (1976). A note on the use of a stochastic mammillary compartmental model as an environmental safety model. *Bulletin of Mathematical Biology*, 38, 467-478.

Mulholland, R. J. and Keener, M. S. (1974). Analysis of linear compartment models for ecosystems. *Journal of Theoretical Biology*, 44, 105-116.

Purdue, P. (1974a). Stochastic theory of compartments. *Bulletin of Mathematical Biology*, 36, 304-309.

Purdue, P. (1974b). Stochastic theory of compartments: one and two compartment systems. *Bulletin of Mathematical Biology*, 36, 577-587.

Purdue, P. (1975). Stochastic theory of compartments: An open, two compartment reversible system with independent Poisson arrivals. *Bulletin of Mathematical Biology*, 37, 269-275.

Raman, S. and Chiang, C. L. (1973). On a solution of the migration process and the application to a problem in epidemiology. *Journal of Applied Probability*, 10, 718-727.

Renshaw, E. (1973). Interconnected population processes. *Journal of Applied Probability*, 10, 1-14.

Ross, S. M. (1970). *Applied Probability Models with Optimization Applications*. Holden-Day, San Francisco.

Soong, T. T. (1971). Pharmacokinetics with uncertainties in rate constants. *Mathematical Biosciences*, 12, 235-243.

Teorell, T. (1937). Kinetics of distribution of substances administered in the body. I and II. *Archives Internationales de Pharmacodynamic et de Therapie*, 57, 205 and 226.

Thakur, A. K., Rescigno, A., and Schafer, D. E. (1972). On the stochastic theory of compartments. I. A single compartment system. *Bulletin of Mathematical Biophysics*, 34, 53-65.

Thakur, A. K., Rescigno, A., and Schafer, D. E. (1973). On the stochastic theory of compartments. II. Multi-compartment systems. *Bulletin of Mathematical Biology*, 35, 263-271.

Thakur, A. K. and Rescigno, A. (1978). On the stochastic theory of compartments. III. General, time-dependent reversible systems. *Bulletin of Mathematical Biology*,

Weiner, D. (1976). *On the stochastic theory of compartments.* Ph.D. thesis, University of Kentucky.

Weiner, D. and Purdue, P. (1977a). A semi-Markov approach to stochastic compartmental models. *Communications in Statistics*, A6, 1231-1243.

Weiner, D. and Purdue, P. (1977b). The stochastic theory of compartments. A mammillary system. *Bulletin of Mathematical Biology*, 39, 533-542.

Wise, M. E., Osborn, S. B., Anderson, J., and Tomlinson, R. W. S. (1968). A stochastic model for turnover of radiocalcium based on the observed power laws. *Mathematical Biosciences*, 2, 199-224.

Wise, M. E. (1974). Interpreting both short and long-term power laws in physiological clearance curves. *Mathematical Biosciences*, 20, 327-337.

Yakubovich, V. A. and Starzhinskii, V. M. (1975). *Linear Differential Equations with Periodic Coefficients.* Wiley, New York.

[*Received June* 1978. *Revised January* 1979]

J. H. Matis, B. C. Patten, and G. C. White, (eds.),
Compartmental Analysis of Ecosystem Models, pp. 261-278. All rights reserved.
Copyright ©1979 by International Co-operative Publishing House, Fairland, Maryland.

SEMI–MARKOV COMPARTMENTAL MODELS IN ECOLOGY AND ENVIRONMENTAL
HEALTH

ALLAN H. MARCUS

Department of Applied Mathematics
Washington State University
Pullman, Washington 99164 USA

SUMMARY. The interactions among distinct compartments in an
ecological or environmental system are often modeled as a Markov
process. In Markov models the residence times in each compartment
are exponentially distributed, and the transition probability
matrix does not depend on the elapsed residence time. The semi-
Markov generalization of such models can be formulated in two
ways: (i) the transition probability matrix is constant, but the
residence time or holding time distributions are arbitrary; (ii)
the transition probability matrix may depend on residence time in
the compartment. In the first example we suggest a semi-Markov
formulation of Horn's Markov chain model for the succession of
species of trees in a forest. In the second model we study the
effect of diffusion out of one or more compartments, leading to
power-law type residence lifetime distributions in those compartments; the metabolism of lead in mammals is an example. The use
of semi-Markov models greatly increases the realism of the model
without greatly increasing the difficulty of the mathematical
analysis of the model.

KEY WORDS. compartmental model, semi-Markov process, forest
succession.

1. INTRODUCTION

A mathematical model is, at best, a caricature of the reality
it is intended to represent. There is always a tradeoff between
building a highly detailed and complex model on the one hand, and
building a simple and easily understood model on the other hand.

Yet, just as a crude cartoon can often be turned into a satisfactory sketch by the addition of a few important details, so
will some simple and appealing Markov chain models in environmental science be made much more realistic by the inclusion of
appropriate waiting time distributions. The use of mathematical
modeling in environmental science and ecology is likely to become
even more important because it is the only means by which the
long-term effects of environmental changes can be predicted. In
both of the examples discussed below, the effects of interest will
emerge only after tens or hundreds of years, so that no timely
observational or experimental verification is possible. Overly
complex models are, in my opinion, more likely to mislead environmental scientists than are overly simple models, for a variety of
reasons: (1) Many people are skeptical about a simple model that
gives a good fit to observations, but are more willing to accept
the results of a highly complicated model -- especially a compu-
terized model. (2) In complicated models it may be almost as
difficult to trace causal relations as it is in the real system
being modeled. It is not always clear which features of a complicated model are essential in fitting the data, and which features
are peripheral if not irrelevant. (3) Most models can be 'tuned'
by adjusting their parameters so as to provide the best possible
fit to the data; but an excellent fit to historical data, which
can usually be achieved by a complicated model, does not guarantee
predictive validity. Since a simple model cannot be so readily
tuned to the data, it is more convincing when it does fit. (4)
Complicated models are not only 'data-hungry,' but also too often
require knowledge of ancillary relations among variables that are
not available.

My first example is that of succession of species of trees
in a forest. Horn (1975a, b) has presented a Markov chain model
for the substitution or replacement of one species of tree by
another (possibly the same) at a given location in the forest.
The state of the Markov chain is the species of tree occupying
the site during a given year, so that time is here a discrete
variable. If a mature forest is removed by a catastrophe such as
a forest fire or clearcutting, a succession of new growths will
occur until a pattern resembling that of the original forest is
achieved. Horn describes as the 'climax community' thus formed
the stationary distribution of the Markov chain, corrected for
longevity of the various species. Horn's model is then implicitly
a semi-Markov model in discrete time. If the semi-Markov machinery
is fully exploited, it can produce useful predictions qualitatively
comparable to those of more complicated computer models. Computer
simulations have been developed by Leak (1970) and by Botkin,
Janak, and Wallis (1972). The Botkin, *et al.* model does not depend
on parameters peculiar to the particular forest they studied
(Hubbard Brook, New Hampshire, USA), and so may be generalizable
to other forests. The parameters for both Leak's birth-and-death

simulation and Horn's semi-Markov model would have to be separately estimated for each forest system. Empirical verification can be obtained by historical retrospection rather than by experiment, since typical time scales for substantial changes are about 50 years.

My second example is the uptake, distribution, and retention of trace contaminants in mammals. For environmental (rather than accidental or occupational) exposure to trace metals such as lead, the problem is one of slow accumulation in certain organs such as bone. The contaminant is not necessarily biologically inactive and progressive damage may occur. For lead and alkaline earths in man, the typical time scales may be 10-50 years. Controlled long-term studies have been carried out only on laboratory animals such as mice, rats, dogs, and baboons. Mathematical models are required to predict the long-term accumulation in man. The states of the Markov chain are usually called 'compartments.' Wise (1974) has raised serious objections to the use of compartmental models for alkaline earth metabolism: (1) The compartments may or may not correspond to specific physiological organs or organ systems. (2) With longer periods of observation it is necessary to add more and more compartments with longer and longer residence times, in order to fit the data. The explicit physiological interpretation of these compartments (much less the identification of biological mechanisms) becomes much more difficult.

Bone-seeking elements apparently leave the bone matrix by a diffusion-type random walk (Marshall and Onkelinx, 1968; Marshall, et al., 1972) rather than by a Poisson random walk appropriate to compartments containing cellular or interstitial fluids. Wise (1974, 1975) and Marcus (1975, 1977) have described random walk mechanisms appropriate to retention of alkaline earth metals by bone. A 3-compartment model that includes diffusion out of bone may give much more accurate estimation of the accumulated body burden of lead over long periods of time than does the more detailed 5-compartment model developed by Bernard (1977). Bernard's model does not include diffusion, but of course provides more accurate modeling of transport among blood and soft tissues. The basic Markovian nature of the transition of lead molecules among compartments is the same in both models, but the usual compartment model does not admit the heavy-tailed residence time distribution in bone due to diffusion, which is included in the semi-Markov model.

2. SEMI-MARKOV PROCESSES: A REVIEW

2.1 Notation.

N = number of states (compartments)(species) in the system

0 = label for an absorbing state, e.g., external environment, if needed

p_{ij} = transition probability from initial state i to next state j

$h_{ij}(t)$ = probability density function (pdf) of time t that a particle just entering state i at time 0 remains there before moving next to j; holding time pdf

$w_i(t) = \sum\limits_{j=0}^{N} p_{ij} h_{ij}(t)$ is the waiting time pdf in state i

$W_i(t) = \int_t^\infty w_i(x)\, dx$ = survival probability in state i at time t

$H_i(t) = w_i(t)/W_i(t)$ = hazard function in state i at time t

$E(T_i) = \int_0^\infty t\, w_i(t)\, dt$ = mean waiting time (longevity) in state i

$c_{ij}(t) = p_{ij} h_{ij}(t)$ = core probability function

$p_{ij}(t)$ = transition probability from initial state i to next state j, given that the waiting time is t

$b_{ij}(t)$ = occupancy probability that a system initially in state i is found in state j at time t

$f_{ij}(t)$ = first passage time pdf of time t when a system initially in i first enters state j (if j ≠ i) or first returns to i (if j = i); cdf $F_{ij}(t)$

u_j = stationary probability of transition to state j

b_j = stationary probability of occupancy of state j

Matrices: $\underset{\sim}{P} = [p_{ij}]$, $\underset{\sim}{w}(t) = \text{Diag}[w_i(t)]$, $\underset{\sim}{W}(t) = \text{Diag}[W_i(t)]$,
$\underset{\sim}{C}(t) = [c_{ij}(t)]$, $\underset{\sim}{B}(t) = [b_{ij}(t)]$, $\underset{\sim}{F}(t) = [f_{ij}(t)]$,
$\underset{\sim}{u} = (u_1,\cdots,u_N)$, $\underset{\sim}{b} = (b_1,\cdots,b_N)$

Fourier transforms: $\underset{\sim}{B}^*(z) = [b_{ij}^*(z)]$, $\underset{\sim}{C}^*(z) = [c_{ij}^*(z)]$,
$\underset{\sim}{F}^*(z) = [f_{ij}^*(z)]$, $\underset{\sim}{W}^*(z) = \text{Diag}[W_i^*(z)]$,

where the Fourier transform of an arbitrary function g(t) is written

$$g^*(z) = \int\limits_{-\infty}^{\infty} \exp(i\, z\, t)\, g(t)\, dt \quad \text{with} \quad i = \sqrt{(-1)}.$$

2.2 Results. A complete and readable review of semi-Markov processes is given by Howard (1971) and a deeper survey by Cinlar (1974). There are two mathematically equivalent specifications of the semi-Markov process (SMP):

$$\text{(i)} \quad \{p_{ij}(t), w_i(t)\}, \qquad \text{(ii)} \quad \{p_{ij}, h_{ij}(t)\}.$$

They are equivalent since $c_{ij}(t) = p_{ij} h_{ij}(t) = p_{ij}(t) w_i(t)$ so:

$$\text{(i)} \quad w_i(t) = \sum_{j=0}^{N} p_{ij} h_{ij}(t), \quad p_{ij}(t) = p_{ij} h_{ij}(t)/w_i(t), \qquad (1)$$

$$\text{(ii)} \quad p_{ij} = \int_0^{\infty} p_{ij}(t) w_i(t) \, dt, \quad h_{ij}(t) = p_{ij}(t) w_i(t)/p_{ij}. \qquad (2)$$

Specification (i) is suitable for the forest succession model, and (ii) for the lead metabolism model. The following results are standard and easily derived:

$$\underset{\sim}{u} = \underset{\sim}{u}\underset{\sim\sim}{P} \quad \text{where} \quad u_j \geq 0, \quad \sum_{j=0}^{N} u_j = 1, \qquad (3)$$

$$b_j = u_j \, E(T_j)/E(T) \quad \text{where} \quad E(T) = \sum_{j=0}^{N} u_j \, E(T_j), \qquad (4)$$

$$\underset{\sim}{B}(t) = \underset{\sim}{W}(t) + \int_0^t \underset{\sim}{C}(x) \, \underset{\sim}{B}(t-x) \, dx, \qquad (5)$$

$$\underset{\sim}{F}(t) = \underset{\sim}{C}(t) + \int_0^t \underset{\sim}{C}(x) \, \{\underset{\sim}{F}(t-x) - \text{Diag}[\underset{\sim}{F}(t-x)]\}dx. \qquad (6)$$

It is useful to note how the semi-Markov process can be specified in terms of the sequence of random variables (J_n, t_n) where J_n is the state of the process (its embedded Markov chain) at step n, and t_n is the time of the *nth* transition. We have $n=0,1,2,\cdots$; $j_n=0,1,2,\cdots,N$; $t_0 < t_1 < t_2 \cdots$. Then

$$\Pr\{J_{n+1} = j \quad \text{and} \quad t_{n+1} \leq t \, | \, J_0, J_1, \cdots, J_n = i; \, t_0, \cdots, t_n\}$$

$$= \Pr\{J_{n+1} = j \quad \text{and} \quad t_{n+1} \leq t \mid J_n = i; \, t_n\}$$

$$= P_{ij} \, G_{ij}(t - t_n) = P_{ij}(t - t_n) \, W_i(t - t_n). \tag{7}$$

Convolution equations such as (5) and (6) lend themselves to transform analysis:

$$\underset{\sim}{B}^*(z) = [\underset{\sim}{I} - \underset{\sim}{C}^*(z)]^{-1} \, \underset{\sim}{W}^*(z), \tag{8}$$

$$\underset{\sim}{F}^*(z) = \underset{\sim}{C}^*(z) \, [\underset{\sim}{I} - \underset{\sim}{C}^*(z)]^{-1} \, \{\text{Diag}[\underset{\sim}{I} - \underset{\sim}{C}^*(z)]^{-1}\}^{-1}. \tag{9}$$

The numerical solution of these Fourier transforms is a straight-forward exercise in symbolic matrix manipulation and numerical integration (Marcus and Becker, 1977); efficient numerical schemes are available.

The Markov process (MP) is a special case of the SMP. Define a constant matrix of transition rates k_{ij} and matrix $\underset{\sim}{K} = (k_{ij})$ such that

$$k_{ii} = k_i > 0, \quad k_{ij} = -k_i P_{ij} \quad \text{if} \quad i \neq j. \tag{10}$$

Then the defining relations may be taken as

$$dB(t)/dt = -\underset{\sim}{B}(t) \, \underset{\sim}{K}. \tag{11}$$

Alternatively, in specification (ii) we have strictly exponential waiting times

$$w_i(t) = k_i \exp(-k_i t) \quad \text{thus} \quad p_{ij}(t) = p_{ij}. \tag{12}$$

In the discrete-time case, the MP has all transitions (including self-transitions) occurring at unit time steps, whereas the SMP will have transitions occurring at intervals that have a random discrete distribution, e.g., geometric.

3. FOREST SUCCESSION

3.1 Markov Chain Models. Horn (1975a, b) has presented the transition probability matrix $\underset{\sim}{P}$ copied in Table 1 for $N = 11$ species of trees recorded over many years in Institute Woods at Princeton, New Jersey, USA. These probabilities were estimated by counting the number n_{ij} of saplings of type j under a tree of species i at a location. The total number of saplings under

TABLE 1: Transition matrix for Institute Woods in Princeton: percent saplings under various species of trees. The number of saplings of each species listed in the row at the top, where the abbreviations are self-explanatory, is expressed as a percentage of the total number of saplings (last column) found under individuals of the species listed in the first column. The entries are interpreted as the percentages of individuals of species listed on the left that will be replaced one generation hence by species listed at the top. A dash implies that no saplings of that species were found beneath that canopy; a zero, that the percentage was less than 0.5%. (From Horn, 1975b.)

Canopy species, i	Sapling species (%), j											Total
	BA	GB	SF	BG	SG	WO	OK	HI	TU	RM	BE	n_j
Big-toothed aspen	3	5	9	6	6	–	2	4	2	60	3	104
Gray birch	–	–	47	12	8	2	8	0	3	17	3	837
Sassafras	3	1	10	3	6	3	10	12	–	37	15	68
Blackgum	1	1	3	20	9	1	7	6	10	25	17	80
Sweetgum	–	–	16	0	31	0	7	7	5	27	7	662
White oak	–	–	6	7	4	10	7	3	14	32	17	71
Red oak	–	–	2	11	7	6	8	8	8	33	17	266
Hickories	–	–	1	3	1	3	13	4	9	49	17	223
Tuliptree	–	–	2	4	4	–	11	7	9	29	34	81
Red maple	–	–	13	10	9	2	8	19	3	13	23	489
Beech	–	–	–	–	1	1	1	1	8	6	80	405
												3286
Stationary distribution (%), u_j	0	0	4	5	5	2	5	6	7	16	50	
Longevity (years), μ_j	80	50	100	150	200	300	200	250	200	150	300	
Age-corrected stationary distribution (%), b_j	0	0	2	3	4	2	4	6	6	10	63	

trees of type i is n_i, so that $p_{ij} = n_{ij}/n_i$. The stationary transition probability vector $\underset{\sim}{u}$ is also shown in Table 1. However, Horn also noted that the long-term occupancies

$$b_{ij}(\infty) = b_j \qquad (13)$$

have to be 'corrected for longevity.' This is precisely what is done in calculating the stationary occupancy probabilities in a semi-Markov process. If the matrix $\underset{\sim}{P}$ is recurrent, and if the mean time μ_{ij} residing in i before going to j is denoted

$$\mu_{ij} = \int_0^\infty t h_{ij}(t) dt, \qquad (14)$$

then the average waiting time in state i is

$$\mu_i = \int_0^\infty t \, dW_i(t) = \sum_{j=1}^N p_{ij} \mu_{ij}. \qquad (15)$$

It is not hard to show that (from (4))

$$b_j = u_j \mu_j / \sum_{i=1}^N u_i \mu_i, \qquad (16)$$

and this is Horn's longevity correction. For a Markov process, $\mu_{ij} = 1/k_{ij} = \mu_i$ for all j, so that $k_i = 1/\mu_i$. I have also copies μ_j and b_j in Table 1.

3.2 *Non-Markovian Properties*. The model developed by Leak (1970) is only superficially a Markov model. Leak develops 'birth' and 'death' rates for trees (see Table 2). These were estimated from replacements at intervals of 8-9 years and 24-25 years after even-aged plantings (second growths) of northern hardwoods in the Bartlett Experimental Forest in New Hampshire, USA, so there are only two species in common with the Princeton forests in Table 1. It is striking that the death rates are typically much higher in the second and third decades of life than in the first decade. This is of course impossible in the context of a purely Markov model.

Time dependent hazard functions are explicit in Leak's work and are possibly an essential feature of the Botkin-Janak-Wallis simulation.

TABLE 2: Death rates per hundred years in second-growth northern hardwoods for three density classes during the intervals (I) *from 0 to 8-9 years and* (II) *from 8-9 to 24-25 years. With obvious abbreviations, the species are Beech, Yellow Birch, Sugar Maple, Red Maple, Paper Birch, White Ash, Red Spruce, and Eastern Hemlock (after Leak, 1970).*

Stand density	Age	Species							
		BE	YB	SM	RM	PB	WA	RE	EH
Low	I	.15	1.32	.01	.01	2.40	.01	.01	.01
	II	.11	1.46	.01	.23	.75	.01	.01	.19
Medium	I	.23	.65	.34	.52	.32	1.91	.01	.01
	II	.42	1.27	.30	.07	1.08	.64	2.97	.01
High	I	.23	1.59	1.23	.23	.24	.45	2.05	.57
	II	.93	2.54	1.37	.60	.49	.45	.01	.09

The Hazard function $H_i(t) = w_i(t)/W_i(t)$ is easily seen to be constant only in the MP model (12). If $H_i(0) > 0$ and if $H_i(\infty)$ is finite and positive, then most of the familiar lifetime distributions such as the exponential, Weibull, Gamma, and lognormal cannot be used. Some less familiar pdf's can be fitted to data such as that given by Leak or by Stephens and Waggoner (1970), however. One family is the *three-parameter* mixed exponential density, and another is the *two-parameter* mixed Gamma.

Let a, k, and m be positive parameters, with a dimensionless and with k, m dimensionally (time)$^{-1}$. The three-parameter mixed exponential is:

$$w(t) = ak \exp(-kt) + (1 - a)m \exp(-mt), \qquad (17)$$

so
$$W(t) = a \exp(-kt) + (1 - a) \exp(-mt). \qquad (17)$$

The pdf (13) has increasing hazard function $H(t) = w(t)/W(t)$ if $a > m/(m-k)$ where $m > k$.

The two-parameter mixed Gamma pdf is given by (assuming n is a known integer)

$$w(t) = \frac{k}{m+k} [ke^{-kt}] + \frac{m}{m+k} \left[\frac{k^n t^{n-1}}{(n-1)!} e^{-kt} \right], \qquad (18)$$

so
$$W(t) = \frac{k}{m+k} e^{-kt} + \frac{m}{m+k} \sum_{j=0}^{n-1} \frac{(kt)^j}{j!} e^{-kt},$$

and also has an increasing hazard function.

Both (17) and (18) can be fit to count data. Suppose that an initial population of N trees begins to grow at time $t = 0$. The trees are revisited at times t_1, t_2, $\cdots$ and the number n_1, n_2, $\cdots$ that die between t_{i-1} and t_i are counted. Then the data (n_i) follow a multinomial distribution with parameters N and (p_i), where $p_i = W(t_i) - W(t_{i-1})$, and the parameters of w can be fit by the method of maximum likelihood. The results appear to be rather messy for the mixtures (17) or (18) and will be discussed elsewhere.

It is evident from Table 2 that the majority of the hazard rates increase from the first decade of life to the next two. For beech trees (a dominant species), the increase in hazard rate with age for stands of medium and high density is striking, suggesting the mixed Gamma pdf is a better model. The usual compartmental (Markov) model with its constant hazard rate is evidently untenable here.

3.3 Transient Analysis. Even the simplest cases may involve a great deal of analysis (for example, see Howard, 1971). For example, even when all of the $h_{ij}(t)$ are exponential, the occupancy probabilities $b_{ij}(t)$ may exhibit considerable sinusoidal oscillations even though they are approaching the stationary limiting value b_j at an exponential rate. If the embedded Markov chain J_n is transient then the use of transform methods appears unavoidable. An important application of this is given in the next section.

4. LEAD METABOLISM

In the usual compartmental models, we are most often interested in the proportion of a trace substance found in compartment j after time t from its original insertion into compartment i. This is of course the occupancy probability $b_{ij}(t)$ calculated from (5) or (8). Although rarely made explicit, the assumption usually adopted is that transitions are independent and Markovian, so that we may use equation (11) as the *defining* relationship.

The solution to (11) may be expressed in terms of the row
vector $b_i(t) = [b_{i0}(t), \cdots, b_{iN}(t)]$ as

$$b_i(t) = b_i(0) \exp(-Kt). \tag{19}$$

The concentration in compartment j must then necessarily be a
mixture of exponential functions of time. We denote the eigenvalues of K by $s_0, \cdots, s_N$ and the corresponding (orthonormal-
ized) eigenvectors by $v_0, \cdots, v_N$ (column form) so

$$Kv_j = s_j v_j \text{ and } v_i^T v_j = 0 \text{ or } 1 \text{ according as } i \neq j \text{ or } i=j. \tag{20}$$

We may then express (16) more explicitly as

$$b_i(t) = b_i(0) \sum_{j=0}^{N} \exp(-s_j t) \, v_j \, v_j^T. \tag{21}$$

In particular, we may define the *retention* function, or total
proportion of the initial dose retained in the body beyond time
t, by the function

$$R(t) = 1 - \sum_{i=1}^{N} b_{i0}(t) = \int_{t}^{\infty} \sum_{i=1}^{N} b_{i0}(0) \, f_{i0}(x) \, dx. \tag{22}$$

With the convenient assumption that all the initial dose is in
compartment 1, so $b_{11}(0) = 1$ then

$$R(t) = \sum_{h=1}^{N} \sum_{j=1}^{N} \exp(-s_j t) \, \bar{v}_{ij} \, v_{hj}. \tag{23}$$

Now, equation (16) or (23) can be fit to data on metabolism of alkaline earth metals such as Ca, Sr, Ba, Ra, or to important radioactives such as U, Am, or to Pb. The resulting models
are not parsimonious, and apparently must be extended as additional data become available. We will consider in detail lead
metabolism. For experiments on baboons up to 300 days, Cohen
(1971) found N=2 terms adequate. For beagle dogs, N=4 terms
were required for times to 1000 days by Hursh (1973), but in
another experiment N=3 terms gave an adequate fit to data collected by Lloyd, *et al.* (1975) for times out to 1600 days.
Bernard (1977) has synthesized elements of these analyses to produce a 5-compartment model for use up to times in excess of
10,000 days.

However, Hursh (1973) also obtained a very good fit to his data by use of a single-term power function for retention at times greater than 10 days,

$$R(t) = 1.96 \; t^{-0.50}, \tag{24}$$

for *in vivo* gamma counting. More complicated but analogous functions have been proposed by Marshall, *et al.* (1972) as the International Commission on Radiological Protection (ICRP) model for Ca, Sr, Ba, and Ra. We may reexpress the ICRP model

$$R(t) = q_1 \exp(-s_1 t) + \sum_{j=2}^{3} q_j \exp(-s_j t) c^a/(t+c)^a. \tag{25}$$

The constant c represents a short initial time delay, much less than a day, and the parameter a is the characteristic exponent of the power function. The ICRP values are 0.1 for Ca, 0.18 for Sr, 0.237 for Ba, and 0.415 for Ra. We may thus adopt the Hursh value of a=0.5 for lead metabolism.

It is clear that if curve-fitting were our only goal, we could just as well adjust N and the parameters s_j, v_{ij} in (23), or adjust a, c, q_j, s_j in (25) to fit the retention data. My opinion is that restricting N to a small number of biologically interpretable compartments may be a more fruitful approach. The necessity of adding long-lived compartments to exponential models such as (23) arises from the nature of the retention of metal molecules in bone. Once bound to the bone matrix, it appears that a metal molecule is most likely to leave the bone by diffusion to a bone surface. It is then necessary to consider the nature of the random walk of a molecule *within* a compartment. Several models have been considered by Marcus (1975) and Wise (1974). Power-law models that are most plausible and also most convenient for the calculation of the inverse Fourier transforms in equations (8) and (9) require either *stable* or *exponentially modified stable* pdf's, $h_{ij}(t)$. These are most conveniently defined by their Fourier transforms or characteristic functions. The stable pdf for holding time has positive parameters, a, m, and

$$h_{ij}^*(z) = \int_{-\infty}^{\infty} h_{ij}(t) \exp(izt) \, dt$$

$$= \exp\{-m|z|^\alpha \; [1 - i\,\mathrm{sgn}(z) \; \tan(\pi\alpha/2)]\}. \tag{26}$$

We also require a < 1. Note that the stable pdf does not have a
finite mean. For this reason, it is convenient to define the EMS
pdf, which for k ≥ 0 is

$$h_{ij}^*(z) = \exp\{m[k^\alpha - (k - iz)^\alpha]/\cos(\pi\alpha/2)\}. \tag{27}$$

The stable pdf (26) is the same as (27) when k = 0. For
k > 0,

$$E(T_{ij}) = \alpha \; m \; k^{\alpha-1}/\cos(\pi\alpha/2). \tag{28}$$

The parameter k can be interpreted as the resorption rate, i.e.,
the rate at which bone material is returned to the blood due to
being worn away. Calcium and strontium in compact (cortical) bone
has a resorption rate of about 4% per year in the ICRP model.

The other parameters in our semi-Markov model are derived
from the Markov assumption and from mass balance relationships
given by Rabinowitz, Wetherill, and Kopple (1976). Their compartmental formulation is shown in Figure 1. Their analysis is a
conventional 2-compartment model, in view of the very long pre-
sumed lifetime of the bone compartment. The compartmental formulation of Bernard's 5-compartment model is shown in Figure 2. In
the face of a time-variable dose rate d(t) (in units of
mg/kg/year, say), the accumulated body burden BB(t) is

$$BB(t) = \int_0^t R(t-x) \; d(x) \; dx, \tag{29}$$

$$BB(\infty) = d \; m_{10} \text{ if } d(x) = d \text{ and } m_{10} = \int_0^\infty t \; f_{10}(t) \; dt. \tag{30}$$

Thus the total body burden reaches equilibrium only if the mean
first passage time from the initial source of ingestion to the
external environment is finite. The accumulated body burden
BB(t) is sketched in Figure 3 for 3-compartment models with,
respectively, exponential, exponentially modified stable, and
stable pdf's for the holding time of lead in bone; Bernard's model
was integrated using (29) and is also shown. It is interesting
that Bernard's model shows a rapid increase then sharp leveling
off as the equilibrium BB is approached after a few years, while
our S and EMS models exhibit an almost linearly increasing
body burden with constant exposure to lead. The parameters of the
models have matched, as nearly as possible. See Marcus (1978a)
for further details and model specifications.

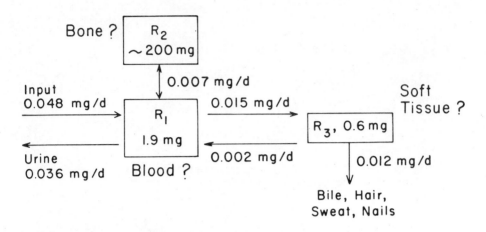

FIG. 1: Rabinowitz-Wetherill-Kopple model (1976) was used by the author to parameterize his model.

In view of relation (30), it is clear that many interesting properties of transient semi-Markov processes can be derived simply from the matrix $\underset{\sim}{P}$ and the mean holding times $E(T_{ij})$. For the 3-compartment model in Figure 1, we obtain upon differentiating (9) or integrating (6)

$$m_{10} = [P_{10}E(T_{10}) + P_{13}P_{30}E(T_{13}+T_{30})$$

$$+ \sum_{j=2}^{3} P_{ij}P_{j1}E(T_{1j}+T_{j1})]/(P_{10}+P_{13}P_{30}). \qquad (31)$$

For a mammillary model such as Bernard's in Figure 2, in which compartment 1 is the central compartment and $2, \cdots, N$ are peripheral, we find that

$$m_{10} = E(T_{10}) + \sum_{j=2}^{N} P_{ij} E(T_{ij} + T_{j1})/P_{10}. \qquad (32)$$

In Bernard's model the compartments are not precisely identified with physiological systems; compartments 2 and 3 are, mainly, cortical and trabecular (or compact and cancellous) bone, and compartments 4 and 5 are mainly liver, kidney, and blood, but compartment 1 includes blood and faster moving parts of the liver

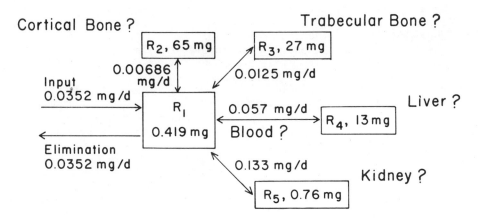

FIG. 2: *Bernard's model (1977).*

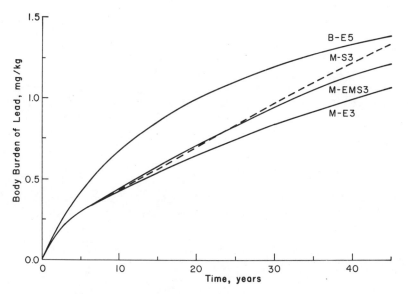

FIG. 3: *Body burden of lead predicted for standard exposure to lead of reference man   B–E5: Bernard's 5-compartment model. M–E3: Marcus 3-compartment model with exponential residence time in bone.   M–EMS3: Marcus 3-compartment model with exponentially modified stable law for bone residence time.   M–S3: Marcus 3-compartment model with stable (power) law residence time in bone.*

and kidney. The same identifiability problem is present in interpreting the compartments in the Rabinowitz-Wetherill-Kopple model,
so that there are serious difficulties in comparing a composite
3-compartment model based on Bernard's model with any other 3-
compartment model. None of the proposed models is completely
satisfactory, in my opinion, but the semi-Markov machinery at
least allows the possibility of a more realistic treatment of the
diffusion of lead molecules out of bone. One consequence of the
semi-Markov model is the predicted increase in blood lead levels
many years after exposure; this is in accord with some observations
(Prerovska and Teisinger, 1970) and is *not* predicted by the more
detailed model of Bernard.

We have also developed semi-Markov compartmental models analogous to the ICRP models for alkaline earth metabolism (Marcus,
1978b). While the analysis of semi-Markov models requires a
computer, e.g., for the inversion of (8), I believe that such
models are not conceptually more complicated than the usual Markov
compartmental models which require use of a computer when the
number N of compartments required to fit the data also grows
large. The results shown here are for N=3.

ACKNOWLEDGMENTS

This work was partially supported by the National Institute
of Environmental Health Sciences and by the Environmental Protection Agency (USA) under Grant 5-Rol-ESO 1236-2. I am particularly grateful to Mr. Arthur Becker for computations, and to Dr.
S. R. Bernard and Dr. J. H. Marshall for comments on related work.

REFERENCES

Bernard, S. R. (1977). Dosimetric data and metabolic model for
    lead. *Health Physics*, 32, 44-46.

Botkin, D. B., Janak, J. F., and Wallis, J. R. (1972). Some
    ecological consequences of a computer model of forest growth.
    Journal of Ecology, 60, 849-872.

Cinlar, E. (1975). *Introduction to Stochastic Processes*.
    Prentice-Hall, Englewood Cliffs, New Jersey.

Horn, H. S. (1975a). Markovian properties of forest succession.
    In *Ecology and Evolution of Communities*, M. L. Cody and J. M.
    Diamond, eds. Belknap Press, Cambridge, Massachusetts.

Horn, H. S. (1975b). Forest succession. *Scientific American*,
    46 (May), 90-98.

Howard, R. A. (1971). *Dynamic Probabilistic Systems, Vol. II.* Wiley, New York.

Hursh, J. B. (1973). Retention of ^{210}Pb in beagle dogs. *Health Physics*, 25, 29–35.

Leak, W. B. (1970). Successional change in northern hardwoods predicted by birth and death simulation. *Ecology*, 51, 794–801.

Lloyd, R. D., Mays, C. W., Atherton, D. R., and Bruenger, F. W. (1975). ^{210}Pb studies in beagles. *Health Physics*, 28, 575–583.

Marcus, A. H. (1975). Power laws in compartmental analysis, I. *Mathematical Biosciences*, 23, 337–350.

Marcus, A. H. and Becker, A. (1977). Power laws in compartmental analysis, II. *Mathematical Biosciences*, 35, 27–45.

Marcus, A. H. (1978a). The body burden of lead: Comparison of mathematical models for accumulation. *Environmental Research*, to appear.

Marcus, A. H. and Becker, A. (1978b). Alkaline earth metabolism: The ICRP model reformulated as a semi-Markov model. Submitted for publication.

Marshall, J. H. and Onkelinx, C. (1968). Radial diffusion and power function retention of alkaline earth radioisotopes. *Nature*, 217, 742–743.

Marshall, J. H., ed. (1972). *Alkaline Earth Metabolism in Adult Man*, Publication 20 of the International Commission on Radiological Protection. Pergamon, Oxford. Reprinted in *Health Physics*, 24, 124–221.

New York University Medical Center. (1971). The retention and distribution of lead 210 in the adult baboon, N. Cohen, McD. Wrenn, M. Eisenbud, eds. Institute of Environmental Medicine Report NYO–3086–10, –11.

Prerovska, I. and Teisinger, J. (1970). Excretion of lead and its biological activity several years after termination of exposure. *British Journal of Industrial Medicine*, 27, 352–355.

Rabinowitz, M. B., Wetherill, G. W., and Kopple, J. D. (1976). Kinetic analysis of lead metabolism in healthy humans. *Journal of Clinical Investigation*, 58, 270–286.

Stephens, G. R. and Waggoner, P. E.   (1970).   The forests anti-
    cipated from 40 years of natural transitions in mixed hard-
    woods. *Bulletin of the Connecticut Agricultural Experiment
    Station* (New Haven), No. 707, 1-58.

Weiner, D. and Purdue, P.   (1977).   A semi-Markov approach to
    stochastic compartmental models. *Communications in Statistics
    - Theory and Methods*, A6, 1231-1243.

Wise, M. E.   (1971).   Discussion on stochastic analysis by Matis
    and Hartley. *Biometrics*, 27, 97-100.

Wise, M. E.   (1974).   Interpreting both short- and long-term power
    laws in physiological clearance curves. *Mathematical Bio-
    sciences*, 20, 327-337.

Wise, M. E.   (1975).   Skew distributions in biomedicine including
    some with negative powers of time. In *Statistical Distri-
    butions in Scientific Work, Vol. 2*, G. P. Patil, S. Kotz,
    and J. K. Ord, eds. D. Reidel, Dordrecht. 241-262.

[Received June 1978. *Revised January* 1979]

J. H. Matis, B. C. Patten, and G. C. White, (eds.),
Compartmental Analysis of Ecosystem Models, pp. 279-293. All rights reserved.
Copyright ©1979 by International Co-operative Publishing House, Fairland, Maryland.

THE NEED FOR RETHINKING ON BOTH COMPARTMENTS AND MODELLING

M. E. WISE*

Physiology Laboratory
Leiden University
Wassenaarseweg 62
Leiden, Netherlands

SUMMARY. Possible mathematical forms of retention, excretion
and clearance curves observed after a single intake of a tracer
are considered briefly. According to models based on homogeneous
compartments these should be sums of exponentials of time, most
of them negative. Over the same period of time in the curve, many
fit equally well or better to functions containing negative powers
of time, or even to a single power function. Numerical illustrations are given, in particular for excretion curves for ^{65}Zn
in some mammals and man, obtained from retention data. The
pdf's for time intervals for the tracer between entering and
leaving a "compartment" should often be changed from negative
exponentials of time to ones for first passage times of drifting
Brownian particles (or Inverse Gaussian functions), so that the
tracer 'drifts' *through* the compartment. Finally, some contrasting quotes on what a model is or should be are given.

KEY WORDS. compartments, modelling, clearance curves, excretion
curves, retention curves.

1. ALTERNATIVE MATHEMATICAL FORMS FOR CLEARANCE, EXCRETION, AND RETENTION CURVES

Compartment modelling has a curious past and a fascinating
present. The future perhaps depends on how much the above title

---

*Permanently seconded from the J. A. Cohen Interuniversity
Institute of Radiopathology and Radiation Protection.

comes to be accepted! Rethinking is most needed in fitting
and interpreting excretion and retention curves of all kinds in
a living organism. Very often what is retained and excreted is a
radioactive isotope. Then its time course clearly needs to be
known in order to assess any radiation damage, but this should
also help us to know more about metabolism, flow and turnover in
particular organs of the substance that is labelled by the
radioactive tracer, or for forecasting what happens to almost
anything that enters the organism. These questions obviously
arise often enough in ecology.

For the curve fitting, nearly everything centers round what
is fitted to a so-called clearance curve $y(t)$, obtained from
a single short intake of the tracer. Then if a proportion
$R(t)$ of this tracer is still in the organism at time t , and
if everything that enters leaves, we have $R(0)=1$. The
excretion curve is defined to be:

$$-R'(t) = dR(t)/dt \tag{1}$$

and we have

$$-\int_{0}^{\infty} R'(t)dt = 1 . \tag{2}$$

Often the concentration $y(t)$ in blood and/or in urine is
measured. With appropriate units, if then η or $\eta(t)$ is the
rate tracee is excreted, then the instantaneous rate that the
tracer is excreted is:

$$-R'(t) = y(t)\eta(t) . \tag{3}$$

So if $\eta(t)$ can be regarded as constant we have

$$-R'(t) = \eta y(t) \tag{4}$$

and this can be regarded as a probability density curve for the
time spent by the tracer in the organism.

There is nothing new in this so far, but it leads straight
to our controversy. If the organism consists of homogeneous
compartments, about the only possible form for $y(t)$ and for
$R(t)$ is a sum of negative exponentials:

$$y(t) \text{ or } R(t) = \sum_{i=1}^{r} A_i \exp -B_i t . \tag{5}$$

Usually $r=2$, 3, or 4. In fitting the observed data, what
is usually done is to try $r=2$, and if this does not work, to
put $r=3$ and go on adding terms until a fit *is* achieved.

Very often, however, over a wide range of values of t ,
y(t) fits either: $At^{-\alpha}$,   (6)

or: $At^{-\alpha}exp-\beta t$,   (7)

or (7) changing into:

$$Ct^{-w}exp-\gamma t \qquad\qquad (8)$$

where $w>\alpha$; then much of a log y v log t plot consists of
one or of two straight lines. But if you try to fit any of
these to (5) you will always succeed, by taking enough terms,
within any reasonable limits of experimental error.

Health physicists and others working on radiation hazards
to humans seem largely to have realized this 15 years ago or more.
Since then, in many publications such as those of the International
Commission of Radiation Protection ICRP (1967) the two approaches
have coexisted.

At the other extreme, most physiologists and pharmacologists
who do compartment analyses of this kind seem still to be
unaware of all this. In relation to ecosystems however, the
position is hopeful in a way. In a number of publications the
authors are clearly convinced that these models ought to be
used, both for data analysis and for long term forecasting, but
there seem to be surprisingly few numerical examples, (Schultz,
Eberhardt, Thomas and Cochran, 1974). Hence, homogeneous
compartmental analyses for ecosystems have not (yet) wasted much
time.

In several other fields I am convinced that many hundreds
of clearance curves fitted by (5) ought to be reanalyzed and re-
interpreted, especially when r=3 or more. For two compartments
and/or when r=2 in eq. (5), the results are more often valid,
but relevant parameters may be estimated very inefficiently.
For example, if the data fit (7) and (5) with r=2 equally
well, the end exponential β or B_1 can be estimated from both
approaches, but that from (7) will have a smaller standard error.
On the other hand, where a one-compartmental model fits the
relevant data this *should* generally be valid.

Other things being equal, the reason why most researchers
have preferred (5) to (6), (7) or (8) is probably that the
resulting interpretations in terms of rate constants, pool sizes
etc., appear to be clear and simple. (But the quantitative
estimates often seem to be physiologically unreasonable.) There
is another reason. If the data really fit (6), (7) and/or (8)

the corresponding exponential terms will be just separate
enough to be capable of numerical estimation. In other words:

$$y(t) = \sum_{i=1}^{r} A_i \exp - B_i t \text{ and } B_1 < B_2 < B_3 \qquad (9)$$

with r=3 for example, it is generally realized that the B's
must be well separated, and that mostly

$$A_1 < A_2 < A_3 . \qquad (10)$$

Sometimes it is found that $A_3 < A_2$, but then this corresponds
to a rapidly disappearing term which is negligible in its
effect after one or two observations. Yet few people, it seems,
have asked why the answers so often come out like this.

2. WHAT COULD BE DONE TO REPLACE HOMOGENEOUS COMPARTMENTAL MODELS

There are already 1000 or more published examples where (5)
has been fitted numerically but there is clear evidence that (6),
(7) and/or (8) would fit as well or better over the same range of
time   (Wise, 1976, 1977, 1979).

The majority of these curves come from clinical studies, but
there are quite a number relating to isotopes of particular
interest to ecologists. In the appendix are given examples from
studies of retention of ^{65}Zn in various animals including man
(Williamson, 1975; Richmond, Furchner, Trafton, and Langham,
1962; Twardock and Crackel, 1969).

Statistical ecologists will, reasonably enough, ask for
general alternative recipes. I have the uncomfortable feeling
that these are both indispensible and unattainable, like policies
for scientific research in general! But what follows should at
least be worth trying.

We can assume that in part of an ecosystem, tracer enters
a biological organism, and that we are interested in the
retention curve or excretion curve only but have inadequate data.

Data on y(t) and R'(t) are generally more accurate than
those on R(t) and are more likely to fit a model. If both
excretion and retention data are available so much the better.

One should always look at a plot of log y(t) or log
{-R'(t)} against log t . If necessary R'(t) is obtained

numerically from R(t) . Very often there will be a considerable
linear portion, and sometimes there are two such straight lines;
if so, it is safe enough to assume that the processes being
observed are different before and after the change from one
straight line to the other.

Often the long term processes will interest ecologists the
most. Direct evidence for these in living organisms comes again
mostly from researches on radiation hazards. A general rule
seems to be that if a second power law is observed, it is observed
for a long time - so long that γ in equation 8 cannot be
estimated directly. This is certainly true for bone seeking
isotopes of Ca, Sr, and Ba for example and also for Radium.
It also seems to hold good for some metabolites. In a recent
study, on clearance and turnover of glycerol, it was possible
to assume that γ was zero and still obtain consistent
quantitative interpretations (Norwich, 1977; Kallai, 1977).

In all cases long term forecasts and extrapolations can differ
widely from those from sums of exponentials, which is not so if
only curve fitting over an observed range is needed.

At zero times, y(t) of course cannot be infinite. The
simplest form of y(t) that meets the obvious requirements seems
to be:

$$y(t) = y_1(t) + y_2(t) \tag{11}$$

where

$$y_i(t) = A_i x^{-w_i} \exp - \phi_i(x+1/x) \tag{12}$$

and

$$x = (t-T_i)/\mu_i, \text{ or simply } t/\mu_i .$$

In other words, there might be a shift in the origin (Wise, 1971).

In practice the factor $\exp(-\phi_i/x)$ may only approximately
describe the increasing part of the curve. On the other hand,
the greater parts of many curves have been fitted to (7) from
a fairly small t onwards; and, as for radiocalcium, (7) can also
fit the transition period.

What then should replace the compartments, or how should we
redefine them? It seems that we can still regard blood (plasma)
as a central homogeneous compartment when interpreting clearance

curves for humans or mammals. But for every other part of the
body, what matters is the distribution of *time intervals* for
individual tracer particles between entering and leaving this
part. In homogeneous compartment theory this can only be a
negative exponential distribution. Even when the tracer is
eaten, it is often assumed that the gut is a homogeneous
compartment. If there is then only one other compartment,
the clearance curve can be of the form:

$$y(t) = A \{e^{-B_1 t} - e^{-B_2 t}\} \tag{13}$$

with $B_2 > B_1$.

Corresponding to this, individual tracer particles must
be moving around randomly *within* the compartments. In fact,
if we are to interpret all these power laws, they must be moving
through each compartment, as if they undergo a random walk
with drift like a Brownian particle. Then the distribution
of time intervals is of the form of equation (12); if it is
a constant drift, with a constant rate of spreading out, then
$w=3/2$ and $\tau=0$; the distribution is that of first passage
times to a fixed boundary (sometimes called the Inverse
Gaussian).

Suppose this holds good in the first 'compartment,' and
that this is the gut. Then (12) will give the pdf of first
arrivals of individual tracer particles in the second compartment. Suppose the probability density distribution of the time
interval, between entering and leaving, is $y_2(t)$. Then the
observed excretion curve will be the convolution of $y_1(t)$
with $y_2(t)$ or

$$y(t) = \int_{\tau=0}^{t} y_2(\tau) \; y_1(t-\tau)d\tau \tag{14}$$

and then $y(t)$ is calculated, either by numerical integrations,
or by adding the cumulants of $y_1(\tau)$ and $y_2(\tau)$. Those for
the Inverse Gaussian distribution form a simple pattern.

For two such passages in series, the convolution has the
same form if ϕ_i/μ_i is the same for $i=1$ and 2, otherwise
this has a pdf approximating (12), with $w>3/2$, and again with
a fairly simple cumulant generating function.

If nothing at all is known about the form of $y(\tau)$ yet we
need to make a long range forecast by a simulation study, I

would recommend assuming y(τ) is of Inverse Gaussian form
instead of that of a negative exponential. There may be many
such passages in series: this will have the effect of smearing
out and yielding a fairly simple form for a multi-convolution
integral. Or if you *must* have compartments, have them in series,
with one way entrances and exits.

3. DISCUSSION

Theoretical forms of clearance curves have been worked
out for radiocalcium in terms of such distributions of time
intervals (Wise, Osborn, Anderson, and Tomlinson, 1968;
Wise, 1974, 1975). In this model there was one central
compartment - blood (plasma) which was regarded as homogeneous;
in it the time interval for any one ^{47}Ca particle between
entering and leaving it was very short - one minute or less,
but *all* transitions to tissue, bone, kidneys, etc. took place
in this 'compartment.' It became clear at Parma that these
are semi-Markov processes, in fact a particular case of those
set out in general terms by Marcus (1979).

There were also many discussions and several papers on what
models are or should be. A few other contrasting views are
worth quoting; at least one author in each of them is associated
with 'ISEP.' "Models based on meaningful and testable hypotheses should be formulated prior to data collection" (Goldstein
and Elwood, 1971). "Biological models are, or at least should
be, evolved by a repeated see-saw process of monitoring against
experimental evidence. Starting from a comparatively simple
model a comparison with experimental evidence will normally
point to modifications of the original model which in turn call
for new experimental evidence to provide a decision between
alternative model modifications and so on" (Matis and Hartley,
1971). "....the construction and use of models involves more
than curve-fitting and prediction of data. Hertz (1900) gives
three criteria for choice between alternative theories which
may be applied to models: (1) logical permissibility,
(2) correctness, and (3) appropriateness. The second of these
is the requirement that the model be confirmed by experience
and one must agree with Hertz that it is a sine qua non. The
third is more subtle and may lead different men to different
choices, but it is a necessary consideration" (Beck and
Rescigno, 1970). "Beck and Rescigno rightly raise the issue
of what a model is or should be. We largely agree with the
desiderata (and the questions!) in the last paragraph, but
only partly with their earlier remarks on mathematical models
and fits for biological data. We think these should be kept

clearly distinguished from any underlying theoretical model; nevertheless they can be very valuable in their own right. ....Among comparably close fits, we would argue that the one with the fewest adjustable parameters is generally the simplest and best" (Anderson, Tomlinson, Osborn, and Wise, 1970).

The first two passages give strongly contrasting views. Each may be right in its own context, but Matis and Hartley are more often right in ours. In ecology, and in physiology, the observations we would really like to make may not be obtainable, and where they are, their choice may depend above all on what is *biologically* important.

If a model yields a quantitative prediction of course it is worth trying this out in every possible way, even if this forecast seems biologically useless.

Apart from this, all the points quoted seem particularly appropriate for rethinking on compartments.

APPENDIX: DETECTING POWER LAWS IN EXCRETION AND RETENTION DATA THAT HAVE BEEN FITTED BY SUMS OF NEGATIVE EXPONENTIALS

When a clearance or retention curve $y(t)$ has been well fitted to a sum of r negative exponentials, but really fits $At^{-\alpha}$ over the same period of time, there is a quick way of detecting this numerically. Obviously we have to consider the shape of $Y(x)$, where $Y=\log y$, $x=\log t$ and $y(t)$ is as in eq. (5), and part of $Y(x)$ will be near to a straight line. This part, if it exists, is almost always picked up by calculating the points of inflexion of $Y(x)$ (where $Y''(x)=0$) and joining successive ones by straight lines; one or more of them then gives good approximations to A and α . There are $0,2...$ up to $2r-2$ points of inflexion when there are r exponentials. Even when there are no points of inflexion in the period of interest, evidence that $y(t)$ really fits (7) can often be obtained by first multiplying $y(t)$ by $\exp(\theta B_1 t)$ with increasing values of θ between 0 and 1, until the logarithm of $y(t)\exp(\theta B_1 t)$ has two more points of inflexion than before. Moreover, in sets of curves with $r=3$ or more, the existence of two power laws in one curve can sometimes be detected.

For illustrating this numerically there are results, published in the last 20 years, for more than 100 sets of

clearance curves to choose from. (Deplorably, in about as
many more sets the authors have reported only model parameters
as distinct from curve fitting parameters.) Of these, many
studies of retention of radioisotopes in mammals that have
appeared in Health Physics are relevant for ecological studies.
Richmond, Furchner, Trafton, and Langham (1962) obtained
retention curves for ^{65}Zn, in mice, rats, dogs and men. They
actually plotted $R(t)$ v t in days, with R on a log scale
but only for averaged retention. Since these are non-linear
functions of t , they are not necessarily of the same form
as the separate retention curves. However, the one for 4 dogs
showed a straight line from about 150 to 500 days. So in this
case a 'final' exponential term was observed. This was not so
in the other sets, but perhaps this was because the observations
could not be continued long enough.

Even for these dogs, and despite the possible distortion
through averaging, there were clear indications that a power
law was present in the clearance curve. After $Zn^{65}Cl_2$ had
been administered orally, its retention was fitted by (t in
days)

$$R(t) = 51.3 \exp -1.2208t + 10.06 \exp -0.0743t + 38.74 \exp -0.0054t .$$

Hence, for $-R'(t)$ the three coefficients A3, A2 and A1 are
0.6303, 0.00747 and 0.00174 respectively. (In general we
write $-R'(t) = \sum_1^r A_i \exp - B_i t$ with $B_1 < B_2 < B_3 ....$)

We now consider $\log -R'(t) = Y$ as a function of $x = \log t$.
Clearly as $x \to -\infty$, $Y \to 0.64$, and when x is large enough,
$Y \to \log A_1 - B_1 e^x$.

There are four points of inflexion at:

	1	2	3	4
$-R'(t)$	.0268	.00559	.00255	.00122
Y	-3.620	-5.186	-5.973	-6.710
t	2.865	8.727	27.046	82.414

The straight lines joining the *ith* to the i+1*th* point of
inflexion, Y_i are:

$$Y_1 = -2.1397 \quad -1.406x \; ,$$

$$Y_2 = -3.6785 \quad -0.6959x \; ,$$

$$Y_3 = -3.9740 \quad -0.6609x \; . \tag{15}$$

The fact that Y_2 and Y_3 represent almost the same line is already more than chance. In the computer print out, $Y-Y_1$, $Y-Y_2$ and $Y-Y_3$ were tabulated. Y_1 appeared to have no particular significance, but for equal intervals of $\log_{10}t$, $Y-Y_2$, and $Y-Y_3$ oscillated about zero from about 5 to 300 days. Table 1 gives values of:

$$\log_e\{-R'(t)/X_2\} \quad \text{and} \quad \log_e\{-R'(t)/X_3\}$$

where $\quad X_2 = 0.02526t^{-0.6959}$

and $\quad X_3 = 0.02251t^{-0.6609} \tag{16}$

from t = 1 to 630 days, by which time the final exponential in $-R'(t)$ has taken over. Then, of course, $Y-Y_2$ and $Y-Y_3$ rapidly become large and negative, which in no way invalidates the evidence for the negative power law in (16) over the given period. What happens before 5 days is less clear in this case; the true form of y(t) could be as in eq. (11), with the power law eq. (16) contained in $y_2(t)$.

The four points of inflexion of Y(x) are almost equally spaced. This is typical when a single power law is fitted by a sum of 3 exponentials over a fairly long period.

Clear indications were found for power laws $-R'(t) = At^{-\alpha}$ in the mean curve for the 6 rats, and for the men, separately, numbers 2, 3, and 4; the results for no. 1 were doubtful. For the mice no power law could be found, but a mean of 12 curves could easily be distorted.

TABLE 1.

$\log_{10}t$	$-R'(t)$	$Y-Y_2$	$Y-Y_3$
0	.194	2.042	2.157
0.1	.144	1.901	2.009
.	.0994	1.691	1.790
.	.0633	1.400	1.491
.	.0373	1.030	1.114
0.5	.0209	0.612	0.687
.	.0122	0.230	0.297
.	.00824	0.002	0.061
.	666	−0.052	−0.001
.	586	−0.018	+0.025
1.0	523	+0.027	+0.061
.	458	+0.055	+0.082
.	393	+0.061	+0.080
.	329	+0.045	+0.056
.	271	+0.012	+0.015
1.5	223	−0.026	−0.031
.	185	−0.053	−0.066
.	157	−0.053	−0.075
.	138	−0.022	−0.052
.	124	+0.030	−0.008
2.0	112	+0.085	+0.039
.	.000990	+0.126	+0.072
.	854	+0.139	+0.077
.	710	+0.114	+0.044
.	563	+0.042	−0.036
2.5	420	−0.091	−0.177
.	291	−0.299	−0.393
.	183	−0.602	−0.705
2.8	102	−1.026	−1.136

TABLE 2

Curve	A	α	between $\log_{10}t$	$-R'(t)$	and $\log_{10}t$	$-R'(t)$
6 rats	.2229	1.378	1.4	.00249	2.5	.000076
man 3	.0119	0.618	1.1	.00234	2.8	.000207
man 4	.0174	0.901	0.8	.00331	1.8	.000442
man 2	.0436	0.753	1.2	.00530	2.5	.000612

In general, the results from excretion or clearance curves, when these are obtained directly, are clearer than those for retention curves. The above is therefore not a particularly favourable case!

According to the referee, it was also not a good choice biologically. He points out that "the time concentration of zinc appears to be governed by a homeostatic (rate adjusting) process." This would be non-linear, whilst fits both by sums of negative exponentials, interpreted in terms of homogeneous compartments, and by power or gamma functions interpreted in terms of random walks with drift, both assume linear steady state processes. Possibly this is still valid as a first approximation, since the form of the clearance curves is not different from so many others. A similar difficulty often arises with clearance curves in pharmacokinetics. I believe we can make progress only when we get the linear models right.

REFERENCES

Anderson, J., Osborn, S. B., Tomlinson, R. W. S., and Wise, M. E. (1969). Clearance curves for radioactive tracers – sums of exponentials or powers of time? *Physics in Medicine and Biology*, 14, 498-501.

Anderson, J., Osborn, S. B., Tomlinson, R. W. S., and Wise, M. E. (1970). Calcium kinetics: The philosophy and practice of science. *Physics in Medicine and Biology*, 15, 567-568.

Beck, J. S. and Rescigno, A. (1970). Calcium kinetics: The philosophy and practice of science. *Physics in Medicine and Biology*, 15, 566-567.

Goldstein, R. A. and Elwood, J. W. (1971). Two compartment, three parameter model for the absorption and retention of ingested elements by animals. *Ecology*, 52, 935–939.

Hertz, H. (1900). *Principles of Mechanics*. Dover (1956), New York, 1–2. (As quoted by Beck and Rescigno.)

ICRP Publication 10 (1967). *Evaluation of Radiation Doses to Body Tissues from Internal Contamination Due to Occupational Exposure*. Pergamon Press, Oxford, New York.

Kallai, Mary-Ann (1977). *A new approach to the measurement of glycerol turnover and preliminary application to the measurement of triglyceride turnover*. Ph.D. thesis, University of Toronto.

Marcus, A. H. (1979). Semi-Markov compartmental models in ecology and environmental health. In *Statistical Distributions in Ecological Work*, J. K. Ord, G. P. Patil, and C. Taillie, eds. Satellite Program in Statistical Ecology, International Co-operative Publishing House, Fairland, Maryland.

Matis, J. H. and Hartley, H. O. (1971). In discussion on their paper: Stochastic compartmental analysis: Model and least squares estimation from time series data. *Biometrics*, 27, 77–101.

Norwich, K. H. (1977). *Molecular Dynamics in Biosystems. The Kinetics of Tracers in Intact Organisms*. Pergamon Press, Oxford, New York.

Richmond, C. R., Furchner, J. E., Trafton, G. A., and Langham, W. H. (1962). Comparative metabolism of radionuclides in mammals. I. Uptake and retention of orally administered Zn^{65} by four mammalian species. *Health Physics*, 8, 481–489.

Schultz, V., Eberhardt, L. L., Thomas, J. M., and Cochran, M. I. (1974). Compartment models. In *A Bibliography of Quantitative Ecology*. Dowden, Hutchinson and Ross, Stroudsburg, PA. 12–17.

Twardock, A. R. and Crackel, W. C. (1969). Cesium 137 retention by cattle, sheep and swine. *Health Physics*, 16, 315–323.

Williamson, P. (1975). Use of ^{65}Zn to determine the field metabolism of the snail *cepaea nemoralis*. *Ecology*, 56, 1185–1192.

Wise, M. E. (1971). Skew probability curves with negative
    powers of time and related to random walks in series.
    Statistica Neerlandica, 25, 159-180.

Wise, M. E. (1974). Interpreting both short and long term
    power laws in physiological clearance curves. *Mathematical
    Biosciences*, 20, 327-337.

Wise, M. E. (1975). Skew distributions in biomedicine
    including some with negative powers of time. In *Statistical
    Distributions in Scientific Work, Vol. 2: Model Building
    and Model Selection*, G. P. Patil, S. Kotz, and J. K. Ord,
    eds. Reidel, Dordrecht, Holland, 241-262.

Wise, M. E. (1977). The form and interpretation of clearance
    curves for injected radioisotopes based on negative power
    laws, especially for [47]Ca and estimation bone accretion
    rate. *Current Topics in Radiation Research Quarterly*,
    12, 63-82.

Wise, M. E. (1976). Interpreting clearance curves in kinetic
    studies and the need for a fresh approach. In *Radio-
    pharmaceutical Dosimetry Symposium, Oak Ridge, Tenn.*
    R. J. Cloutier, J. L. Coffey, W. S. Snyder, and E. E. Watson,
    eds. U. S. Department of Health, Education and Welfare,
    Rockville, Maryland, 41-65.

Wise, M. E. (1979). Fitting and modelling excretion data in
    radiobioassay. Section 6.8, in *Handbook of Measurement
    and Radiation Protection, Section A, Vol. II: Biological and
    Mathematical*, Allen Brodsky, ed. CRC Press, Boca Raton,
    Florida.

See also: Ecological data and sources of information by
Gertz, S. M., Section 5.2, and other items in this volume.

The general recipes (Wise, 1976, 1977, 1979) should be
applicable to very many systems relating to absorption,
retention, and excretion in biological systems. The 1979
reference includes numerical information on about 100 sets
of excretion and clearance curves in the literature.

[*Received June* 1978. *Revised December* 1978]

MATHEMATICAL ANALYSIS OF COMPARTMENTAL STRUCTURES

INTRODUCTION TO SECTION IV

This section, dealing with the mathematical analysis of compartmental systems, is more abstract than previous sections. The papers contain several mathematical advances which clarify basic concepts and hold promise for future practical applications in ecosystem compartmental analysis. Indeed, practical applicability of the Walter paper is clear as results derived therein were used in his paper in Section I. Walter develops the relationship between differential equations and Markov chains, and exploits this to deduce properties of equilibrium solutions of the differential equations. The Gerald and Matis paper extends the usual compartment model formulation to include births and a mixed recipient and donor controlled transfer mechanism. The paper also illustrates the fitting of several moments simultaneously to observed data, thereby enhancing the identifiability of parameters and the precision of their estimators. The Rescigno paper extends previous research of this author on operational calculus from the one variable to the two variable case. This problem is of particular interest in compartmental systems involving both time and age varying coefficients, often due to extrinsic and intrinsic forces, respectively, which have not yet been amenable to applied analysis. This papers should serve to illustrate the production of pure theory under the stimulus of applied problems in the synergistic theory, practice, theory, practice,···, cycles characteristic of scientific development.

J. H. Matis, B. C. Patten, and G. C. White, (eds.),
Compartmental Analysis of Ecosystem Models, pp. 295-310. All rights reserved.
Copyright ©1979 by International Co-operative Publishing House, Fairland, Maryland.

COMPARTMENTAL MODELS, DIGRAPHS, AND MARKOV CHAINS[*]

G. G. WALTER

Department of Mathematics
The University of Wisconsin
Milwaukee, Wisconsin 53201 USA

SUMMARY. The relationship between the structure and the differential equation of compartmental models of ecosystems is explored. The structure is given by a directed graph whose properties are used, together with Markov chains, to deduce properties of the equilibrium solutions to the differential equation. Two simple examples corresponding respectively to energy and nutrient flows are shown, because of their different structure, to have different asymptotic behavior. In general, it is shown that strong connectivity is necessary and sufficient for existence of a feasible equilibrium solution of a closed system. A non strongly connected system may be decomposed into strong components whose inputs and outputs determine whether a feasible equilibrium solution exists.

KEY WORDS. compartmental model, ecosystem, Markov chain, graph.

1. INTRODUCTION

Compartmental models, by which the flow of energy or nutrients through an ecosystem can be studied, have a dual aspect: one structural and the other dynamic. Both are useful in the analysis of the behavior of the ecosystem. The structural aspect leads to a directed graph in which compartments are vertices and flows are arcs. It is independent of the functional form of the flows. The dynamic aspect on the other hand leads to a system of differential equations:

---

[*] Center for Great Lakes Studies contribution no. 182.

$$\frac{dX}{dt} = F(X) \quad . \tag{1.1}$$

Its study requires knowledge both of the functional form of the flows and their magnitude. In the case of linear donor controlled flows the equation has the form

$$\frac{dX}{dt} = AX \quad , \tag{1.2}$$

where $X^T = [x_1, \cdots, x_n]$ is the vector of levels in each component and $A = [a_{ij}]$ is matrix of flow rates a_{ij} from compartment j to compartment i when $i \neq j$, and

$$a_{ii} = -\sum_{\substack{j=1 \\ j \neq i}}^{n} a_{ij} - a_{oi} \quad .$$

Here a_{oi} is the flow from compartment i to the outside of the system.

The structural aspects seem related to the intuitive concepts of complexity and trophic level. The dynamic aspects seem more closely related to stability and diversity. Yet that the two aspects are intimately related was shown by Cobelli and Romanin-Jacur (1975) who showed that strong connectivity (a structural concept) implied controllability (a dynamic concept).

Both aspects are related as well to certain properties of associated *Markov chains*. The differential equation (1.2) may be converted into a Markov chain provided the outputs are all 0. This is done by first approximating dX/dt by the difference quotient $[X(t+h) - X(t)]/h$ where the time step h is taken sufficiently small to make the diagonal element of hA less than 1 in magnitude. Then equation (1.2) is approximately

$$X^T(t+h) = (X(t) + hAX(t))^T = X^T(t)(I + hA)^T \quad . \tag{1.3}$$

Since the choice of h forces all the diagonal elements and hence all elements of $(I + hA)^T = P$ to be positive, and since the columns of A add up to 0, P is the transition matrix of a Markov chain.

Such a chain has associated with it a directed graph obtained by letting the vertices correspond to the states and the arcs to the positive transition probabilities.

The associated digraph (i.e., directed graph) has, except for loops, the same structure as the original compartmental model. In cases when the outputs are not all zero, another compartment consisting of the outside universe could be added and then the Markov chain constructed.

While compartmental models in their differential equation form have become a standard tool of ecologists, Markov chains and graph theory have not. The concepts from the latter two subjects that are needed in this work may be found in Roberts (1976), or in more concise form in Kemeny and Snell (1962).

The same ecosystem, of course, can give rise to different compartmental models depending on whether the flow of energy or which of several nutrients is being traced. Typically energy flow can be modelled by an open hierarchical system while nutrient flows often correspond to closed systems. In this work we shall first examine the behavior of two simple compartmental models. They will correspond to the same ecosystem but will model energy and nutrient flows respectively and hence have a different structure. We shall generalize the latter model to more general systems, and show that the strong connectivity is necessary and sufficient for the existence of a "real" equilibrium solution to the equation. We shall also consider food webs, and finally the decomposition of a general compartmental model into strong components. We shall show in this case that the location of inputs and outputs determines whether or not the equilibrium solution will be positive.

2. TWO SIMPLE EXAMPLES

Consider the following simple system consisting of producers (x_1) , herbivores (x_2) , decomposers (x_3) , and abiotic elements (x_4) . The energy flow begins with an input f_{10} to the primary producers. There is a flow from the producers to the herbivors and from the herbivores to the decomposers. The losses from each compartment due primarily to metabolism and respiration are represented by flows to the outside. This model of energy flow is show in Figure 1.

Nutrient flow on the other hand could be given by the model shown in Figure 2. Some of the nutrient in the herbivore compartment is recycled directly to the abiotic portion, whereas some is cycled through the decomposers. There is also a direct interchange between the abiotic elements and the primary producers.

These, except for the absence of carnivores, are similar to the systems discussed by M. Usher (1973), and represent perhaps,

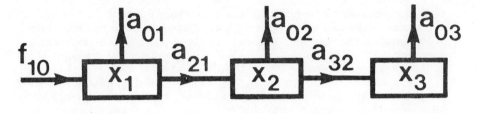

FIG. 1: *Energy flow in a simple compartmental model of an eco-system consisting of producers* (x_1), *herbivores* (x_2), *decomposers* (x_3), *and abiotic environment* (x_4).

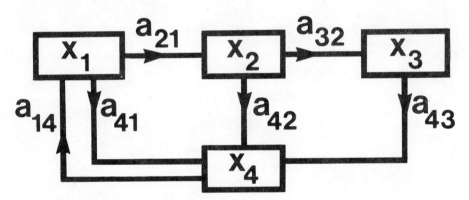

FIG. 2: *Typical nutrient flow in the ecosystem model with the same compartments as in Figure 1.*

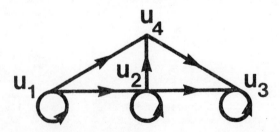

FIG. 3: *The directed graph corresponding to the Markov chain model of the energy flow of Figure 1.*

the simplest compartmental models which can properly be termed
ecosystem models. Yet even in these simple cases, the distinction between the energy flow and nutrient flow model becomes
apparent.

The associated digraph in the first case is not connected,
whereas in the second case it is strongly connected (i.e., there
is a closed path which includes all vertices). The associated
differential equations are:

$$\frac{dX}{dt} = \begin{bmatrix} -a_{01}-a_{21} & 0 & 0 & 0 \\ a_{21} & -a_{02}-a_{32} & 0 & 0 \\ 0 & a_{32} & -a_{03} & 0 \\ 0 & 0 & 0 & 0 \end{bmatrix} X + \begin{bmatrix} f_{10} \\ 0 \\ 0 \\ 0 \end{bmatrix}$$

and

$$\frac{dX}{dt} = \begin{bmatrix} -a_{21}-a_{41} & 0 & 0 & a_{14} \\ a_{21} & -a_{42}-a_{32} & 0 & 0 \\ 0 & a_{32} & -a_{43} & 0 \\ a_{41} & a_{42} & a_{43} & -a_{14} \end{bmatrix} X . \quad (2.2)$$

In the first case the eigenvalues are all negative or zero,
while in the second they all have negative real parts or are zero.

In the first case one must ignore the input and must add
another compartment for the outputs in order to construct the
corresponding Markov chain. However, since it has no inputs or
outputs, compartment 4 may be ignored. The transition matrix is
then given by

$$P = \begin{bmatrix} 1-h(a_{21}+a_{01}) & ha_{21} & 0 & ha_{01} \\ 0 & 1-h(a_{02}+a_{32}) & ha_{32} & ha_{02} \\ 0 & 0 & 1-ha_{03} & ha_{03} \\ 0 & 0 & 0 & 1 \end{bmatrix} \quad (2.3)$$

which is the transition matrix of an absorbing Markov chain. The
last row corresponds to an absorbing state and the others to transient states. That is, there is a positive probability that a
unit of energy in any of the states except the last will leave it.
Ultimately all energy will be absorbed by the last state which
corresponds to the outside of the system. This, of course, is
not much of a surprise and is already obvious from the digraph
of Figure 3, where u_1, u_2, u_3, correspond to the first three
compartments and u_4 to the outside. The transition matrix
(2.3) may be put in the form

$$P = \begin{bmatrix} Q & R \\ 0 & I \end{bmatrix}.$$

(2.5)

From the theory of absorbing chains (see Roberts, 1976,
p. 282) the submatrix Q has the property that $Q^n \to 0$ and
$(I-Q)$ is invertible. This is also clear in this particular case.
In fact the inverse is given by

$$(I-Q)^{-1} = \frac{1}{h} \begin{bmatrix} 1/b_1 & a_{21}/(b_1 b_2) & a_{21}a_{32}/(b_1 b_2 a_{03}) \\ 0 & 1/b_2 & a_{32}/b_2 a_{03} \\ 0 & 0 & 1/a_{03} \end{bmatrix}$$

(2.6)

where $b_1 = a_{01} + a_{21}$ and $b_2 = a_{02} + a_{32}$. Again from the
theory of absorbing Markov chains, the elements in the first row
are the expected time in each compartment before absorption (in
units of h steps). In the original time units it is the same
with the quotient h omitted. The total expected time to absorption is

$$E = \frac{1}{b_1} + \frac{a_{21}}{b_1 b_2} + \frac{a_{21}a_{32}}{b_1 b_2 a_{03}} \ .$$

(2.7)

The differential equation (2.1) may be easily solved since
the eigenvalues λ are $-b_1$, $-b_2$, $-a_{03}$, and $\cdot 0$. It is

$$X(t) = Ke^{\lambda t}K^{-1}X(0) + \int_0^t Ke^{\lambda(t-s)}K^{-1}Fds \ ,$$

(2.8)

where $F = [f_{10},0,0,0]^T$, K is the matrix eigenvectors and $\exp(\lambda t)$ is the matrix with $\exp(\lambda t)$'s on the diagonal and 0's elsewhere. In this case we may omit the fourth compartment again and obtain for the first three compartments the asymptotic levels:

$$\lim_{t\to\infty} X_1(t) = 0 + \lim_{t\to\infty} K_{11} \begin{bmatrix} \dfrac{e^{\lambda_1 t}-1}{\lambda_1} & 0 & 0 \\[2ex] 0 & \dfrac{e^{\lambda_2 t}-1}{\lambda_2} & 0 \\[2ex] 0 & 0 & \dfrac{e^{\lambda_3 t}-1}{\lambda_3} \end{bmatrix} K_{11}^{-1} F_1$$

$$= 0 - K_{11} \Lambda_{11}^{-1} K_{11}^{-1} F_1 = -A_{11}^{-1} F_1 , \qquad (2.9)$$

where K_{11}, Λ_{11}, and A_{11} are partitions (3×3) of the matrices K, Λ, and A. Since $F_1^T = [f_{10},0,0]$, the value in (2.9) is $\bar{f}_{10}$ times the first column of $-A_{11}^{-1}$, and since $-A_{11} = (I-Q)/h$, this first column is the same as the first row of 2.6). Thus the energy distribution with a constant rate of input is the same as the expected time in each of the compartments before absorption.

The behavior of the Markov chain and differential equation for the nutrient flow is completely different. Since the digraph is strongly connected the Markov chain is ergodic and if h is sufficiently small the diagonal of the transition matrix has no zero elements. Such chains are regular as well (see Roberts, 1976, p. 305), and hence P^n converges as $n \to \infty$ to a matrix W all of whose rows are the same. Each row is an eigenvector of P corresponding to the eigenvalue 1.

This follows from the fact that

$$WP = (\lim P^n)P = \lim P^{n+1} = P \lim P^n = PW \qquad (2.10)$$

and hence if w is a row of W , wP = w. Moreover each component of w is positive. This can easily be seen in this case since

$$0 = w(I-P) = w(I - hA^T - I) = -hwA^T \qquad (2.11)$$

or w^T is an eigenvector of A belonging to 0:

$$Aw^T = \begin{bmatrix} -a_{21}-a_{41} & 0 & 0 & a_{14} \\ a_{21} & -a_{42}-a_{32} & 0 & 0 \\ 0 & a_{32} & -a_{43} & 0 \\ a_{41} & a_{42} & a_{43} & -a_{14} \end{bmatrix} \begin{bmatrix} w_1 \\ w_2 \\ w_3 \\ w_4 \end{bmatrix} = 0. \quad (2.12)$$

From the first equation it is seen that w_1 and w_4 have the same sign, from the second that w_1 and w_2 have the same sign, and from the third that w_2 and w_3 do. Hence all components have the same sign and if one is zero all are zero.

This same vector w arises when the differential equation is considered. Indeed the matrix A has exactly one zero eigenvalue and three with negative real parts. Hence the general solution is of the form

$$X(t) = Ke^{\Lambda t} K^{-1}X(0)$$

$$= c_1 e^{\lambda_1 t} K_1 + c_2 e^{\lambda_2 t} K_2 + c_3 e^{\lambda_3 t} K_3 + c_4 K_4. \quad (2.13)$$

As $t \to \infty$, $X(t)$ approaches $c_4 K_4$, which is also an equilibrium solution to equation (2.2), since K_4 is an eigenvector of A belonging to the eigenvalue 0. Since w^T is also, and since the multiplicity of 0 is 1, $c_4 K_4$ is just some multiple of w^T , say a. To evaluate a we sum all the components of both sides of (2.13).

$$x_1(t) + x_2(t) + x_3(t) + x_4(t) = [1,1,1,1]X(t)$$

$$= [1,1,1,1] \sum_{i=1}^{3} c_i e^{\lambda_i t} K_i + a[1,1,1,1]w^T. \quad (2.14)$$

However, each of the columns of A adds up to 0; hence since $\lambda_i K_i = AK_i$, the sum of all the components of K_i is zero. Since w^T is a probability vector its components all add up to one, and we have the result

$$a = \sum_{i=1}^{4} x_i(t) \quad .$$

(2.15)

This is not altogether too surprising, since the system is closed and always contains the same total amount of nutrient a . What is surprising is that this nutrient will be redistributed in such a way that each compartment will ultimately end up with a positive share.

In the next section we show that some of the results obtained here are perfectly general and do not depend on the particular compartmental model.

3. STRONGLY CONNECTED COMPARTMENTAL MODELS

In this section we generalize the results of the example of nutrient flow. We shall suppose that the compartmental model is closed and donor controlled, and shall investigate the relation between strong connectivity and equilibrium solutions to the differential equations.

Since A is singular in the case of closed systems, the equation $AX = 0$ has a non-zero solution. However, this solution is not necessarily feasible, in that it may have some components which are zero or have opposite signs. If the rank of A is $n-1$ these equilibrium solutions are also stable, since all the non-zero eigenvalues have negative real parts (see Hearon, 1963). *Theorem 1. Let* D *be the digraph and* $dX/dt = AX$ *the differential equation associated with a closed donor controlled compartmental model. If* D *is strongly connected then any solution* X(t) *approaches an equilibrium solution of* $dX/dt = AX$ *which is stable, feasible, and unique (except for magnitude). If* D *is not strongly connected there is no unique, feasible equilibrium solution.*

The proof again makes use of a Markov chain with transition matrix $P = (I + hA)^T$ and emulates the arguments in the last section. Since D is strongly connected, the chain must be ergodic. But P has non-zero elements on the diagonal. Hence the chain is regular as well, and the iterates of P^m approach a stochastic matrix W all of whose rows are the same and all of whose elements are positive. (See Roberts, 1976, p. 292.)

Now suppose X is an equilibrium solution. Then $(I + hA)^m X = X$ and hence

$$P = \begin{bmatrix} Q & R \\ 0 & 1 \end{bmatrix} \qquad (4.2)$$

where $Q = (I + hA_{11})^T$ is an $(n-1) \times (n-1)$ submatrix. From the theory of such chains the expected time to absorption is given by the sum of those rows of $(I-Q)^{-1} = N$ corresponding to inputs. This time is units of h by which N must be multiplied to obtain the time in the original units.

The matrix hN may be found directly by calculating the inverst of $I - Q = -hA_{11}^T$. In the case of a simple food chain, it is given by

$$hN = h(I-Q)^{-1} = -A_{11}^{-1T} = \begin{bmatrix} \dfrac{1}{a_1} & \dfrac{1}{a_2} & \cdots & \dfrac{1}{a_{n-1}} \\ 0 & \dfrac{1}{a_2} & \cdots & \dfrac{1}{a_{n-1}} \\ \cdots & \cdots & \cdots & \cdots \\ 0 & 0 & \cdots & \dfrac{1}{a_{n-1}} \end{bmatrix}$$

where a_i is the flow rate from compartment i to compartment $i + 1$. The expected time to absorption is thus

$$E = \sum_{i=1}^{n-1} a_i^{-1} .$$

This, of course, is exactly what one would suspect.

5.   THE GENERAL CASE

In general, whether energy or nutrients are being traced, the compartmental model is neither a food web nor is it strongly connected. Of course, as with any digraph, it can be decomposed into strongly connected components. The 'condensation' or digraph consisting of the such components is acyclic and hence has associated with it a measure of trophic level. We assume that the components $K_1, K_2, \cdots, K_m$ are ordered according to trophic level. The differential equation governing the system has the form

$$\frac{dX}{dt} = AX + F - DX \qquad (5.1)$$

where

$$A = \begin{bmatrix} A_{11} & 0 & 0 & 0 \\ A_{21} & A_{22} & 0 & 0 \\ \vdots & \vdots & \vdots & \vdots \\ A_{m1} & A_{m2} & A_{m3} & A_{mm} \end{bmatrix} ,$$

F is the vector of inputs, and DX of outputs. Here the diagonal submatrix, except for its diagonal elements, consists of flow rates within the ith compartment, while A_{ji}, $i < j$, consists of those between compartments.

The term DX may be omitted from (5.1) if an additional compartment corresponding to the 'outside' is adjoined to the model as was done in the last section. However for most of the analysis in this section we retain the form of (5.1).

Each component K_i of our model is strongly connected and may have flows in and out of it. It is governed by a differential equation of the form (5.1) again except that the matrix is of the form

$$A_i = \begin{bmatrix} -a_{11} & a_{12} & \cdots & a_{1k} \\ a_{21} & -a_{22} & \cdots & a_{2k} \\ \vdots & \vdots & & \vdots \\ a_{k1} & a_{k2} & & -a_{kk} \end{bmatrix} .$$

This differs from the matrix A_{ii} in that the flows out of the component are not included in the diagonal elements, but rather appear in the matrix D_i .

In Section 3 we saw that in strongly connected compartmental models with no flows in or out, the nutrients distributed them-

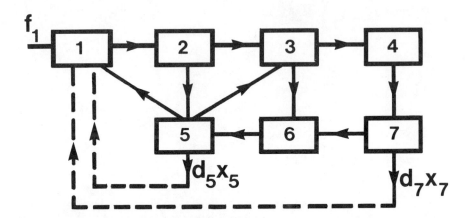

*FIG. 4: A strongly connected component of a compartmental model.
The first compartment has an input and compartments 5 and 7 have
outputs. The dotted lines refer to the rerouting of the output
to the input compartments.*

selves in such a way that each compartment got a positive share.
It's not obvious that the same would be true if there were inputs
and outputs. However we shall show that it is.

Accordingly let us assume that our component has an input
flow f_1 to the first compartment, and has output rates d_1,
$d_2, \cdots, d_k$ from the corresponding compartments. An example is
given by Figure 4. We could as easily assume the input is to any
other compartment. Any result we get can be extended to the case
of general input by superposition.

Let us modify our model by routing our outputs back to the
first compartment and for the moment ignoring the input. Then
the matrix of the resulting system is

$$
A'_i =
\begin{bmatrix}
-a_{11} & a_{12} + d_2 & \cdots & a_{1k} + d_k \\
a_{21} & -a_{22} - d_2 & \cdots & a_{2k} \\
\cdots & \cdots & \cdots & \cdots \\
a_{k1} & a_{k2} & \cdots & -a_{kk} - d_k
\end{bmatrix}
\tag{5.3}
$$

and the digraph is strongly connected with no inputs or outputs.

By theorem in Section 3 there is an equilibrium solution X_i' with all positive components such that

$$A_i' X_i' = 0 . \tag{5.4}$$

Any multiple of X_i' is again a solution. Returning to our original equation we see that an equilibrium solution must satisfy the equation

$$
\begin{bmatrix}
-a_{11}-d_1 & a_{12} & \cdots & a_{1k} \\
a_{21} & -a_{12}-d_2 & \cdots & a_{2k} \\
\cdots & \cdots & \cdots & \cdots \\
a_{k1} & a_{k2} & & -a_{kk}-d_k
\end{bmatrix}
\begin{bmatrix} x_1 \\ x_2 \\ \vdots \\ x_k \end{bmatrix}
= -
\begin{bmatrix} f_1 \\ 0 \\ \vdots \\ 0 \end{bmatrix}
\tag{5.5}
$$

The components of the vector X_i' satisfy the second through kth equations in (5.5). Moreover a multiple $c_1 X_i'$ of X_i' satisfies the first equation since

$$(-a_{11}-d_1)c_1 x_1' + a_{12}c_1 x_1' + \cdots + a_{1k}c_1 x_1'$$

$$= -a_{11}c_1 x_1' + (a_{12}+d_2)c_1 x_2' + \cdots (a_{1k}+d_k)c_1 x_k'$$

$$- c_1[d_1 x_1' + d_2 x_2' + \cdots + d_k x_k']$$

$$= 0 - c_1[d_1 x_1' + d_2 x_2' + \cdots + d_k x_k'] = -f_1 . \tag{5.6}$$

This is satisfied for

$$c_1 = (d_1 x_1' + d_2 x_2' + \cdots + d_k x_k')^{-1} f_1 . \tag{5.7}$$

Since all the components of X_i' are positive and all the d_j's are non negative, the denominator will be positive if there is any nonzero flow from any compartment in K_i . If there is no flow from K_i then intuitively we would expect a continuing build-up with no finite equilibrium.

The same procedure may be repeated with the flow going into compartment 2. Now the outputs are routed back to compartment 2 instead of 1, the resulting equilibrium solution X_i'' has a multiple $c_2 X_i''$ which satisfies an equation similar to (5.5) with f_2 instead of f_1 . The constant is given by

$$c_2 = (\sum_{p=1}^{k} d_p x_p'')^{-1} f_2 . \qquad (5.8)$$

We repeat this procedure k times to find the equilibrium solution for a general input. It is

$$X_i = \sum_{j=1}^{k} c_j X_i^{(j)} = \sum_{j=1}^{k} (\sum_{p=1}^{k} d_p x_p^{(j)})^{-1} f_j X_i^{(j)} .$$

Hence we have, for every strongly connected component with a constant input to any of its compartments and with an output to any other component, a unique stable positive equilibrium solution to its differential equation. Components with zero inputs have only the zero equilibrium solution unless there is no output as well in which case the analysis of Section 3 applies. Components with inputs and no outputs have no equilibrium solution. Restated in a slightly different form is the

Theorem 2. Let D *be the digraph of a compartmental model,* D* *its condensation consisting of the strongly connected components of* D; *let* B *be the vertex basis and* C *the vertex contrabasis of* D*. *If each component in* B *has an input and each component in* C *an output, then the equation* (5.1) *has a unique stable positive equilibrium solution.*

REFERENCES

Cobelli, C. and Romanin-Jacur, G. (1975). Structural identifiability of strongly connected biological compartmental systems. *Medical and Biological Engineering,* 831–837.

Hearon, J. Z. (1963). Theorems on linear systems. *Annals, New York Academy of Sciences,* 108, 36–68.

Kemeny, J. G. and Snell, J. L. (1962). *Mathematical Models in the Social Sciences.* MIT Press, Cambridge.

Roberts, F. S. (1976). *Discrete Mathematical Models.* Prentice Hall, Englewood Cliffs, New Jersey.

Usher, M. B. (1973). *Biological Management and Conservation.* Chapman and Hall, London.

[*Received June* 1978. *Revised December* 1978]

J. H. Matis, B. C. Patten, and G. C. White, (eds.),
Compartmental Analysis of Ecosystem Models, pp. 311-334. All rights reserved.
Copyright ©1979 by International Co-operative Publishing House, Fairland, Maryland.

ON THE CUMULANTS OF SOME STOCHASTIC COMPARTMENTAL MODELS APPLIED
TO ECOLOGICAL SYSTEMS

K. B. GERALD[*] AND J. H. MATIS

Institute of Statistics
Texas A & M University
College Station, Texas 77843 USA

SUMMARY. This research develops the first and second cumulant
functions for a broad stochastic compartmental model. In addition
to the usual stochastic formulation which includes immigrations,
deaths, and a donor-controlled transfer mechanism, the present
model structure also incorporates births and a recipient-controlled
transfer mechanism. The present paper derives the differential
equations for the first two cumulants of a n-compartment irreversible system, and illustrates their solution for a two-compartment
model. These first two cumulants provide a basis for statistical
inference concerning the model and the feasibility of several
approaches to parameter estimation is illustrated with simulated
data. In particular it is shown that several parameters, which
are not identifiable from the mean value function alone, are
identifiable when the mean and variance functions are fitted
simultaneously. It is also shown how the present work may be
generalized in many ways.

KEY WORDS. stochastic compartmental model, donor-controlled model,
recipient-controlled model.

1. INTRODUCTION

Compartmental analysis has had a significant impact on the
biological modeling of the past few decades. Although originally
developed for drug and tracer kinetics, the compartmental approach

---
[*]Present address: Rockwell International, P. O. Box 464, Golden,
Colorado 80401 USA.

has more recently been applied to ecosystem modeling (see, e.g., review in O'Neill, 1979). Schultz, *et al.* (1976) contains a bibliography of compartmental models in ecology and Jacquez (1972) and Atkins (1969) survey the applications to other fields.

Most of the development in classical compartmental analysis has been concerned with a deterministic formulation only and has ignored the stochastic formulation. Gold (1977) defines a system in which the output is uncertain to be a stochastic (as opposed to deterministic) system. While many systems have been adequately modeled deterministically, in other cases the deterministic model has come under criticism. Cornfield, *et al.* (1960) state that no real biological system could strictly satisfy all of the assumptions incorporated in the formulation of the deterministic model and Kodell (1974) has found instances where deterministic considerations have given unrealistic or disappointing interval estimates of the parameters. Explaining the desirability of stochastic models, Gold (1977) notes

Every real system must be considered to be subject to uncertainties of one type or another, all of which are ignored in the formulation of a deterministic model. As a result, deterministic models generally present fewer mathematical difficulties, but can only be considered to describe system behavior in some average sense. Stochastic models are required whenever it is necessary to explicitly account for the randomness of underlying events.

One should not conclude from the preceding criticisms that the stochastic model should always be used in place of the deterministic model. Indeed, Gold (1977) and Poole (1979) give circumstances where the deterministic formulation would be preferable. However, the foregoing discussion does point out the need for further research in the stochastic formulation of compartmental analysis, which is the focus of this paper.

In the classical linear compartmental model, the rate at which material leaves a compartment is proportional to the amount of substance in that compartment. In an ecosystem context, this model is referred to as a donor controlled model. By way of contrast, a recipient controlled model refers to a system in which the rate at which material leaves a compartment is proportional to the amount of material in the receiving compartment. Work on the recipient controlled model has not received much attention in the literature, yet it is very useful in the modeling of certain ecosystems.

Section 2 of this paper illustrates the use of a recipient controlled model. The system of differential equations for the model will be derived for certain n-compartment models in Section 3. Section 4 illustrates the solution for the special n=2 compartment case; however, the more general solution is lengthy and is given in Gerald (1978). The estimation aspects of the models are discussed in Section 5 and some general conclusions are contained in Section 6.

2. AN EXAMPLE OF A RECIPIENT CONTROLLED COMPARTMENTAL SYSTEM: THE PLANT-HERBIVORE SUBSYSTEM

While there is a prolific amount of applications and theory on the linear donor controlled model, there is virtually no work in the literature on the applications of the theory of recipient controlled models. However, many ecological situations can be postulated that would require a recipient controlled transfer. W. E. Grant of Texas A&M University has suggested the generalized model of the plant-herbivore subsystem illustrated in Figure 1 as a possible example. In this example it is desired to determine the rate of exchange of energy from the sun to the vegetation (by photosynthesis) and from the plants to the herbivores. Since the sun can be considered to be an infinite energy source, the rate of energy transferred to the plants will not be proportional to the amount of energy in the sun, solely, but will also be a function of the amount of vegetation present to carry out photosynthesis. Thus it would be incorrect to model the first transfer, k_{21}, as a donor controlled transfer only. Additionally,

if the amount of plants available to the herbivores is very plentiful (i.e., no competition for vegetation), then the rate of energy transfer from the plants to the herbivores would be a recipient controlled transfer (at least until vegetation were reduced to a scarcity in which competition among herbivores for plants was present).

3. THE COMBINED DONOR AND RECIPIENT CONTROLLED, GENERAL IRREVERSIBLE MODEL

3.1 A Rationale for the Combined Model. The example of the generalized model of the plant-herbivore subsystem suggests that a recipient controlled transfer may occur when there is either an infinite source in the donor compartment or a sufficiently greater number of units in the donor compartment over the recipient compartment so that the transfer rate would not depend on the number of units in the donor compartment. This suggests that an exclusive recipient controlled transfer mechanism is likely to occur in real-life applications only when a donor compartment exceeds

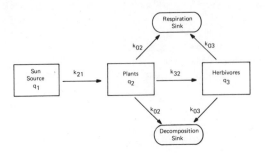

FIG. 1: *Generalized model of the plant-herbivore subsystem.*

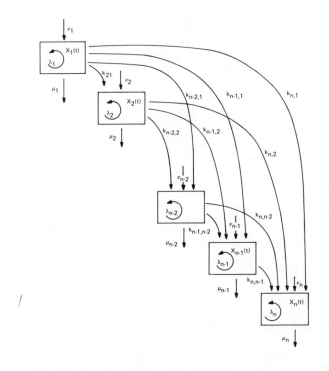

FIG. 2: *The general n-compartment irreversible system with birth, death, and immigration in each compartment.*

some 'threshold' value. In many other cases, however, it is likely to occur in combination with donor controlled transfer.

In this paper, a simple, combined donor and recipient control mechanism will be incorporated into a model framework wherein ecological resources are considered particulate. The notion of constructing stochastic models based on resource particles, or 'units,' has been recently suggested by Barber (see, e.g., Barber, 1978a, b; and Barber, Patten, and Finn, 1979) and has long been used successfully in practice to model BOD and DO in streams (see, e.g., Thayer and Krutchkoff, 1967; also Padgett, 1979, for a review). For present simplicity, the usual assumption of the homogeneity of all units is also made; however the present work may be generalized along the lines of Matis and Wehrly (1979) if this assumption proves untenable.

3.2 Development of the Combined Model. Given such homogeneous units, consider now the general n-compartment irreversible system with birth, death, and immigration in each compartment. The schematic of this compartmental model is illustrated in Figure 2. The notation used refers to the rates of possible transitions. Transitions may occur in any one of four ways; a unit may immigrate to compartment i from outside the system with transition rate ν_i, a unit in compartment i may produce (or give 'birth' to) another unit with transition rate λ_i, a unit in compartment i may leave the system ('die') with transition rate μ_i, or a unit in compartment i may migrate to compartment j with transition rate k_{ji}.

Let $X_i(t)$ (i=1,$\cdots$,n) be random variables denoting the number of particles or units in the *ith* compartment at time t. Now the probability of any of the four possible transitions occurring are assumed as follows:

Pr{a single unit immigrates to compartment i from outside the system in the interval (t, t + Δt)}
= $\nu_i \Delta t + o(\Delta t)$ (i=1,$\cdots$,n),

Pr{a single birth in compartment i occurs in the interval (t, t + Δt) given $X_i(t)$}
= $X_i(t)\lambda_i \Delta t + o(\Delta t)$ (i=1,$\cdots$,n),

Pr{a single death in compartment i occurs in the interval (t, t + Δt) given $X_i(t)$}
= $X_i(t)\mu_i \Delta t + o(\Delta t)$ (i=1,$\cdots$,n),

Pr{a single unit moves from compartment i to compartment j
in the interval (t, t + Δt) given $X_i(t)$ and $X_j(t)$}

$= k_{ji}[p_{j,i,1}X_i(t) + p_{j,i,2}X_j(t)]\Delta t + o(\Delta t)$ $(1 \leq i < j \leq n)$,

where the probability of two or more transitions occurring in the
interval (t, t + Δt) is o(Δt). These assumptions differ from
other work in the last statement. Note that the p's can be used
in two possible ways; either (i) as zero-one variables indicating
whether a recipient controlled transfer or a donor controlled
transfer should be in effect (e.g., whether the threshold value
has been exceeded or not) or (ii) to create a linear combination
of donor and recipient controlled transfers.

It is assumed that all of the units in the system act independently of each other and that immigration of units from outside
the system does not alter the behavior of the system. Also, the
initial conditions are assumed known for the system.

It is assumed that all transition rate coefficients (λ_i, μ_i,
k_{ij}: i, j=1,$\cdots$,n) are constant and that all transitions
involved occur at random. There are $(n^2+5n)/2$ transitions with
probability of first order magnitude of Δt and other (multiple)
transitions can be ignored since they have a probability of o(Δt).

The first, second, and first cross cumulants (i.e., the mean,
variance, and covariance) for each compartment will be determined
for the combined donor and recipient controlled, general irreversible model. These cumulants can then be used to predict the
behavior of the system and to obtain estimates of the parameters.
The procedure for finding the cumulants is described in Bailey
(1964). The technique consists of first finding a partial differential equation for a generating function of the distribution of
the number of units, or individuals, in each compartment of the
system. There are various methods for finding this desired partial
differential equation. One method is to construct the Kolmogorov
forward equation associated with all possible single event probabilities. This method derives the partial differential equation
from the definition of a partial derivative. While this approach
is intuitively appealing, the simplest technique of determining
the partial differential equation of the generating function is
given by Bartlett (1949). The value of this technique is that the
appropriate partial differential equation for the cumulant generating function can be written down with relative ease. The partial differential equation is then expanded in series notation
and appropriate terms are collected to give a set of differential
equations of the first, second, and first cross cumulants for
each compartment. Solving these differential equations gives the
desired solutions.

Applying Bartlett's 'random variable' technique for obtaining the cumulant generating function (see Bailey, 1964), the partial differential equation for the cumulant generating function of the distribution of the number of units in each compartment of the system becomes

$$\frac{\partial K(\underset{\sim}{\theta},t)}{\partial t} = \sum_{i=1}^{n} (e^{\theta_i} - 1) \left\{ \lambda_i \frac{\partial K(\underset{\sim}{\theta},t)}{\partial \theta_i} + \nu_i \right\}$$

$$+ \sum_{i=1}^{n} (e^{-\theta_i} - 1) \mu_i \frac{\partial K(\underset{\sim}{\theta},t)}{\partial \theta_i} \tag{1}$$

$$+ \sum_{i=1}^{n-1} \sum_{j>i}^{n} k_{j,i} (e^{-\theta_i + \theta_j} - 1)$$

$$\left\{ P_{j,i,1} \frac{\partial K(\underset{\sim}{\theta},t)}{\partial \theta_i} + P_{j,i,2} \frac{\partial K(\underset{\sim}{\theta},t)}{\partial \theta_j} \right\} \, ,$$

where $K(\underset{\sim}{\theta},t)$ denotes the joint cumulant generating function for $X_1(t), \cdots, X_n(t)$ and the vector $\underset{\sim}{\theta}$ is defined as $\underset{\sim}{\theta} = (\theta_1, \cdots, \theta_n)$.

Expanding $(e^{\theta_i} - 1)$ and $\partial K / \partial \theta_i$ for all i, equation (1) becomes

$$\frac{\partial K(\underset{\sim}{\theta},t)}{\partial t} = \sum_{i=1}^{n} \nu_i (\theta_i + \frac{\theta_i^2}{2} + \cdots)$$

$$+ \sum_{i=1}^{n} \lambda_i (\theta_i + \frac{\theta_i^2}{2} + \cdots)$$

$$[_i\kappa_1(t) + {}_i\kappa_2(t)\theta_i + \sum_{j\neq i}^{n} {}_{i,j}\kappa_{11}(t)\theta_j + \cdots]$$

$$+ \sum_{i=1}^{n} \mu_i (-\theta_i + \frac{\theta_i^2}{2} + \cdots)$$

$$[_i\kappa_1(t) + {}_i\kappa_2(t)\theta_i + \sum_{\substack{j\neq i}}^{n} {}_{i,j}\kappa_{11}(\theta_j + \cdots]$$

$$+ \sum_{i=1}^{n-1} \sum_{j>i}^{n} k_{j,i}P_{j,i,1}(-\theta_i + \theta_j + \frac{(\theta_i - \theta_j)^2}{2} + \cdots)$$

$$[_i\kappa_1(t) + {}_i\kappa_2(t)\theta_i + \sum_{\substack{\ell\neq i}}^{n} {}_{i,\ell}\kappa_{11}\theta_\ell + \cdots]$$

$$+ \sum_{i=1}^{n-1} \sum_{j>i}^{n} k_{j,i}P_{j,i,2}(-\theta_i + \theta_j + \frac{(\theta_i - \theta_j)^2}{2} + \cdots)$$

$$[_j\kappa_1(t) + {}_j\kappa_2(t)\theta_j + \sum_{\substack{\ell\neq j}}^{n} {}_{j,\ell}\kappa_{11}(t)\theta_\ell + \cdots],$$

$$(2)$$

where the κ are defined as follows:

$_i\kappa_1(t)$ = first cumulant (mean) for compartment i,

$_i\kappa_2(t)$ = second cumulant (variance) for compartment i,

$_{i,j}\kappa_{11}(t)$ = first cross cumulant (covariance) between the ith and jth compartment.

Note also that taking the partial derivative of the series expansion of $K(\underset{\sim}{\theta},t)$ with respect to t, i.e., the left-hand side of (1), gives the following:

$$\frac{\partial K(\underset{\sim}{\theta},t)}{\partial t} = \sum_{i=1}^{n} {}_i\dot{\kappa}_1(t)\theta_i + \sum_{i=1}^{n} {}_i\dot{\kappa}_2(t)\frac{\theta_i^2}{2}$$

$$+ \sum_{i=1}^{n} \sum_{j>i}^{n} {}_{i,j}\dot{\kappa}_{11}(t)\theta_i\theta_j + \cdots, \qquad (3)$$

where the dot denotes differentiation with respect to t.

3.3 Differential Equations of the First and Second Cumulants.
After rearranging the terms of (2) and equating coefficients of
θ_i, θ_i^2, and $\theta_i\theta_j$ ($i=1,\cdots,n$; $j=i+1,\cdots,n$) to those on the
right hand side of (3), one has the following set of differential
equations:

$$_i\dot{\kappa}_1(t) = \xi_i[_i\kappa_1(t)] + \sum_{\ell=1}^{i-1} k_{i,\ell,1}^*[_{\ell+1}\kappa_1(t)]$$

$$- \sum_{\ell=i+1}^{n} k_{\ell,i,2}^*[_{i+1}\kappa_1(t)]$$

$$+ \sum_{\ell=1}^{i-2} k_{i,\ell,2}^*[_{\ell+1}\kappa_1(t)] + \nu_i \qquad\qquad (i=1\cdots,n)$$

$$_i\dot{\kappa}_2(t) = 2\xi_i[_i\kappa_2(t)] + (2\lambda_i - \xi_i)[_i\kappa_1(t)]$$

$$+ \sum_{\ell=1}^{i-1} k_{i,\ell,1}^* \left\{ 2[_{\ell,i}\kappa_{11}(t)] + _\ell\kappa_1(t) \right\}$$

$$+ \sum_{\ell=i+1}^{n} k_{\ell,i,2}^* \left\{ 2[_{i,i+1}\kappa_{11}(t)] + _{i+1}\kappa_1(t) \right\}$$

$$+ \sum_{\ell=1}^{i-2} k_{i,\ell,2}^* \left\{ 2[_{\ell+1,i}\kappa_{11}(t)] + _{\ell+1}\kappa_1(t) \right\} + \nu_i$$

$$(i=1,\cdots,n)$$

$$_{i,i+1}\dot{\kappa}_{11}(t) = (\xi_i + \xi_{i+1})[_{i,i+1}\kappa_{11}(t)] + k_{i+1,i,1}^*[_i\kappa_2(t) - _i\kappa_1(t)]$$

$$- \sum_{\ell=i+1}^{n} k_{\ell,i,2}^*[_{i+1}\kappa_2(t)] + \sum_{\ell=1}^{i-1} \left\{ k_{i,\ell,1}^*[_{\ell,i+1}\kappa_{11}(t)] \right.$$

$$+ \overset{*}{k}_{i+1,\ell,1}[_{\ell,i}\kappa_{11}(t)]\Big\} - \sum_{\ell=i+2}^{n} \overset{*}{k}_{\ell,\ell+1,2}[_{i,i+2}\kappa_{11}(t)]$$

$$- \overset{*}{k}_{i+1,i,2}[_{i+1}\kappa_1(t)] + \overset{*}{k}_{i+1,i-1,2}[_i\kappa_2(t)]$$

$$+ \sum_{\ell=1}^{i-2}\Big\{\overset{*}{k}_{i,\ell,2}[_{\ell+1,i+1}\kappa_{11}(t)] + \overset{*}{k}_{i+1,\ell,2}[_{\ell+1,i}\kappa_{11}(t)]\Big\}$$

$$(i=1,\cdots,n-1)$$

$$_{i,j}\overset{\bullet}{\kappa}_{11}(t) = (\xi_i + \xi_j)[_{i,j}\kappa_{11}(t)] + \overset{*}{k}_{j,i,1}[_i\kappa_2(t) - {}_i\kappa_1(t)]$$

$$+ \overset{*}{k}_{j,i,2}[_{i,i+1}\kappa_{11}(t) - {}_{i+1}\kappa_1(t)] + \sum_{\ell=1}^{i-1}\overset{*}{k}_{i,\ell,1}[_{\ell,j}\kappa_{11}(t)]$$

$$+ \sum_{\substack{\ell=1\\ \ell\neq i}}^{j-1}\overset{*}{k}_{j,\ell,1}[_{\ell,i}\kappa_{11}(t)] - \sum_{\ell=i+1}^{n}\overset{*}{k}_{\ell,i,2}[_{i+1,j}\kappa_{11}(t)]$$

$$+ \overset{*}{k}_{j,i-1,2}[_i\kappa_2(t)] - \sum_{\ell=j+1}^{n}\overset{*}{k}_{\ell,j,2}[_{i,j+1}\kappa_{11}(t)]$$

$$\sum_{\ell=1}^{j-2}\overset{*}{k}_{j,\ell,2}[_{i,\ell+1}\kappa_{11}(t)] + \sum_{\ell=1}^{i-2}\overset{*}{k}_{i,\ell,2}[_{\ell+1,j}\kappa_{11}(t)]$$

$$(i=1,\cdots,n-2; \quad j=i+2,\cdots,n) \qquad\qquad (4)$$

with the following definitions

$$\xi_i = \lambda_i - \mu_i + \overset{*}{k}_{i,i-1,2} - \sum_{\ell=i+1}^{n}\overset{*}{k}_{\ell,i,1},$$

$$k^*_{i,j,m} = k_{i,j}P_{i,j,m} \quad (1 \le j \le i \le n; \ m=1,2),$$

and

$$\sum_{\ell=a}^{b} f_{i,j,\ell} \equiv 0 \quad (b<a),$$

where $f_{i,j,\ell}$ is any function of the cumulants.

This system of equations may be solved as follows. First, the set of differential equations involving the first cumulants of (4) may be solved simultaneously. Then the differential equations involving the second order cumulants of (4) may be solved simultaneously with the solutions to the first cumulants appropriately substituted into the differential equations.

4. A SPECIAL CASE OF THE GENERAL IRREVERSIBLE MODEL: THE TWO-COMPARTMENT MODEL

Consider now the solution of the model for the two compartment case. The set of differential equations (4) when $n=2$ becomes

$$\dot{\kappa}_{10}(t) = \xi_1 \kappa_{10}(t) - k^*_2 \kappa_{01}(t) + \nu_1$$

$$\dot{\kappa}_{01}(t) = \xi_2 \kappa_{01}(t) + k^*_1 \kappa_{10}(t) + \nu_2$$

$$\dot{\kappa}_{20}(t) = 2\xi_1 \kappa_{20}(t) + (2\lambda_1 - \xi_1)\kappa_{10}(t) + 2k^*_1 \kappa_{11}(t) + k^*_2 \kappa_{01}(t) + \nu_1$$

$$\dot{\kappa}_{02}(t) = 2\xi_2 \kappa_{02}(t) + (2\lambda_2 - \xi_2)\kappa_{01}(t) + 2k^*_1 \kappa_{11}(t) + k^*_1 \kappa_{10}(t) + \nu_2$$

$$\dot{\kappa}_{11}(t) = (\xi_1 + \xi_2)\kappa_{11}(t) + k^*_1 [\kappa_{20}(t) - \kappa_{10}(t)]$$
$$- k^*_2 [\kappa_{02}(t) + \kappa_{01}(t)] \qquad (5)$$

where $\xi_1 = \lambda_1 - \mu_1 - k_1^*$, $\xi_2 = \lambda_2 - \mu_2 + k_2^*$, $k_1^* = k_{21}P_{211}$, and $k_2^* = k_{21}P_{212}$.

The solution for the first and second cumulants to the set of differential equations   (5)   is:

$$K_{10}(t) = \sum_{i=1}^{2} c_{1i} \exp\{-\omega_i t\} + \delta_1$$

$$K_{01}(t) = \sum_{i=1}^{2} c_{2i} \exp\{-\omega_i t\} + \delta_2$$

$$K_{20}(t) = \sum_{i=1}^{5} c_{3i} \exp\{-\omega_i t\} + \delta_3$$

$$K_{02}(t) = \sum_{i=1}^{5} c_{4i} \exp\{-\omega_i t\} + \delta_4 \tag{6}$$

where

$$c_{1i} = \frac{[g_2 - (-1)^{i+1} g_1]a_2 - (-1)^i 2k_2^* a_3}{2g_2} \qquad (i=1,2)$$

$$c_{2i} = \frac{(-1)^i c_{1i}[g_1 - (-1)^i g_2]}{2k_2^*} \qquad (i=1,2)$$

$$c_{ij} = \frac{2k_1^* k_2^*}{g_2^2 k_{i-2}^*} \left[- \frac{[g_1 - (-1)^i g_2](c_{1,j}r_{11} + c_{2,j}r_{21})}{2(\omega_3 - \omega_j)} \right.$$

$$\left. - \frac{[g_1 - (-1)^{i+1} g_2](c_{1,j}r_{12} + c_{2,j}r_{22})}{2(\omega_4 - \omega_j)} + \frac{g_1(c_{1,j}r_4 + c_{2,j}r_5)}{\omega_5 - \omega_j} \right]$$

$$(i=3,4; \; j=1,2)$$

$$c_{3,i} = \frac{k_2^*[g_1 - (-1)^i g_2]}{g_2^2} \left[\frac{c_{11}r_{1,i} + c_{21}r_{2,i}}{\omega_i - \omega_2} + \frac{c_{12}r_{1,i} + c_{22}r_{2,i}}{\omega_i - \omega_2} \right.$$

$$\left. + \frac{\delta_1 r_{1i} + \delta_2 r_{2i} + r_{3i}}{\omega_i} \right] \qquad (i=3,4)$$

$$c_{4,i} = \frac{k_1^*[g_1 - (-1)^{i+1}g_2]}{k_3^*[g_1 - (-1)^i g_2]} c_{3,i} \qquad (i=3,4)$$

$$c_{i,5} = \frac{-2g_1 k_1^* k_2^*}{g_2^2 k_{i-2}^*} \left[\frac{c_{11}r_4 + c_{21}r_5}{\omega_5 - \omega_1} + \frac{c_{12}r_4 + c_{22}r_5}{\omega_5 - \omega_2} + \frac{\delta_1 r_4 + \delta_2 r_5 + r_6}{\omega_5} \right]$$

$$(i=3,4)$$

$$\omega_1 = \frac{-(\xi_1 + \xi_2) + g_2}{2} \quad , \quad \omega_2 = \frac{-(\xi_1 + \xi_2) - g_2}{2}$$

$$\omega_3 = 2\omega_2 \; , \qquad \omega_4 = 2\omega_1 \; , \qquad \omega_5 = -(\xi_1 + \xi_2)$$

$$\delta_1 = \frac{\omega_1[\nu_1(g_1 + g_2) - 2k_2^* \nu_2] - \omega_2[\nu_1(g_1 - g_2) - 2k_2^* \nu_2]}{2g_2 \omega_1 \omega_2}$$

$$\delta_2 = \frac{\omega_1[2k_1^* \nu_1 - \nu_2(g_1 - g_2)] - \omega_2[2k_1^* \nu_1 - \nu_2(g_1 + g_2)]}{2g_2 \omega_1 \omega_2}$$

$$\delta_i = \frac{2k_1^* k_2^*}{g_2^2 k_{i-2}^*} \left[- \frac{[g_1 - (-1)^i g_2](\delta_1 r_{11} + \delta_2 r_{21} + r_{31}}{2\omega_3} \right.$$

$$-\frac{[g_1 - (-1)^{i+1}g_2](\delta_1 r_{12} + \delta_2 r_{22} + r_{32})}{2\omega_4} + \frac{g_1(\delta_1 r_4 + \delta_2 r_5 + r_6)}{2\omega_5}\Bigg]$$

$$(i=3,4)$$

$$g_1 = \xi_1 - \xi_2 , \qquad g_2 = \sqrt{g_1^2 - 4k_1^* k_2^*}$$

$$r_{1,i} = \frac{(2\lambda_1 - \xi_1)[g_1 - (-1)^i g_2] + k_2^*[g_1 - (-1)^{i+1}g_2]}{4k_2^*} \qquad (i=1,2)$$

$$r_{2,i} = \frac{k_1^*[g_1 - (-1)^i g_2] + (2\lambda_2 - \xi_2)[g_1 - (-1)^{i+1}g_2] + 4k_1^* k_2^*}{4k_1^*}$$

$$(i=1,2)$$

$$r_4 = \frac{-k_1^*(2\lambda_1 - \xi_1 + k_2^*)}{g_1} , \qquad r_5 = \frac{-k_2^*(2\lambda_2 - \xi_2 + k_1^* + g_1)}{g_1}$$

$$r_6 = \frac{-k_1^* \nu_1 + k_2^* \nu_2}{g_1}$$

The above solution (6) may be verified by substitution into the set of differential equations (5). In particular, it is not difficult to verify these solutions for any assumed numerical values of the coefficients.

5. ESTIMATION RESULTS WITH SIMULATED DATA

To investigate the precision of certain proposed parameter estimators, data was simulated for the following assumed systems: 1) a single realization of a two-compartment donor controlled model with constant transition rates and initial conditions: $\nu_1 = \nu_2 = 0$, $\lambda_1 = .03$, $\mu_1 = .02$, $k_2 = .03$, $\lambda_2 = .02$, $\mu_2 = .01$, $X_1(0) = 1000$ and $X_2(0) = 100$; 2) a mean of ten replicates from system 1;   3) a single realization of a two-compartment recipient controlled model with constant transition rates and initial conditions: $\nu_1 = \nu_2 = 0$, $\lambda_1 = .03$, $\mu_1 = .02$, $k_2 = .005$, $\lambda_2 = .01$, $\mu_2 = .02$, $X_1(0) = 500$ and $X_2(0) = 500$; and 4) a mean of ten replicates from system 3. The observations consisted of either the number or the mean number of units in each compartment at times t=1,···,40. All the data are given explicitly in Gerald (1978). The sample means and variances of the ten replicates of the donor controlled model are plotted in Figure 3.

A modified Gauss-Newton estimation procedure was used to fit the first cumulants to the data for each of the four systems described above. It is easy to verify that the partial derivatives of the first cumulants with respect to the five unknown parameters are not linearly independent; therefore, it is not possible to obtain separate estimates for each of the five parameters. However it is possible to obtain for each compartment an estimate of the difference between the birth and death rates, $\lambda_i - \mu_i$, and also an estimate on the migration rate, k_{21}. A summary of the results is given in Table 1 for the donor and the recipient controlled models. The fitted and theoretical values for the means and variances of the ten simulations of the donor controlled model are plotted in Figure 3.

The modified Gauss-Newton estimation procedure was also used to simultaneously fit the first and second cumulants for each compartment of the donor controlled model to the sample means and variances of the ten replicates. This is a novel application of the estimation procedure which solves the identifiability problem of the parameter set. When fitted simultaneously, the partial derivatives of the first and second cumulants with respect to each of the five unknown parameters are linearly independent and hence it is possible, in principle, to obtain separate estimates of all five parameters. Table 2 summarizes the results for the donor controlled model. Note that although each parameter is identifiable, the precision of the individual birth and death rates is rather low for this simulation.

The comparative precision of the various estimates is illustrated for the donor controlled model by Figure 4 which contains the residuals for the means of compartments 1 and 2, respectively. The following three cases are illustrated:

  i) fitting the first cumulants to the data consisting of only one replicate;

  ii) fitting the first cumulants to the means of the ten replicates; and

TABLE 1: *Parameter estimates, asymptotic confidence limits, and asymptotic correlations from fitting the first cumulants of the donor controlled and the recipient controlled models.*

	Donor controlled			Recipient controlled		
	$\lambda_1-\mu_1$	$\lambda_2-\mu_2$	k_{21}	$\lambda_1-\mu_1$	$\lambda_2-\mu_2$	k_{21}
True value	.01	.01	.03	.01	−.01	.005
One replicate						
Estimated value	.0109	.0073	.0340	.0131	−.0170	.0133
95% Confidence limits: lower	.0100	.0062	.0331	.0045	−.0182	.0119
upper	.0118	.0084	.0348	.0316	.0158	.0147
Correlations						
$\lambda_1 - \mu_1$	1.0000	−.9487	.9279	1.0000	.9088	−.7624
$\lambda_2 - \mu_2$		1.0000	−.9534		1.0000	−.8566
k_{21}			1.0000			1.0000
Ten replicates						
Estimated value	.0097	.0095	.0030	.0099	−.0089	.0042
95% Confidence limits: lower	.0094	.0090	.0297	.0068	−.0108	.0024
upper	.0101	.0100	.0303	.0130	−.0071	.0061
Correlations						
$\lambda_1 - \mu_1$	1.0000	−.9488	.9335	1.0000	−.9961	.9959
$\lambda_2 - \mu_2$		1.0000	−.9597		1.0000	−.9955
k_{21}			1.0000			1.0000

    iii) fitting the first and second cumulants simultaneously to the means and variances of the ten replicates.

Figure 5 illustrates the residuals for the first two cases of the recipient controlled model. As expected, the residuals tend to be much smaller for the aggregate data than for the single realization. Also, as expected, the residual (simulated minus predicted) values were lower for the case ii) than for case iii), however the mean square errors (theoretical minus predicted) were higher for ii) than for iii).

TABLE 2: Parameter estimates, asymptotic confidence limits, and asymptotic correlations from fitting the first and second cumulants of the donor controlled model.

	λ_1	μ_1	k_{21}	λ_2	μ_2
True value	.03	.02	.03	.02	.01
Estimated value	.0192	.0072	.0292	.0345	.0251
95% Confidence limits: lower	.0140	-.0000	.0262	.0324	.0228
upper	.0244	.0144	.0323	.0365	.0274
Correlations					
λ_1	1.0000	.9163	-.3349	.2779	-.5291
μ_1			-.6817	.2359	-.3616
k_{21}				-.0517	-.1098
λ_2					.5966
μ_2					

This limited simulation study is intended only to indicate the feasibility of the estimation procedure with such data. A full simulation study is under present investigation.

6.   DISCUSSION

This research has derived the first and second cumulants for a broad stochastic compartmental model which incorporates birth and recipient-controlled transfer mechanisms in addition to the usual death and donor-controlled transfer mechanisms. These added mechanisms, although very important in certain ecological systems, have not been previously developed in the literature of stochastic compartmental models. Such a separate development is not necessary in a deterministic model with births since the deterministic model may be parameterized in terms of the difference between the birth and death rates. However, the second cumulants of the stochastic model are functions of both the birth and the death rates and this in turn necessitates the separate development of the present stochastic model.

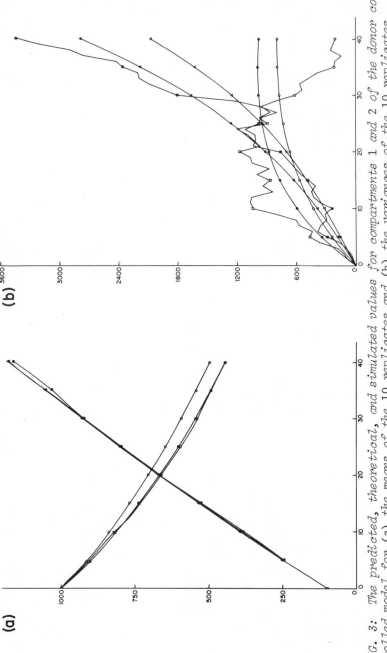

FIG. 3: The predicted, theoretical, and simulated values for compartments 1 and 2 of the donor controlled model for (a) the means of the 10 replicates and (b) the variances of the 10 replicates.
Legend: Open and Closed Circles - Predicted and Theoretical Values for Comp.1; Open and Closed
Squares - Simulated Values for Compartments 1 and 2; Open and Closed Triangles - Theoretical and Predicted Values for Comp. 2.

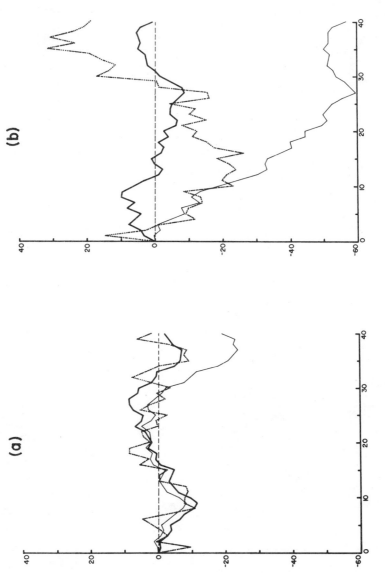

FIG. 4: Plots of the residuals for the means of the donor controlled model for (a) comp. 1 and (b) comp. 2. Legend: Heavy Line – Fitting mean of 10 replicates; Dotted Line – Fitting single replicate; Thin Line – Simultaneous fitting of mean and variance of 10 replicates.

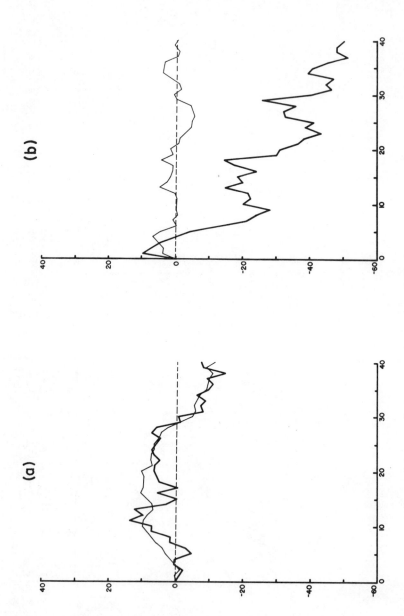

FIG. 5: Plots of the residuals for the means of the recipient controlled model for (a) comp. 1 and (b) comp. 2. Legend: Thin Line - Fitting mean of 10 replicates; Heavy Line - Fitting single replicate.

The first and second cumulants also provide a basis for some statistical inference, and an approach to parameter estimation was illustrated with simulated data. As expected, there is a high degree of correlation between the parameter estimates resulting from the elongated and narrow joint confidence regions of the estimates. Thus it may be difficult to obtain good point and interval estimates on certain individual parameters, however all parameters are identifiable in this stochastic model.

The solutions to these cumulants of the general irreversible model could be generalized in several ways. Firstly, one could assume that the initial number of units in the system is not fixed but random. This would introduce the means, variances, and covariances of the distribution of the initial conditions into the solutions of the cumulants. Secondly, the transition rates could be generalized to the time-dependent case. Gerald (1978) gives the solution for the first cumulants of the n-compartment donor and recipient controlled models with time-dependent transition rates.

Thakur, *et al.* (1973) note that the coefficient of variation (i.e., standard deviation divided by the mean) in many stochastic compartmental models has a limit of 0 as the number of particles approaches infinity. In practice, this means that the relative variability is negligible in the stochastic formulation of many models since the particle count in many applications is astronomical. However, this is not necessarily the case in stochastic models with a birth mechanism, as in the present models, since the variances may be much larger than the means.

Further work is now in progress in fitting these broad models to data and in identifying the individual mechanisms.

ACKNOWLEDGMENT

The authors are indebted to W. E. Grant of Texas A&M University for many helpful discussions concerning these problems.

REFERENCES

Atkins, G. L. (1969). *Multicompartment Models for Biological Systems*. Wiley, New York.

Bailey, N. T. J. (1964). *The Elements of Stochastic Processes with Applications to the Natural Sciences*. Wiley, New York.

Barber, M. C. (1978a). A retrospective Markovian model for ecosystem resource flow. *Ecological Modelling*, 5, 125-135.

Barber, M. C. (1978b). A Markovian model for ecosystem flow analysis. *Ecological Modelling*, 5, 193-206.

Barber, M. C., Patten, B. C., and Finn, J. T. (1979). Review and evaluation of input-output flow analysis for ecological applications. In *Compartmental Analysis of Ecosystem Models*, J. H. Matis, B. C. Patten, and G. C. White, eds. Satellite Program in Statistical Ecology, International Co-operative Publishing House, Fairland, Maryland.

Bartlett, M. S. (1949). Some evolutionary stochastic processes. *Journal of the Royal Statistical Society, Series B*, 2, 211-229.

Cornfield, J., Steinfield, J., and Greenhouse, S. W. (1960). Models for the interpretations of experiments using tracer compounds. *Biometrics*, 16, 212-234.

Gerald, K. B. (1978). *On the theory of some stochastic compartmental systems in ecological work*. Ph.D. dissertation, Texas A&M University.

Gold, H. J. (1977). *Mathematical Modeling of Biological Systems - An Introductory Guidebook*. Wiley, New York.

Jacquez, J. A. (1972). *Compartmental Analysis in Biology and Medicine*. Elsevier, New York.

Kodell, R. L. (1974). *Nonlinear estimation with a known covariance structure over time*. Ph.D. dissertation, Texas A&M University.

Matis, J. H. and Wehrly, T. E. (1979). An approach to a compartmental model with multiple sources of stochasticity for modeling ecological systems. In *Compartmental Analysis of Ecosystem Models*, J. H. Matis, B. C. Patten, and G. C. White, eds. Satellite Program in Statistical Ecology, International Co-operative Publishing House, Fairland, Maryland.

O'Neill, R. V. (1979). A review of compartmental analysis in ecosystem science. In *Compartmental Analysis of Ecosystem Models*, J. H. Matis, B. C. Patten, and G. C. White, eds. Satellite Program in Statistical Ecology, International Co-operative Publishing House, Fairland, Maryland.

Padgett, W. J. (1979). A survey of stochastic models for BOD and DO in streams. Manuscript.

Poole, R. W. (1979). Ecological models and the stochastic - deterministic questions. In *Scientific Modeling and Quantitative Thinking with Examples in Ecology*, G. P. Patil, D. Simberloff, and D. Solomon, eds. Satellite Program in Statistical Ecology, International Co-operative Publishing House, Fairland, Maryland.

Schultz, V., Eberhardt, L. L., Thomas, J. M., and Cochran, M. I. (1976). *A Bibliography of Quantitative Ecology*. Dowden, Hutchinson, and Ross,

Thakur, A. K., Rescigno, A., and Schafer, D. E. (1973). On the stochastic theory of compartments: II. Multicompartment systems. *Bulletin of Mathematical Biology*, 35, 263-271.

Thayer, R. P. and Krutchkoff, R. G. (1967). Stochastic model for BOD and DO in streams. *Journal of the Sanitary Engineering Division, ASCE*, (SA3), 93, 59-72.

[*Received July* 1978. *Revised February* 1979]

J. H. Matis, B. C. Patten, and G. C. White, (eds.),
Compartmental Analysis of Ecosystem Models, pp. 335-358. All rights reserved.
Copyright ©1979 by International Co-operative Publishing House, Fairland, Maryland.

THE TWO–VARIABLE OPERATIONAL CALCULUS IN THE CONSTRUCTION OF COMPARTMENTAL ECOLOGICAL MODELS

ALDO RESCIGNO

The Bragg Creek Institute for Natural Philosophy
Bragg Creek, Alberta, Canada

SUMMARY. The one variable operational calculus has been used extensively in compartmental analysis; its extension to ecological modeling is straightforward and it lends itself to many interesting applications. When building time–dependent models, i.e., when considering species whose age composition is relevant to the problem, the two-variable operational calculus is a very powerful tool and can easily be integrated with graph-theoretical methods.

KEY WORDS. operational calculus in two dimensions, compartmental ecological models, two-variable operational calculus.

1. INTRODUCTION

The operational calculus in one variable, introduced by Mikusinski (1959), has been used extensively in the construction of compartmental models; see for instance Rescigno (1963a), Rescigno (1969), Rescigno and Beck (1972).

Alternative presentations of Mikusinski's calculus were made, for instance, by Erdelyi (1962), using a strictly algebraic approach, and by Berg (1967), using the Duhamel product in lieu of the convolution integral.

In higher dimensions, an operational calculus has been developed by Ditkin and Prudnikov (1962), but it was completely based on the Laplace transform. Here I want to show how easily the Mikusinski operational calculus can be extended to two dimensions, and by implication to any higher dimensions.

where B is the domain ε, $\rho \leqq T$; $\varepsilon + \rho \geqq 0$; η, $\sigma \leqq T$;
$\eta + \sigma \geqq 0$.   Call C the domain $-T \leqq \varepsilon$, ρ; $\varepsilon + \rho \leqq 0$;
$-T \leqq \eta$, σ; $\eta + \sigma \leqq 0$; then B+C is the domain $-T \leqq \varepsilon$, ρ, η,
$\sigma \leqq +T$ and

$$\iiiint_{B+C} = \iiiint_{B} + \iiiint_{C} \;;$$

but I=0, therefore

$$\int_{-T}^{+T} \int_{-T}^{+T} e^{n(\varepsilon+\eta)} \; f(T-\varepsilon,T-\eta) \; [\int_{-T}^{+T} \int_{-T}^{+T} e^{n(\rho+\sigma)} \; f(T-\rho,T-\sigma) d\rho d\sigma] d\varepsilon d\eta$$

$$= \iiiint_{C} e^{n(\varepsilon+\rho)+n(\eta+\sigma)} \; f(T-\varepsilon,T-\eta) \; f(T-\rho,T-\sigma) d\rho \; d\sigma \; d\varepsilon \; d\eta \;.$$

Call M the maximum value of $|f|$ in C; in the same domain

$$\exp[n(\varepsilon+\rho) + n(\eta+\sigma)] < 1;$$

therefore

$$\left| \int_{-T}^{+T} \int_{-T}^{+T} e^{n(\varepsilon+\eta)} \; f(T-\varepsilon,T-\eta) \; [\int_{-T}^{+T} \int_{-T}^{+T} e^{n(\rho+\sigma)} \; f(T-\rho,T-\sigma) d\rho d\sigma] d\varepsilon d\eta \right| \leqq$$

$$\iiiint_{C} M^2 \; d\rho \; d\sigma \; d\varepsilon \; d\eta = 4T^4 M^2,$$

and

$$\int_{-T}^{+T} \int_{-T}^{+T} e^{n(\varepsilon+\eta)} \; f(T-\varepsilon,T-\eta) d\varepsilon \; d\eta = 2 \; T^2 \; M,$$

for any n > 0 and $T \geqq \varepsilon$, η.   We have also

$$\int_{0}^{+T} e^{n\varepsilon} \int_{-T}^{+T} e^{n\eta} \; f(T-\varepsilon,T-\eta) d\eta \; d\varepsilon + \int_{0}^{+T} e^{n\eta} \int_{-T}^{0} e^{n\varepsilon} \; f(T-\varepsilon,T-\eta) d\varepsilon \; d\eta$$

$$= \int_{-T}^{+T} \int_{-T}^{+T} e^{n(\varepsilon+\eta)} \; f(T-\varepsilon,T-\eta) d\varepsilon \; d\eta - \int_{-T}^{0} e^{n\varepsilon} \int_{-T}^{0} e^{n\eta} \; f(T-\varepsilon,T-\eta) d\eta \; d\varepsilon$$

$$= \int_{0}^{+T} e^{nx} \; [\int_{-T}^{+T} e^{n\eta} \; f(T-x,T-\eta) d\eta + \int_{-T}^{0} e^{n\varepsilon} \; f(T-\varepsilon,T-x) d\varepsilon] dx$$

$$\left| \int_0^{+T} e^{nx} \left[\int_{-T}^{+T} e^{n\eta} f(T-x,T-\eta)\,d\eta + \int_{-T}^{0} e^{n\varepsilon} f(T-\varepsilon,T-x)\,d\varepsilon \right] dx \right|$$

$$\leqq 2\, T^2\, M + \left| \int_{-T}^{0} e^{n\varepsilon} \int_{-T}^{0} e^{n\eta} f(T-\varepsilon,T-\eta)\,d\eta\; d\varepsilon \right|$$

$$\leqq 2\, T^2\, M + T^2\, M = 3\, T^2\, M.$$

From the theorem on moments (Mikusinski, 1951),

$$\int_{-T}^{+T} e^{n\eta} f(T-x,T-\eta)\,d\eta + \int_{-T}^{0} e^{n\varepsilon} f(T-\varepsilon,T-x)\,d\varepsilon = 0$$

for any x in the interval $-T, +T$. It follows

$$\left| \int_{-T}^{+T} e^{n\eta} f(T-x,T-\eta)\,d\eta \right| = \left| \int_{-T}^{0} e^{n\varepsilon} f(T-\varepsilon,T-x)\,d\varepsilon \right| \leqq T\, M,$$

and again from the theorem on moments,

$$f(T-x,T-\eta) = 0$$

for any x in the interval $-T, +T$, and any η in the interval $-T, +T$; as T is arbitrary, the proof is now complete.

Theorem. If

$$\int_0^u \int_0^v f(u-x,v-y)\; g(x,y)\; dx\; dy = 0, \quad \forall\, u,v \geqq 0,$$

then $f(u,v) = 0$ or $g(u,v) = 0$, $\forall\, u,v \geqq 0$.

Proof. Put $f_n(u,v) = u^n f(u,v)$ and $g_n(u,v) = u^n g(u,v)$; from the hypothesis,

$$\int_0^u \int_0^v (u-x)\; f(u-x,v-y)\; g(x,y)\; dx\; dy$$

$$+ \int_0^u \int_0^v f(u-x,v-y)\; x\; g(x,y)\; dx\; dy$$

$$= u \cdot \int_0^u \int_0^v f(u-x,v-y)\; g(x,y)\; dx\; dy = 0,$$

i.e.,

$$\{f_1\} \cdot \{g\} + \{f\} \cdot \{g_1\} = 0;$$

also,

$$\{f\} \cdot \{g_1\} \cdot (\{f_1\} \cdot \{g\} + \{f\} \cdot \{g_1\}) = 0.$$

Therefore

$$(\{f\} \cdot \{g_1\})^2 = 0,$$

and by induction

$$(\{f\} \cdot \{g_n\})^2 = 0$$

for any $n \geqq 1$.

From the lemma it follows that

$$\int_0^u \int_0^v f(u-x, v-y) \, x^n \, g(x,y) \, dx \, dy = 0,$$

or $$\int_0^u x^n \left[\int_0^v f(u-x, v-y) \, g(x,y) \, dy \right] dx = 0 \quad (n=1,2,\cdots);$$

and by Lerch theorem (Lerch, 1903),

$$\int_0^v f(u-x, v-y) \, g(x,y) \, dy = 0$$

for any $0 \leqq x \leqq u$. This is the convolution, in one dimension, of the two functions $f(u-x, v)$ and $g(x,v)$; for the theorem of Titchmarsh (1926) either function must be zero, for any v, any u, and any $x \leqq u$.

3. THE QUOTIENT FIELD OF OPERATORS

We can define the *quotient* $\{h\}$ of two function $\{f\}$ and $\{g\}$, with $\{g\} \neq \{0\}$, and write

$$\{f\}/\{g\} = \{h\},$$

if a function $\{h\}$ exists, such that $\{f\} = \{g\} \cdot \{h\}$.

The theorem guarantees that, if $\{h\}$ exists, it is unique. In fact if another function $\{h_1\}$ exists with the property

$\{f\} = \{g\} \cdot \{h_1\}$,   then

$$\{0\} = \{f\} - \{f\} = \{g\} \cdot (\{h\} - \{h_1\}),$$

whence either $\{g\} = \{0\}$ against the hypothesis, or $\{h\} = \{h_1\}$.

It is easy to see that for many pairs of functions, no quotient exists; for instance with $\{f\} = \{g\} = \{1\}$, we cannot find a function $\{h\}$ such that $\{1\} = \{1\} \cdot \{h\}$, for that would imply

$$\int_o^t h(\tau) \, d\tau = 1$$

for any $t \geqq 0$, a clear impossibility.

To make the operation of quotient $\{f\}/\{g\}$ always possible, with the only condition $\{g\} \neq \{0\}$, we introduce a new entity, called *operator*, with the same algorithm used to define rational numbers:

$\{f\}/\{g\} = \{\phi\}/\{\psi\}$ iff $\{f\} \cdot \{\psi\} = \{g\} \cdot \{\phi\}$

(Euclid, Book VII, Proposition 19),

$(\{f\}/\{g\}) + (\{\phi\}/\{\psi\}) = (\{f\} \cdot \{\psi\} + \{g\} \cdot \{\phi\})/(\{g\} \cdot \{\psi\})$

(Peano, 1908, III, 3.2),

$\{f\}/\{g\} \cdot \{\phi\}/\{\psi\} = (\{f\} \cdot \{\phi\})/(\{g\} \cdot \{\psi\})$

(Euclid, Book VIII, Proposition 5).

Addition and multiplication thus defined are associative and commutative; multiplication is distributive with respect to addition; the proofs of these theorems are elementary. In the language of abstract algebra we have a quotient field.

From the definition of equality of operators we have the important property

$$(\{f\} \cdot \{\phi\})/(\{g\} \cdot \{\phi\}) = \{f\}/\{g\},$$

with $\{\phi\} \neq \{0\}$.

From the definition of quotient of functions we have the property

$$\{f\} = (\{f\} \cdot \{\phi\})/\{\phi\},$$

where the expression at the right hand side of the equal sign is a function; looking at the same expression as an operator, we can say that to each function corresponds an operator, or that the ring of functions is *embedded* in the field of operators. Thus the identity above can be seen as the definition of a special class of operators. From now on we shall look at the symbol $\{f\}$ as representing an operator of the kind defined above.

4. NUMERICAL OPERATORS

Another important class of operators, called *numerical operators* is formed by operators of the type $\{\alpha\}/\{1\}$, where α is a constant. It is easy to prove that

$$\{\alpha\}/\{1\} = \{\beta\}/\{1\} \quad \text{iff} \quad \alpha = \beta,$$

$$\{\alpha\}/\{1\} + \{\beta\}/\{1\} = \{\alpha + \beta\}/\{1\},$$

$$\{\alpha\}/\{1\} \cdot \{\beta\}/\{1\} = \{\alpha \cdot \beta\}/\{1\};$$

thus for a numerical operator the rules of arithmetic apply, and we can introduce the notation $\alpha = \{\alpha\}/\{1\}$; in other words, the real numbers are embedded in the operators.

An important identity is $\alpha \cdot \{f\} = \{\alpha f\}$; in fact, writing α and $\{f\}$ in their operational form,

$$\frac{\{\alpha\}}{\{1\}} \cdot \frac{\{f\} \cdot \{\phi\}}{\{\phi\}} = \frac{\{\alpha\} \cdot \{f\} \cdot \{\phi\}}{\{1\} \cdot \{\phi\}}$$

$$= \frac{\{\int_o^t \alpha\, f(\tau)\, d\tau\}}{\{1\}}$$

$$= \frac{\{1\} \cdot \{\alpha f\}}{\{1\}} = \{\alpha f\}.$$

The numerical operator 0 has the unique property $0 = \{0\}$; in fact

$$0 = \{0\}/\{1\} = (\{0\} \cdot \{1\})/\{1\} = \{0\}.$$

5. INTEGRAL OPERATORS

The operator {1} can be called an *integral operator* because

$$\{1\} \cdot \{f\} = \{\textstyle\int_0^u \int_0^v f(x,y)\ dx\ dy\};$$

two more integral operators p and q can be defined with

$$p^\lambda \cdot q = \{u^{\lambda-1}/\Gamma(\lambda)\},\ \lambda > 0;$$

$$q^\mu \cdot p = \{v^{\mu-1}/\Gamma(\mu)\},\ \mu > 0.$$

As a consequence

$$p^\lambda \cdot q^\mu = \{u^{\lambda-2}/\Gamma(\lambda-1)\}\{v^{\mu-2}/\Gamma(\mu-1)\}$$

$$= \{\textstyle\int_0^u \int_0^v [x^{\lambda-2}/\Gamma(\lambda-1)][y^{\mu-2}/\Gamma(\mu-1)]dy\ dx\}$$

$$= \{[u^{\lambda-1}/\Gamma(\lambda)][v^{\mu-1}/\Gamma(\mu)]\}$$

and, in particular, $p \cdot q = \{1\}$.

We can prove the theorem

$$p\{f\} = \{\textstyle\int_0^u f(x,v)\ dx\},\quad q\{f\} = \{\textstyle\int_0^v f(u,y)\ dy\},$$

that justifies the name of integral operators given to p and q.
In fact, according to the definition of p,

$$p^2 q\{f\} = \{u\}\{f\} = \{\textstyle\int_0^u \int_0^v (u-x)f(x,y)\ dx\ dy\}$$

$$= \{\textstyle\int_0^v \int_0^u f(x,y) \int_x^u d\sigma\ dx\ dy\}$$

$$= \{\textstyle\int_0^v \int_0^u \int_0^\sigma f(x,y)\ dx\ d\sigma\ dy\}$$

$$= \{1\} \cdot \{\textstyle\int_0^u f(x,v)\ dx\},$$

whence $p\{f\} = \{\int_0^u f(x,v)\ dx\}$. Similarly for q{f}.

If a(u) and b(u) are two functions of the variable u alone, then

$$(\{a\}/q) + (\{b\}/q) = (q \cdot \{a\} + q \cdot \{b\})/(q \cdot q) = \{a+b\}/q,$$

$$(\{a\}/q) \cdot (\{b\}/q) = (\{a\} \cdot \{b\})/(q \cdot q) = \{\int_0^u a(u-x)\ b(x)\ dx\}/q;$$

therefore for an element as $\{a\}/q,$ the rules of the one-variable operational calculus apply. We can introduce the notation

$$[a(u)] = \{a(u)\}/q, \quad [a(v)] = \{a(v)\}/p,$$

as previously suggested (Rescigno, 1963b), and regard the functions in square brackets as elements of the one-variable operational calculus embedded in the two-variable operational calculus.

An important property of these operators is

$$[a(u)] \cdot [b(v)] = \{a(u) \cdot b(v)\};$$

in fact, by definition,

$$[a(u)] \cdot [b(v)] = (\{a\}/q) \cdot (\{b\}/p) = \{\int_0^u \int_0^v a(u-x)\ b(y)\ dx\ dy\}/\{1\};$$

hence the statement above follows.

The different products between operators can be summarized in this list:

$$\alpha \cdot \beta = \text{arithmetic product}$$

$$\alpha \cdot [f] = [\alpha f]$$

$$\alpha \cdot \{f\} = \{\alpha f\}$$

$$[f(u)] \cdot [g(u)] = [\int_0^u f(u-x)\ g(x)\ dx]$$

$$[f(u)] \cdot [g(v)] = \{f \cdot g\}$$

$$[f(u)] \cdot \{g\} = \{\int_0^u f(u-x)\ g(x,v)\ dx\}$$

$$\{f\} \cdot \{g\} = \{\int_0^u \int_0^v f(u-x,v-y)\ g(x,y)\ dy\ dx\}.$$

The last product is a definition; all others can be proven by putting $\alpha = \{\alpha\}/\{1\}, \quad [f(u)] = \{f\}/q,$ and using the definition.

It should be clear that, though the above products have been written in different forms, the multiplication, as defined in Section 3, is unique.

It is useful now to state a result without proof.

Lemma. For a very large class of functions f(u,v),

$$\lim_{r\to\infty} \{f\} = \lim_{s\to\infty} \{f\} = 0,$$

where the limits above don't have the usual meaning, as r and s
are operators, not variables, but have most of its formal properties.

6. DIFFERENTIAL OPERATORS

Given an arbitrary operator h = {f}/{g}, with h ≠ 0, we
define the inverse of h, denoted by 1/h, as the operator
{g}/{f}; obviously h · (1/h) = 1; in fact

$$h \cdot (1/h) = (\{f\}/\{g\}) \cdot (\{g\}/\{f\})$$

$$= (\{f\} \cdot \{g\})/(\{f\} \cdot \{g\}) = 1.$$

Call r and s the inverses of the integral operators p
and q introduced in the previous section: r = 1/p, s = 1/q.
If f(u,v) has all the derivatives we are going to consider, then

$$\{\int_{0}^{u} \partial f/\partial u \cdot du\} = \{f\} - \{f(0,v)\}$$

$$p \cdot \{\partial f/\partial u\} = \{f\} - \{f(0,v)\}$$

$$\{\partial f/\partial u\} = r\{f\} - [f(0,v)] \tag{1}$$

or

$$r\{f\} = \{\partial f/\partial u\} + [f(0,v)]. \tag{2}$$

Similarly we can compute

$$\{\partial f/\partial v\} = s\{f\} - [f(u,0)] \tag{3}$$

or

$$s\{f\} = \{\partial f/\partial v\} + [f(u,0)].$$

These identities justify the names of *differential operators* to r
and s.

Proceeding in the same way,

$$\{\partial^2 f/\partial u^2\} = r^2\{f\} - r[f(0,v)] - [\partial f/\partial u\big|_{u=0}],$$

$$\{\partial^2 f/\partial u\partial v\} = r\ s\ \{f\} - r[f(u,0)] - s[f(0,v)] + f(0,0),$$

$$\{\partial^2 f/\partial v^2\} = s^2\{f\} - s[f(u,0)] - [\partial f/\partial v\big|_{v=0}],$$

and so forth.

The formulas above are valid provided $f(u,v)$ has the derivatives indicated. If this is not the case, then they can be interpreted as an extension of the concept of derivative. For instance we have seen that the expression $r\{f\} - [f(0,v)]$ is equal to $\{\partial f/\partial u\}$ if this last derivative exists; if it does not, then the above expression is still an operator, though it does not correspond to any function, and we can call it 'derivative of f with respect to u'.

We can now state the

Initial Value Theorem. For any function $f(u,v)$ whose derivative $\partial f/\partial u$ satisfies the condition of the lemma of Section 5,

$$[f(0,v)] = \lim_{r\to\infty} r\{f\}.$$

To prove this, observe that the left hand side of (1) becomes zero when making r infinity, in the sense of the lemma.

In a similar way, if $\partial f/\partial v$ satisfies the condition of the lemma,

$$[f(u,0)] = \lim_{s\to\infty} s\{f\};$$

if both first derivatives satisfy the above condition,

$$f(0,0) = \lim_{\substack{r\to\infty \\ s\to\infty}} r\ s\ \{f\}.$$

7.   FUNCTIONAL CORRELATES

From the definitions of p, q, r, s, we get $\{1\} = 1/(r{\cdot}s)$.
From formula (2) we get $r\{u\} = \{1\}$, whence

$$\{u\} = 1/(r^2{\cdot}s),$$

and by induction

$$\{u^i\} = i!/(r^{i+1}{\cdot}s);$$

similarly

$$\{v^j\} = j!/(r{\cdot}s^{j+1}),$$

and in general

$$\{u^i{\cdot}v^j\} = \Gamma(i{+}1){\cdot}\Gamma(j{+}1)/(r^{i+1}s^{j+1}) \qquad (i > 0,\ j > 0).$$

From the identities

$$\frac{\partial}{\partial u}\, e^{\alpha u + \beta v} = \alpha{\cdot}e^{\alpha u + \beta v}, \qquad \frac{\partial}{\partial v}\, e^{\beta v} = \beta{\cdot}e^{\beta v},$$

using formulae (1) and (3),

$$\alpha\{e^{\alpha u + \beta v}\} = r\{e^{\alpha u + \beta v}\} - [e^{\beta v}],$$

$$\beta\{e^{\beta v}\} = s\{e^{\beta v}\} - p;$$

whence

$$[e^{\beta v}] = 1/(s{-}\beta) \qquad \text{and} \qquad \{e^{\alpha u + \beta v}\} = (r{-}\alpha)^{-1}(s{-}\beta)^{-1}.$$

From the identities

$$\frac{\partial}{\partial u}\, \cos(\alpha u + \beta v) = -\alpha\,\sin(\alpha u + \beta v),$$

$$\frac{\partial}{\partial v}\, \cos(\alpha u + \beta v) = -\beta\,\sin(\alpha u + \beta v),$$

using (1) and (3) we get

$$-\alpha\{\sin(\alpha u + \beta v)\} = r\{\cos(\alpha u + \beta v)\} - [\cos\beta v],$$

$$-\beta\{\sin(\alpha u + \beta v)\} = s\{\cos(\alpha u + \beta v)\} - [\cos\alpha u].$$

But from the results of the one-dimensional operational calculus we have

$$[\cos\beta v] = s/(s^2+\beta^2), \quad [\cos\alpha u] = r/(r^2+\alpha^2);$$

therefore

$$\{\sin(\alpha u+\beta v)\} = (\alpha s+\beta r)(r^2+\alpha^2)^{-1}(s^2+\beta^2)^{-1},$$

$$\{\cos(\alpha u+\beta v)\} = (rs-\alpha\beta)(r^2+\alpha^2)^{-1}(s^2+\beta^2)^{-1}.$$

Many more functions can be represented in the same way as operators in terms of r and s.

8. DISCONTINUOUS FUNCTIONS

In Section 2 we considered the ring of functions of class C; now we are able to include in the field of operators a much larger class of functions. Call K the class of functions f(u,v) defined for any u, v $\geq$ 0, such that $\int_0^u \int_0^v f(x,y)\ dx\ dy$ is continuous for all u, v $\geq$ 0; class K includes all functions that have a finite number of discontinuities on any finite set of dimension one contained in their domain of definition, and such that $\int_0^u \int_0^v |f(x,y)|dx\ dy$ is finite for any u, v $\geq$ 0.

For any function of class K,

$$\{f\} = \{\int_0^u \int_0^v f(x,y)\ dx\ dy\}/\{1\};$$

i.e., a function of class K corresponds to an operator defined with two functions of class C.

As an example of discontinuous function consider the *jump function*

$$H_{a,b}(u,v) = \begin{cases} 0 & \text{for } u < a \text{ or } v < b, \\ 1 & \text{for } u \geq a \text{ and } v \geq b, \end{cases}$$

i.e., the function which vanishes everywhere except on the angle with vertex at the point (a,b) and sides parallel to and oriented with the axes. Define the operator

$$h_{a,b} = r\ s\{H_{a,b}\};$$

for any function $\{f\}$,

$$h_{a,b}\{f\} = r \ s \ \{\int_o^u \int_o^v H_{a,b}(u-x,v-y) \ f(x,y) \ dx \ dy\};$$

but the above integral vanishes for $u < a$ or $v < b$, while for $u \geq a$ and $v \geq b$ the integrand is zero when $x > u - a$ and $y > v - b$; therefore

$$h_{a,b}\{f\} = \begin{cases} r \ s \ \{\int_o^{u-a} \int_o^{v-b} f(x,y) \ dx \ dy\} = \{f(u-a,v-b)\}, \\ \qquad\qquad\qquad\qquad\qquad u \geq a \quad \text{and} \quad v \geq b, \\ 0, \ u < a \ \text{or} \ v < b. \end{cases}$$

This result shows that the operator $h_{a,b}$ shifts a function $f(u,v)$ by a quantity a along the u axis, and a quantity b along the v axis, and makes it zero for all $u < a$ and $v < b$. With $b=0$ we have the well known *translation operator* of the one-dimensional operational calculus.

The jump function is therefore represented by the operator

$$\{H_{a,b}(u,v)\} = h_{a,b}/(r{\cdot}s).$$

Many more functions can be represented with the operator $h_{a,b}$. For instance define the 'gate' function

$$G_{a,b;c,d}(u,v) = \begin{cases} 1 \ \text{for} \ a \leq u \leq b, \ c \leq v \leq d, \\ 0 \ \text{everywhere else}; \end{cases}$$

it is easy to verify that

$$\{G_{a,b;c,d}\} = (h_{a,0} - h_{b,0}){\cdot}(h_{0,c} - h_{0,d})/(r{\cdot}s).$$

9. OPERATIONS A AND B

Define the operation A that transforms a function into another function thus

$$A\{f(u,v)\} = \{f(u/a,v)\}.$$

It is easy to prove that

$$A(\{f\} + \{g\}) = A\{f\} + A\{g\},$$

$$A(\{f\} \cdot \{g\}) = A\{f\} \cdot A\{g\},$$

$$\{f\}/\{g\} = \{h\} \Rightarrow A(\{f\}/\{g\}) = A\{h\}.$$

Define the operation A that transforms an operator into another operator thus

$$h = \{f\}/\{g\} \Rightarrow Ah = A\{f\}/A\{g\}.$$

It is easy to prove that this operation is distributive with respect to addition, multiplication, and division, as the operation A on functions.

Theorem. $A1 = 1$.

In fact, $A1 = A\{1\}/A\{1\} = 1$.

Theorem. $As = s$.

In fact, $As = A\{f(u,v)\}/A\{\int_0^v f(u,y)\ dy\}$

$$= \{f(u/a,v)\}/q\{f(u/a,v)\} = s.$$

Theorem. $Ar = a \cdot r$.

In fact, $Ar = A\{f(u,v)\}/A\{\int_0^u f(x,v)\ dx\}$

$$= \{f(u/a,v)\}/\{\int_0^{u/a} f(x,v)\ dx\}$$

$$= \{f(u/a,v)\}/\{1/a \cdot \int_0^u f(x/a,v)\ dx\}$$

$$= a\{f(u/a,v)\}/p\{f(u/a,v)\} = a \cdot r.$$

Symmetrical properties are valid for the operation

$$B\{f(u,v)\} = \{f(u,v/b)\}.$$

In general,

$$\{f(u,v)\} = \phi(r,s) \Rightarrow \{f(u/a,v/b)\} = \phi(ar,bs)$$

for any positive a, b.

10. OPERATIONS R_α AND S_β

Define the operations for functions

$$R_\alpha S_\beta \{f\} = \{e^{-\alpha u - \beta v} f(u,v)\},$$

and for operators

$$R_\alpha S_\beta (\{f\}/\{g\}) = (R_\alpha S_\beta \{f\})/(R_\alpha S_\beta \{g\}).$$

Theorem. $R_\alpha S_\beta \phi(r,s) = \phi(r+\alpha, s+\beta)$ where $\phi(r,s)$ is a rational expression of the operators r, s.

The proof is analogous to the one in the operational calculus in one variable.

11. SPECIAL OPERATORS

While p, q, r, s are two—variable operators analogous to corresponding one—variable operators, some important operators have no counterpart in the one—variable operational calculus.

Define the operation

$$I\{f\} = \{\int_0^{\min(u,v)} f(u-x, v-x) \; dx\} \tag{4}$$

that transforms the function $f(u,v)$ into the integral indicated, where its upper limit is the smaller of u and v.

It is easy to verify that

$$r \cdot I\{f\} = \begin{cases} \int_0^v f_u(u-x, v-x) \; dx & (u \geqq v) \\ \\ \int_0^u f_u(u-x, v-x) \; dx + f(0, v-u) & (u < v), \end{cases}$$

where f_u represents the partial derivative of f with respect to u; similarly,

$$s \cdot I\{f\} = \begin{cases} \int_0^v f_v(u-x,v-x) \; dx + f(u-v,0) & (u \geqq v) \\[4mm] \int_0^u f_v(u-x,v-x) \; dx & (u < v); \end{cases}$$

it follows

$$(r+s) \cdot I\{f\} = \begin{cases} f(u-v,0) + \int_0^v (f_u + f_v) \; dx & (u \geqq v) \\[4mm] f(0,v-u) + \int_0^u (f_u + f_v) \; dx & (u < v); \end{cases}$$

therefore $I\{f\} = \{f\}/(r + s)$. In other words the operator $1/(r + s)$ transforms a function of two variables into the integral defined by (4).

Define now the operation

$$L[f] = \begin{cases} f(u-v) & (u \geqq v) \\[3mm] 0 & (u < v) \end{cases} \tag{5}$$

that transforms the one-variable function $f(u)$ into the two-variable function shown. It is easy to verify that

$$p \cdot L[f] = \left\{ \begin{array}{c} \int_0^u f(x-v) \; dx \\[4mm] 0 \end{array} \right\} = \left\{ \begin{array}{c} \int_0^{u-v} f(y) \; dy \\[4mm] 0 \end{array} \right\},$$

where the upper line inside the brackets corresponds to $u \geqq v$, and the lower line to $u < v$; similarly,

$$q \cdot L[f] = \left\{ \begin{array}{c} \int_0^v f(u-x) \; dx \\[4mm] \int_0^u f(u-x) \; dx \end{array} \right\} = \left\{ \begin{array}{c} \int_{a-v}^u f(y) \; dy \\[4mm] \int_0^u f(y) \; dy \end{array} \right\};$$

it follows that

$$(p+q) \cdot L[f] = \{\int_0^u f(y)\} dy = p \cdot \{f\} = p.q.[f];$$

therefore L[f] = p q/(p+q) [f] = [f]/(r+s). In other words, the
same operator 1/(r+s) transforms a one-variable function into
the two-variable function defined by (5).

Many more special operators can be defined in a similar way.

Define now the operation on a one-variable function

$$K[f(u)] = s\begin{Bmatrix} f \\ 0 \end{Bmatrix} \quad \begin{matrix} u < v \\ u \geq v \end{matrix} \, .$$

Theorem. K([f] + [g]) = K[f] + K[g]

$$K([f] \cdot [g]) = K[f] \cdot K[g].$$

The first identity is trivial. For the second,

$$K([f] \cdot [g]) = K[\int_0^u f(u-x) \, g(x) \, dx]$$

$$= s \begin{Bmatrix} \int_0^u f(u-x) \, g(x) \, dx \\ 0 \end{Bmatrix}$$

$$= s^2 \begin{Bmatrix} \int_u^v \int_0^u f(u-x) \, g(x) \, dx \, dy \\ 0 \end{Bmatrix}$$

$$= s^2 \begin{Bmatrix} (v-u) \int_0^u f(u-x) \, g(x) \, dx \\ 0 \end{Bmatrix} \quad ;$$

also

$$K[f] \cdot K[g] = s^2 \begin{Bmatrix} f \\ 0 \end{Bmatrix} \cdot \begin{Bmatrix} g \\ 0 \end{Bmatrix}$$

$$= s^2 \left\{ \int_0^v \int_0^u f(u-x,v-y) \, g(x,y) \, dx \, dy \right\},$$

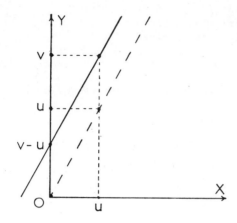

where

$$f(u-x,v-y) = \begin{cases} f(u-x) & \text{for } u-x < v-y \\ 0 & \text{elsewhere,} \end{cases}$$

$$g(x,y) = \begin{cases} g(x) & \text{for } x < y \\ 0 & \text{elsewhere;} \end{cases}$$

$$K[f]\cdot K[g] = s^2\{\iint_A f(u-x)\ g(x)\ dx\},$$

where A is the parallelogram with vertices $(0,0)$; $(0,v-u)$; (u,v); (u,u).

Finally

$$K[f]\cdot K[g] = s^2 \begin{Bmatrix} (v-u) \int_o^u f(u-x)\ g(x)\ dx \\ \\ 0 \end{Bmatrix}.$$

Theorem. $[f]\cdot[g] = [h] \Rightarrow K[f] = K[h]/K[g].$

Definition.   If h is an operator not containing s, i.e., of the type

$$h = [f(u)]/[g(u)], \quad \text{then} \quad Kh = K[f]/K[g].$$

Scholium. The operation K cannot be defined for a function of v or for an operator containing r.

Theorem. K1 = 1.

Theorem. Kr = r + s.

In fact,

$$r = r^2/r = \frac{1/r}{1/r^2} = [1]/[u]; \quad so$$

$$Kr = K[1]/K[u] = \frac{1/(r+s)}{1/(r+s)^2} = r + s.$$

Theorem. $K\phi(r) = \phi(r+s)$, where $\phi(r)$ is a rational expression in r.

Symmetrical properties hold for the operation

$$K^*[f(v)] = r\begin{Bmatrix} 0 \\ f \end{Bmatrix} \begin{matrix} u \leq v \\ u > v \end{matrix} \quad .$$

Define the operation

$$M\{f(u,v)\} = \{\int_0^u f(u-x,v+x) \; dx\}.$$

Theorem. $M\{f\} = \{f\}/(r-s)$.

In fact,

$$r \cdot M\{f\} = \{\int_0^u f_u(u-x,v+x) \; dx + f(0,u+x)\},$$

$$s \cdot M\{f\} = \{\int_0^u f_v(u-x,v+x) \; dx\},$$

where $f_u = \partial f/\partial u$, $f_v = \partial f/\partial v$. Thus

$$(r - s) \cdot M\{f\} = \{f(0,u+v) + \int_0^u (f_u - f_v) \, dx\}$$

$$= \{f(u,v)\}.$$

Define the operation $N[f(u)] = \{f(u+v)\}$.

Theorem. $N[f(u)] = \dfrac{[f(v)] - [f(u)]}{r - s}$.

In fact,

$$p \cdot N[f] = \{\int_0^u f(x+v) \, dx\} = \{\int_v^{u+v} f(x) \, dx\},$$

$$q \cdot N[f] = \{\int_0^v f(u+x) \, dx\} = \{\int_u^{u+v} f(x) \, dx\}; \quad so$$

$$(q-p) \cdot N[f] = \{\int_u^v f(x) \, dx\} = \{\int_0^v f\} - \{\int_0^u f\}$$

$$= q\{f(v)\} - p\{f(u)\}$$

$$= p \cdot q[f(v)] - p \cdot q[f(u)],$$

therefore

$$N[f] = p \cdot q([f(v)] - [f(u)])/(q - p)$$

$$= \dfrac{[f(v)] - [f(u)]}{r - s} \quad .$$

12. APPLICATIONS

As an example of the use of this calculus consider this simple ecological problem.

$N(t,\tau) \, d\tau$ is the number of individuals of a species at time t with age between τ and $\tau + d\tau$; $Z(\tau)$ is the death rate of individuals of that species of age τ. Then

$$N(t,\tau) \, d\tau - N(t+dt,\tau+dt) \, d\tau = Z(\tau) \cdot N(t,\tau) \, d\tau \, dt$$

is the number of individuals dying in the interval of time from t
to t + dt. Consequently we can write the differential equation

$$\partial N/\partial t + \partial N/\partial \tau = -Z(\tau)\cdot N(t,\tau),$$

or

$$\frac{\partial}{\partial \tau} \ln N + \frac{\partial}{\partial \tau} \ln N = -Z;$$

using (1) and (3),

$$r\{\ln N\} - [\ln N(0,\tau)] + s\{\ln N\} - [\ln N(t,0)] = -\{Z\},$$

$$(r + s)\ \{\ln N\} = -\{Z\} + [\ln N(0,\tau)] + [\ln N(t,0)].$$

With the results of Section 11, we have now

$$\{\ln N\} = -I\{Z\} + L[\ln N(0,\tau)] + L[\ln N(t,0)]$$

$$= -\begin{cases} \int_0^\tau Z(\tau-x)\ dx + \ln N(t-\tau,0) & (t \geqq \tau) \\[3mm] \int_0^t Z(\tau-x)\ dx + \ln N(0,\tau-t) & (t < \tau) \end{cases}$$

hence

$$N(t,\tau) = \begin{cases} N(t-\tau,0)\cdot\exp(-\int_0^\tau Z(\tau-x)\ dx) & (t \geqq \tau) \\[3mm] N(0,\tau-t)\cdot\exp(-\int_0^t Z(\tau-x)\ dx) & (t < \tau); \end{cases}$$

and finally, with a change of variable,

$$N(t,\tau) = \begin{cases} N(t-\tau,0)\cdot\exp(-\int_0^\tau Z(x)\ dx) & (t \geqq \tau) \\[3mm] N(0,\tau-t)\cdot\exp(-\int_{\tau-t}^\tau Z(x)\ dx) & (t < \tau). \end{cases}$$

The advantages of the calculus introduced here are not as
much in the solution of partial differential equations as in providing a heuristic language for the manipulation of linear systems
of many-variable functions.

REFERENCES

Berg, L. (1967). *Introduction to the Operational Calculus.*
    North-Holland Publishing Company, Amsterdam.

Ditkin, V. A. and Prudnikov, A. P. (1962). *Operational Calculus*
    in Two Variables and Its Applications. Pergamon, London.

Erdelyi, A. (1962). *Operational Calculus and Generalized Functions.*
    Holt, Rinehart and Winston, New York.

Lerch, M. (1903). Sur un point de la théorie des fonctions
    génératrices d'Abel. *Acta Mathematica,* 27, 339-352.

Mikusinski, J. (1951). Remarks on the moment problem and a
    theorem of Picone. *Colloquium Mathematicum II,* 2, 138-141.

Mikusinski, J. (1959). *Operational Calculus.* Pergamon, London.

Peano, G. (1908). *Formulario Mathematico,* Editio V. Bocca,
    Torino.

Rescigno, A. (1963a). Flow diagrams of multi-compartment systems.
    Annals of the New York Academy of Sciences, 108, 204-216.

Rescigno, A. (1963b). *Operational calculus in two variables.*
    Lawrence Radiation Laboratory Report UCRL-11033. 65-68.

Rescigno, A. (1969). An introduction to linear systems analysis.
    In *Systems Analysis Approach to Neurophysiological Problems,*
    University of Minnesota Laboratory of Neurophysiology, Minnea-
    polis. 1-30.

Rescigno, A. and Beck, J. S. (1972). Compartments. In *Foundations*
    of Mathematical Biology, Vol. 2, Chapter 5, R. Rosen, ed.
    Academic Press, New York.

Titchmarsh, E. C. (1926). The zeros of certain integral functions.
    Proceedings of the London Mathematical Society, 25, 283-302.

[*Received August* 1978. *Revised January* 1979]

AUTHOR INDEX

SUBJECT INDEX